D0857103

SYSTEMATIC AND REGIONAL BIOGEOGRAPHY

Systematic and Regional Biogeography

Stanley A. Morain
University of New Mexico

VNR VAN NOSTRAND REINHOLD COMPANY

Library of Congress Catalog Card Number: 84-3728
ISBN: 0-442-26186-1

Manufactured in the United States of America.

Published by Van Nostrand Reinhold Company Inc.
135 West 50th Street
New York, New York 10020

Van Nostrand Reinhold Company Limited
Molly Millars Lane
Wokingham, Berkshire RG11 2PY, England

Van Nostrand Reinhold
480 Latrobe Street
Melbourne, Victoria 3000, Australia

Macmillan of Canada
Division of Gage Publishing Limited
164 Commander Boulevard
Agincourt, Ontario MIS 3C7, Canada

15 14 13 12 11 10 9 8 7 6 5 4 3 2 1

Library of Congress Cataloging in Publication Data
Morain, Stanley A.
Systematic and regional biogeography.
Includes bibliographical references and index.
1. Biogeography. I. Title.
QH84.M66 1984 574.5 84-3728
ISBN 0-442-26186-1

Contents

Preface

Biogeography is defined by Webster as "a branch of biology that deals with the geographical distribution of plants and animals." The definition is incomplete to the extent that it fails to mention the origins of those distributions. Although it is true that biologists, writing under the rubrics of many subdisciplines, have pursued the topic most vigorously, its treatment in a holistic sense involves the synthesis of information from virtually all the natural sciences. As a geographer, my intention here is to view the field as a dichotomy: first, the book will examine the biological processes that govern the rise and fall of species; and second, it will review the regional developments of biotic realms to discover how the distributions arose. Part I of the book is aimed primarily at introducing intermediate undergraduate geographers to the processes, concepts, and terminology associated with the broader questions of speciation, adaptation, and extinction. Part II reverses this role by addressing the geographical content of biogeography and is aimed at intermediate biology students. My experience in the classroom convinces me that such a marriage is mutually beneficial to both sides of the house.

No book this size can completely describe the intricacies of historical biogeography in the Cenozoic era, or portray in detail the rapid advancement being made in genetics and theoretical ecology. Other authors, for example, MacArthur (*Geographical Ecology: Patterns in the Distribution of Species*), Pielou (*Biogeography*) and Cox, Healey, and Moore (*Biogeography: An Ecological and Evolutionary Approach*), provide alternative approaches to the one adopted here, but until now there has been no single volume that provides a global perspective on the outcome of geological, ecological, and human time in relation to speciation, dispersal, and extinction.

If the book serves in any capacity to arouse the interest of students and to stimulate their pursuit of the literature my aims will have been met.

ACKNOWLEDGMENTS

Since the first compilation of notes over a decade ago, dozens of students, colleagues, and friends have assisted and encouraged me to complete this effort. I have tried to thank them along the way, but most have scattered and are no longer in contact. I take this opportunity to express my deepest appreciation for all their time and support, and hope that if they should read this they will smile and recall our once closer ties. I must expecially thank my wife, Marilyn, who not only encouraged me to continue, but whose hand and skills are directly visible in many of the illustrations. Ms. Julie Trollinger's artistry and meticulous attention to detail in both illustrating and typing also deserve special applause. Finally, I must thank all biogeographers for creating such an exciting and varied literature as to make this a personally rewarding and enjoyable project.

STANLEY A. MORAIN

SYSTEMATIC AND REGIONAL BIOGEOGRAPHY

PART I

Systematic Biogeography

Chapter 1

Biogeography Scope and Content

> Explanatory biogeography also treats of the origins of particular groups of organisms in specified regions and their changes in geographic distribution through time. The approach is historical; the primary question is 'How come?'; and explanations are evolutionary.
>
> G. G. Simpson, *This View of Life*

OVERVIEW

What is Biogeography?

The history of earth's terrestrial flora and fauna is the story of progressive adaptations to changing conditions of the environment. There can be little argument that the ancestors of today's modern species emerged from the sea and began their differentiation in the incubator of low-lying, swampy, tropical, humid regions. As their succeeding generations became better able to survive seasonal drying, new forms originated that were adapted to terra firma; and each such adaptation provided other plants and animals with niches for their own further development. In this fashion, symbiotic, coevolutionary associations and communities have developed. Throughout the long history of land plants dating from the Permo-Carboniferous period to the present, these progressive adaptations have made survival possible, first in an aerial rather than hydrologic medium, then in seasonal drought, seasonal cold, and finally, combinations of drought and cold. Today even the most inhospitable environments, from bleak arctic tundras and felsenmeers to arid playa lake beds, support plant life. The enduring forces of evolution and adaptation have thus permitted a thin biological skin of life to spread out of the tropical incubator to high polar latitudes and Himalayan altitudes.

The basic premises upon which studies in evolutionary biogeography rest are:

Environments are not uniform in space or in time.
To survive, species must constantly be adapting to gradual environmental changes. These changes involve concurrent time-space functions if the land mass on which they are located is physically moving beneath more or less constant climatic belts.
To expand their range, species must adapt to microgeographic differences associated with ecological settings, habitats, competition, and predators.
As part of the process of adapting, species undergo genetic change, most of which leads to phyletic or evolutionary differentiation, but not necessarily to the origin of new species.

However, to state that the evolution and dissemination of life rests on the intricacies of genetics and environment seems too simple. The idea is so easily expressed that its full impact is often lost on those hearing it for the first time. Without carefully relating facts and concepts from geology, botany, geography, genetics, hydrology, zoology, paleontology, climatology, and many other disciplines, the meaning of life from patterns observed today cannot be known. Only to the extent that given regions have been thoroughly studied from these diverse points of view can distributions be described with certainty; and the number of such localities is incredibly few.

For the laws of nature to increase the diversity of life, viable reproducing populations must exist long enough to become species. The elements of this process of speciation and dispersal are themselves varied and complex. Among the many processes affecting the end result, one can list genetic and nongenetic variation, selection, adaptation, evolution, extinction, dispersal mechanisms, and isolating mechanisms. Since none of these processes is independent in time or space, the general advance of species can be compared to a stream of interwoven smaller processes acting and reacting upon each other. There is, consequently, no obvious or best beginning point for describing the mystery and fascination of organic increase. Nor is there an absolute point at which a species can be said to exist beyond its artificial binomial identifier. Any entry into the stream of processes inevitably requires consideration of all the channels defining that particular thread of life.

In the treatment that follows, this process stream will be entered by looking first at variation and isolation, those processes that result in individuals or populations that are genetically and/or spatially marginal to their predecessors. Chapter 3 will expand upon these basic ideas by considering the processes of selection, adaptation, and evolutionary rates, which heavily influence the viability of peripheral or isolated populations. Chapter 4 will focus directly on the various models of speciation and the question of extinction. Chapter 5 begins a review of geographic considerations including active dispersal mechanisms, migration, the rise of refugia, and the nature of barriers. The last chapter in Part I summarizes the processes of passive dispersal and their impact on regional biological development. On the far shore of this process stream one should have an introductory understanding of how organisms continually emerge and how those of obviously different historical and geographic backgrounds assemble into communities.

Throughout these chapters there is one fundamental question directing the discussion; namely, how do plant and animal species arise and disperse into their observed geographical patterns? Since this question involves movement and a spatial dimension, it is clear that geographic methodology, as well as the pure biology of the matter, must be considered. In addition to the fundamental question of "how," the discussion in Part I will explore several related questions, the summary answers to which are given below.

1. What is the evidence for evolution and how do we measure it? *Answer: Phylogeny* as recorded in the fossil record. In part, at least, one can recreate the past by arranging fossil remains into logical sequences showing morphological change. Darwin himself was afraid that too few fossils would ever be unearthed to permit total reconstruction of the past, and this "missing link" problem has resulted in a more modern theory of "punctuated equilibrium."
2. What is the goal of evolution? *Answer: Adaptation.* Through the process of adaptive radiation, variation becomes fixed in a dynamic, rather than static, sense.
3. How is adaptation achieved? *Answer: Genetic mutation* in individuals; *recombination* of gene sets during sexual reproduction; and *gene flow* throughout a population.
4. Why are all species not adapted to all environments? *Answer:* Because of differences in their homeostatic limits (*tolerances*), *selection pressures,* and rates of *gene flow.* It is in this context that growth models, predation theory, competition theory, and general population dynamics become critical fields of inquiry.
5. Where does evolution end? *Answer: Extinction,* which occurs when all populations of a species fail or are unable to adapt to environmental changes.
6. If species are not adapted to all environments, and if extinction is inevitable, how has life persisted for all these thousands and millions of years? *Answer: Speciation.* This process differs from gradual

evolutionary change in the sense that new forms are created that are better able to cope with new environmental situations.

7. What are the requirements for speciation? *Answer: Active dispersal* and *isolation* at the individual and population levels and eventual reproductive isolation from sister populations of the same species. There are also passive dispersal mechanisms, and it is in this context that avenues of migration, land bridges, and continental drift are important topics of discussion.
8. How does speciation work? *Answer:* One holistic mechanism is given by the *theory of geographic speciation,* which hinges on geographic segregation and reproductive isolation of populations. In this context it is essential to review *peripheral isolates, founder populations,* and the adjunct question of how, in the face of selection against genetic deviations, sufficient variability accumulates to promote speciation.
9. Are there limits to species diversification? *Answer:* Yes, according to the theory of island biogeography. Area/species curves lead to the notion of species equilibrium, or carrying capacity, for a given area. Size is used as a surrogate for habitat diversity.
10. If the foregoing questions represent the foundations of biological increase and the geographic spread of life forms, how are regional developmental histories compiled? *Answer:* By aggregation of a wide range of data. By its nature, historical or evolutionary biogeography is usually led by the data available, rather than by testable hypotheses. Large continental land masses are less well known than smaller localities, especially islands. Moreover, the kind and quality of information available differs between localities. Despite these problems, it is possible to piece together sufficiently consistent data to at least outline basic patterns of development.

Why Study Biogeography?

It is no longer a matter for debate that *Homo sapiens* is explosively and, in many instances, irreversibly altering earth's habitable environments (see Eckholm, 1978; Brown, McGrath, and Stokes, 1976; Inskipp and Wells, 1979). No one can dispute that the chemical impact of our industrial and nuclear age is being felt throughout the trophic levels of virtually every known food chain, or that the elements of those food chains are being victimized and systematically annihilated. If biological "success" is measured by competitive ability, habitat breadth, niche breadth, and adaptability, then man represents the apex of evolutionary success.

Lest our species become complacent, however, we should realize that nature plays no favorites. The parade of life over the past half-billion years or more testifies that while we pursue our cultural destiny, we are also subject to our genetic destiny; and the two are ever more entwined. We will have to adapt to the environments we create at least as quickly as do our current best competitors, the insects. If we do not, then we should start immediately to write the final chapters for "The Age of Humanity." A stratigraphic episode will record our relatively short burst of dominance at the tag end of the mammalian era. The message will be that we strutted on life's stage for a mere six million years, one-twentieth of the time of the dinosaurs and less than one-tenth of the recorded history of mammals. In the wings, nature is quietly providing our successors. It is worth our time therefore to study this parade of life and where we fit into it.

When turned toward the service of man, biogeography forms part of the core of our agricultural, medical, and general environmental knowledge (Cole, 1965). People are concerned over the fate of the sperm whale and encouraged to learn that a little known desert plant called the jojoba (*Simmondsia chinensis*) produces a liquid similar to sperm oil. We capitalize on the knowledge that an obscure beetle produces more cortexone than the adrenal glands of a thousand cattle (Evans, 1973). How could we ever learn these things that improve our own lives if we were not committed to discovering, describing, and preserving even the most inconspicuous forms of plant and animal life? But the quest itself presents us with an incredible job of cataloging. There are an estimated 10 million living species of plants and animals, of which a scant 15 percent have been described, and fewer studied (Raven, Berlin, and Breedlove, 1971). There are about 300,000 species of flowering plants (Angiosperms) and approximately 1 million known animals. And what about the molds, bacteria, and algae that play increasingly important roles in dietary, medical, and economic areas? Where these organisms came from, and where they are located today, form a cornerstone for their protection and future use. Unfortunately, "in view of the available taxonomic manpower and the enormous rate of extinction that will characterize the next century, it is doubtful that even five percent more of the world's organisms can be added to our inventory before the remaining 80 percent become extinct" (Raven, Berlin, and Breedlove, 1971, 1210).

One can see, therefore, that there are fundamental questions in biogeography that go beyond mere excitement over whether there have ever been land bridges or drifting continents. Is humankind doomed to succumb to an age of insects, or can we adapt rapidly enough to compete with these heirs apparent? Are the extinctions we observe today necessarily bad, or does humanity merely represent another in a long series of extreme environmental changes from which nature will recover? Can evolutionary biogeography teach us anything about the relationships between humans and their environment? Put another way, are air and water pollution part of our species' inevitable phylogeny? Or, having experienced them, can we alter or reduce these toxic by-products of civilization? Is it possible that the urban ecosystem represents a cultural evolutionary step in environmental change that we, the creator of it, cannot adapt to? Having successfully occupied every terrestrial environment, have we found the natural limits to our tolerance in the form of urban environments? Granted that these are highly philosophical questions (and perhaps not the best ones that could have been posed), it is abundantly clear that the study of evolutionary biogeography is a worthwhile undertaking.

SYSTEMS OF REFERENCE

There are three systems of reference pertinent to evolutionary biogeography: *taxonomic, temporal,* and *geographic.* The one least familiar to most geographers, and indeed, one that often leads to heated discussions at biological meetings, is the taxonomic. Temporal scales can be confusing if "ecological time" and "geologic time" are not kept distinctly separate; and the geographic reference can be muddled if ecological settings and habitats (favored words in biology) are not recognized for what they are—microgeographic locales. Problems of semantics and syntax can arise in the very best literature, and these in turn may lead to misinterpretations by readers, especially those brave enough to cross disciplinary lines to synthesize information. The reference systems used in this book, while traditional in most respects, must be briefly described.

Taxonomic System

Children are taught to ask whether an object is animal, vegetable, or mineral. Most systems of biological classification recognize only the two kingdoms (plant and animal) at the top of the hierarchy. This view, of course, has resulted from a relatively limited knowledge of the world at the time each classification was proposed. Modern understanding of the diversity of living organisms, however, indicates that a two-kingdom system not only imposes an unnatural division of one-celled organisms, but also treats bacteria and higher fungi inadequately. Fifteen years ago,

Whittaker (1969) proposed a five-kingdom classification of life based on three modes of nutrition: photosynthesis, absorption, and ingestion.

Figure 1.1 compares the traditional two-kingdom organization of life with that proposed by Whittaker. In the plant kingdom attention is called to the phylum Tracheophyta, and especially to the subphylum Spermatophyta which includes the Gymnosperms (e.g., pine, spruce) and Angiosperms (flowering plants). These will be the elements of focus throughout subsequent discussions. For the animal kingdom attention will rest on the phylum Chordata, subphylum Vertebrata (the mammals, reptiles, birds, and fishes). A quick glance at Figure 1.1 indicates that this book generalizes evolutionary biogeography on the basis of a very small segment of earth's overall range of life forms and, indeed, on disparate information pertaining even to that small segment.

Major levels below the subphylum are given as class, order, family, genus, and species. The latter two represent the Linnean system of dichotomous classification upon which most of the discussion in Part I will rest. The basic unit of reference is the species and its scattered populations, and the focus of interest will be on how these populations cope with their environments and change through time. In Part II, sheer weight of numbers forces us to focus on generic and familial levels of taxonomy. In some cases monographic revisions of genera and families have added new species, and in others condensed them into fewer taxa. Either way, much that was thought to be known about the distributions of those taxa is left questionable. These problems are magnified even more if a taxonomic revision is performed on a fossil collection, because both the associated geographical and temporal systems of reference must be correlated.

The practical application of the current taxonomic systems appears to be limited to pigeonholing. As Raven, Berlin, and Breedlove (1971) point out, such systems were never designed for information retrieval whereby one can quickly learn whether a species has valuable properties or is endangered. "In dealing with the vast number of organisms that exist, we tend to overemphasize the process of classification, and the decisions it involves, at the expense of the information about the organisms that we are supposedly accumulating. Frequent changes in names exacerbate the difficulties of the system." (Raven, Berlin, and Breedlove, 1971, 1213). Nevertheless, some system must be employed; otherwise we could not proceed. For most purposes there is no viable alternative because the literature is written using the universal scheme shown in Table 1.1. One simply must be aware that species regarded today as being distinct may be refined or regrouped tomorrow, thus changing our views about their origins. As one climbs the ladder of abstraction to genera and

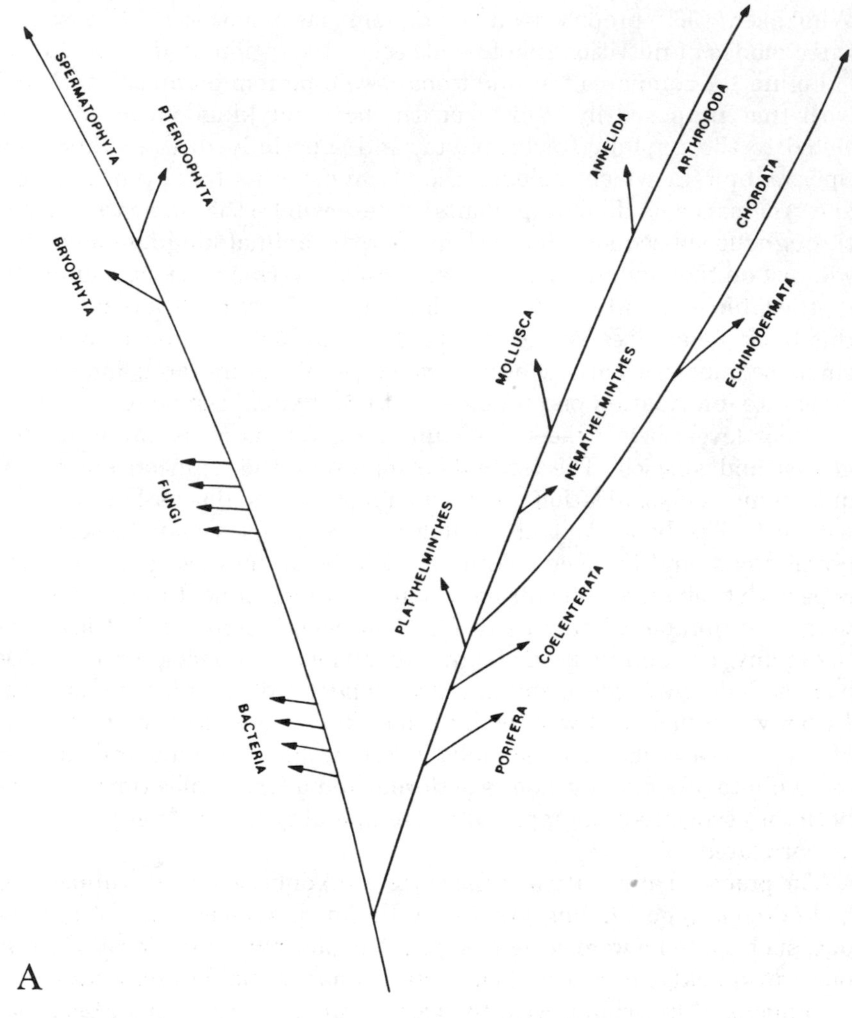

Figure 1.1 (A) A simplified two-kingdom system of taxonomic classification. **(B)** A modern five-kingdom system of classification recognizing unicellular and multicellular plants, fungi, and animals. Each of the five kingdoms can be subdivided into photosynthetic, absorptive, and ingestive forms. [**(B)** *From J. W. Valentine, 1978, Evolution of Multicellular Plants and Animals, Scientific American* **239***(3):140; copyright © 1978 by Scientific American, Inc. All rights reserved.*]

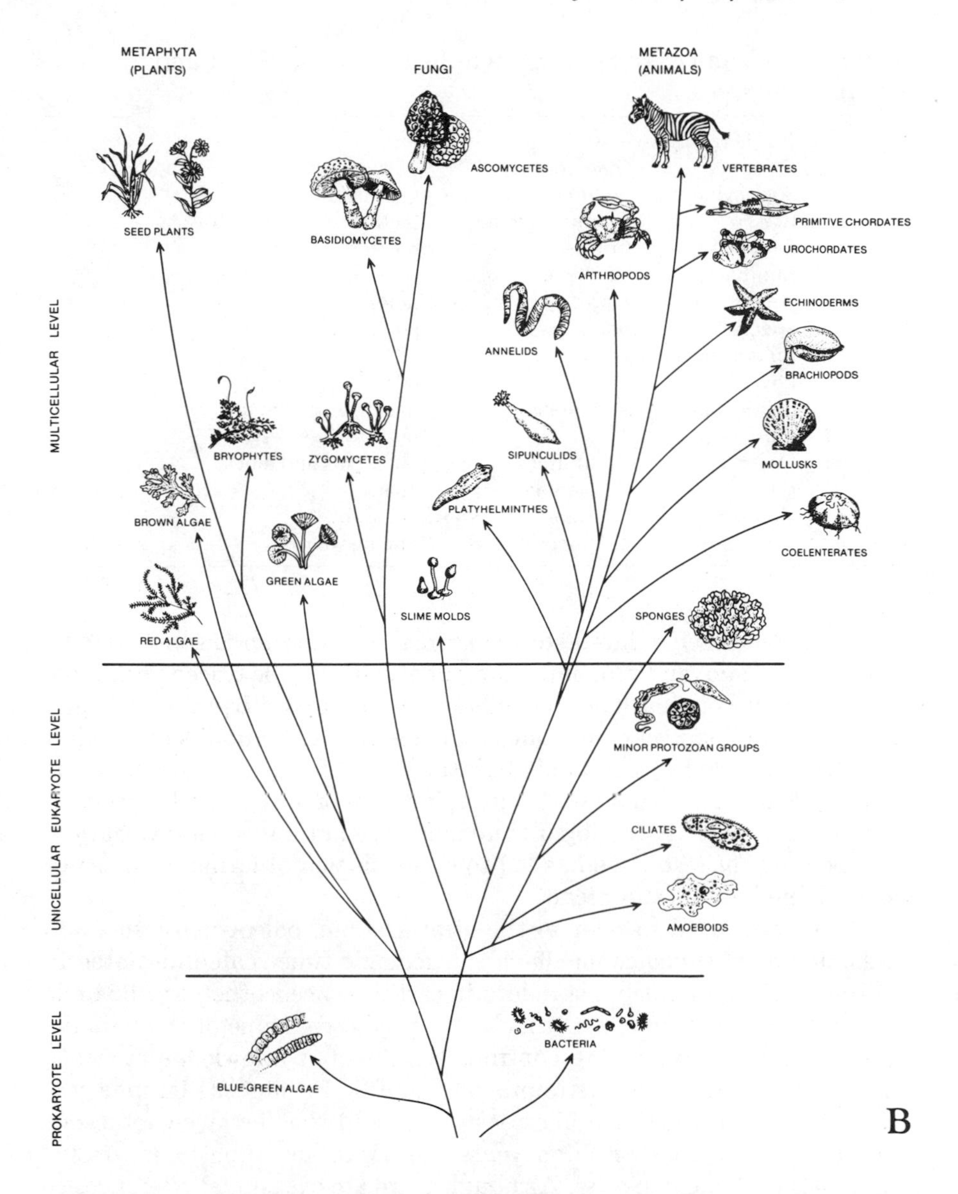

families, there is less controversy over names, primarily because the identifying attributes are more clearly defined.

Temporal System

The majority of angiosperm and mammalian life forms existing today are the result of explosive speciation during the last 1 million years. They are

Table 1.1. Examples of Species Identification in the Plant and Animal Kingdoms.

A. *Ponderosa pine*		
Phylum	Tracheophyta	(vascular plants)
Subphylum	Spermatophyta	(seed bearing)
Class	Gymnospermae	(seed not enclosed in ovary)
Order	Coniferales	(conifer)
Family	Coniferae	(conifer)
Genus	*Pinus*	(pines)
Species	*ponderosa*	(ponderosa)
B. *Ring-necked pheasant*		
Phylum	Chordata	(bilateral symmetry)
Subphylum	Vertebrata	(having a backbone)
Class	Aves	(birds)
Order	Galliformes	(pheasants and allies)
Family	Phasianidae	(quail)
Genus	*Phasianus*	(pheasant)
Species	*colchicus*	(ring-necked)

derived from closely related stock that had its beginnings some 70 to 90 million years ago; and progenitors for all of them can be traced with more or less clarity as far back as 350 million years. Life, itself, probably began more than a billion years ago, although actual fossil remains of the first invertebrates can be traced only half that far. When considered in terms of geologic history, therefore, the diversification of life presently observed on earth is the result of a logarithmic increase in total genetic variation. Competition for resources has led to the evolution of forms to fill every conceivable habitat and niche.

The system used to refer to the stratigraphic, paleogeographic, and paleontological evidence for life is the geologic time scale illustrated in Figure 1.2. It is as much a standard in earth science as the periodic table of elements is in chemistry. It lacks the predictive value of the periodic table, however, and is being continually refined as knowledge of stratigraphy and earth processes improves. The Plio-Pleistocene boundary is currently set at about 1 million years ago and the Tertiary-Cretaceous boundary at 65 to 70 million years; both are key time lines for the mammals and angiosperms. Although there are occasional references to pre-Tertiary environments and taxa, the focus of attention, especially in the regional descriptions of Part II, will be on the development of a narrow segment of life since the demise of dinosaurs.

Geologic time provides a means of tracing geneologies of terrestrial life forms, but the closer one gets to the present, the less valuable it is as a system of reference. To know the history of the United States, for example, is not necessarily to know what will happen in West Virginia today.

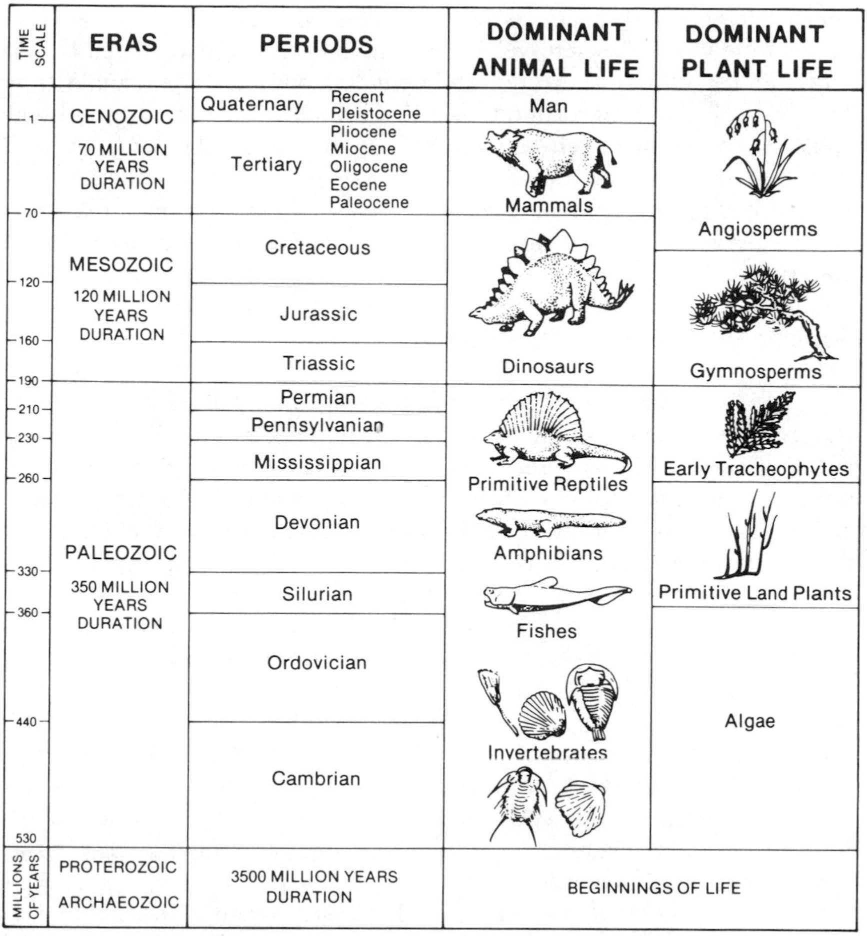

TIME SCALE (MILLIONS OF YEARS)	ERAS	PERIODS		DOMINANT ANIMAL LIFE	DOMINANT PLANT LIFE
1	CENOZOIC 70 MILLION YEARS DURATION	Quaternary	Recent Pleistocene	Man	Angiosperms
70		Tertiary	Pliocene Miocene Oligocene Eocene Paleocene	Mammals	
	MESOZOIC 120 MILLION YEARS DURATION	Cretaceous		Dinosaurs	
120		Jurassic			Gymnosperms
160		Triassic			
190	PALEOZOIC 350 MILLION YEARS DURATION	Permian		Primitive Reptiles	Early Tracheophytes
210		Pennsylvanian			
230		Mississippian			
260		Devonian		Amphibians	Primitive Land Plants
330		Silurian		Fishes	
360		Ordovician		Invertebrates	Algae
440		Cambrian			
530	PROTEROZOIC ARCHAEOZOIC	3500 MILLION YEARS DURATION		BEGINNINGS OF LIFE	

Figure 1.2. The geologic time scale showing dominant plant and animal life at various times in history.

Certainly it would give some predictive ability in the sense that one could state what would not happen, or what might develop, in the next few years; but the moment-to-moment details would have to be observed and recorded. In biological terms this present-oriented time frame is referred to as ecological time (see Simberloff, 1976).

The scope of ecological time permits a glimpse of such phenomena as predator/prey relationships, population growth, colonization rates, and even somewhat longer-term events like plant migrations and rates of dispersal. All of these processes impact on the viability of a species and provide insight into whether its range is expanding or contracting, or whether it should be declared endangered. For general purposes, ecological time is roughly equivalent to human cultural history. All that is known about living forms, and can be predicted about them, comes from the accumulated observations of their ecological status. As might be expected, the following chapters in Part I are concerned essentially with this time frame, punctuated for emphasis by references to geologic history. In Part II, emphasis will be placed on geologic time, except in those areas where man has accelerated the pace of change to an ecological time frame.

Geographic System

The drama of life is performed on stages called habitats, which are defined by a set of autecological and synecological factors. Except in rare and often violent circumstances, the action of life-form survival is passive, and several generations of a life form usually pass before humans are able to perceive change. The events of ecological time are also usually local or, at best, regional, and it is within these more or less restricted settings that new species originate and the ill-adapted perish.

If the earth were a uniform terrestrial plane characterized by shallow environmental gradients, the scope for biotic diversification would be severely limited. There would be fewer, and certainly less-well defined, habitats where pioneers could adapt and flourish. The excitement of life's parade is that earth is not uniform and does exhibit rather steep gradients. For this reason, the prospect for unique genetic events to multiply and disperse is greatly enhanced. Environmental diversity promotes biotic diversity and, equally important, ensures against the uniform distribution of all species in all localities.

During the review of systematics in Part I, reference is directed mostly at processes that occur on a microgeographic scale. On a more abstract but nonetheless measurable scale, plants and animals can then be grouped into a variety of broader units for study on a world basis. The review of regional biogeography in Part II is organized around six biogeographic realms. The term *realm* was used originally to define areas of faunal similarity. In botanical circles the words *biome* and *formation* are roughly equivalent, but refer to broad "climax" vegetational types.

The map of world biogeographical provinces in Figure 1.3 utilizes the faunal realms as basic units that can be further subdivided into provinces based on plant and animal distributions. The areas delineated differ considerably both in their animal and plant species and in the character of their vegetation. Although Udvardy (1975) recognized 193 biogeographical provinces, knowledge of the evolutionary histories of them is quite uneven. In any case, a systematic catalog of information on a

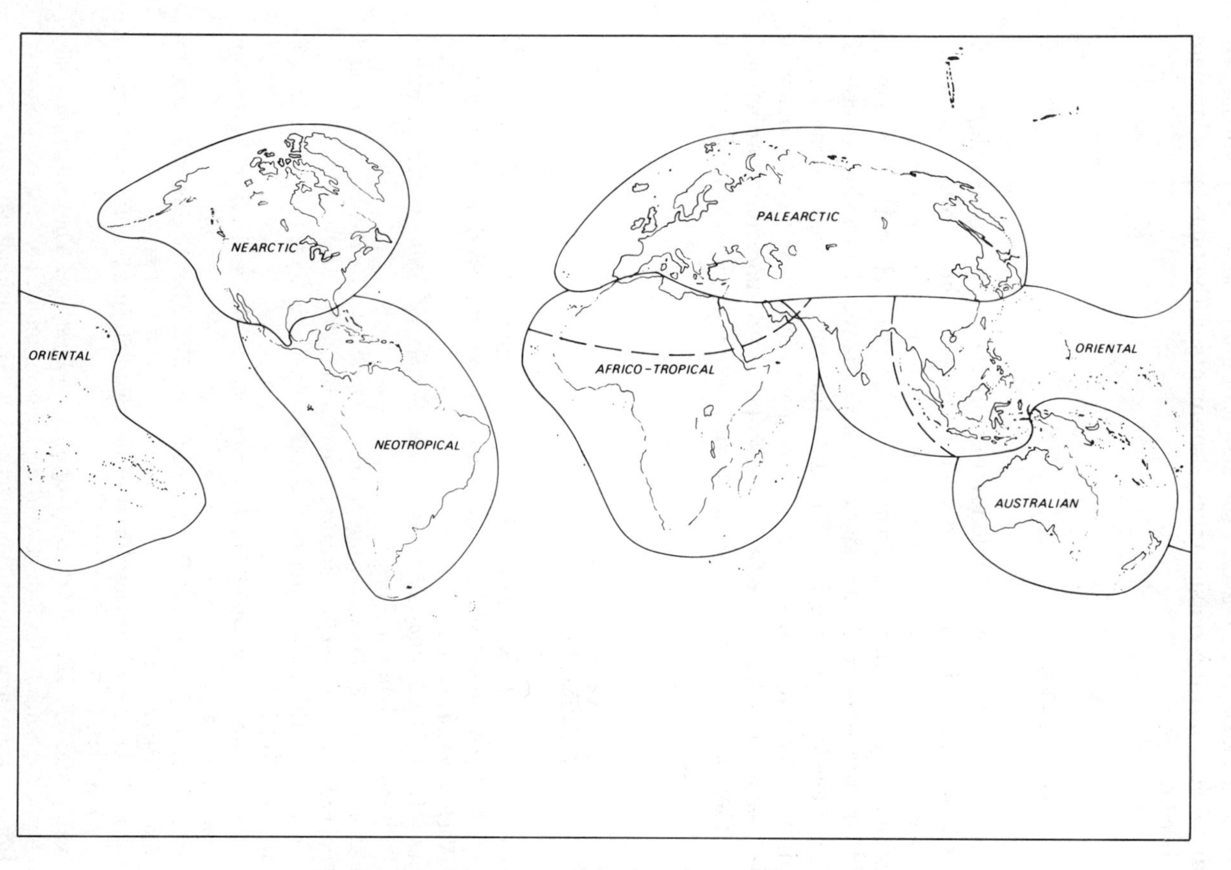

Figure 1.3. Biogeographical realms of the world.

provincial basis is far beyond the scope of effort here. As a substitute, Part II concentrates on summary statements of the biotic development in each realm.

SUMMARY

The discussions that follow explore the processes affecting speciation and extinction first, and then proceed to the factors influencing taxonomic distribution. Most of the attention in Part I is on populations and species occupying habitats in ecological time. In Part II the extension of these processes to describe global patterns requires a shift to several higher levels of abstraction. The distributions of species, genera, and families within and between realms in geologic time forms the basis for discussion. Finally, to reduce the scope to manageable proportions, attention is essentially limited to the subphyla Spermatophyta and Vertebrata.

REFERENCES

Brown, L. R., P. L. McGrath, and B. Stokes, 1976, *Twenty-two Dimensions of the Population Problem,* Worldwatch Paper #5, Worldwatch Institute, Washington, D.C., 83p.

Cole, M. M., 1965, *Biogeography in the Service of Man,* Bedford College, University of London, 63p.

Eckholm, E., 1978, *Disappearing Species: The Social Challenge,* Worldwatch Paper #22, Worldwatch Institute, Washington, D.C., 38p.

Evans, H. E., 1973, Taxonomists' Curiosity May Help Save the World, *Smithsonian* **4**(6):36-45.

Inskipp, T., and S. Wells, 1979, *International Trade in Wildlife,* Earthscan, London, 104p.

Raven, P. H., B. Berlin, and D. E. Breedlove, 1971, The Origins of Taxonomy, *Science* **174**:1210-1213.

Simberloff, D., 1976, Species Turnover and Equilibrium Island Biogeography, *Science* **194**:572-578.

Udvardy, M. D. F., 1975, *World Biogeographical Provinces,* The Co-Evolutionary Quarterly, Sausalito, Calif. (Based on International Union for the Conservation of Nature and Natural Resources. [IUCN] occasional paper no. 18.)

Valentine, J. W., 1978, The Evolution of Multicellular Plants and Animals, *Scientific American* **239**(3):140-159.

Whittaker, R. H., 1969, New Concepts of Kingdoms of Organisms, *Science* **163**:150-160.

FURTHER READINGS

Briggs, D., and S. M. Walters, 1969, *Plant Variation and Evolution,* McGraw-Hill, New York, 256p.

Brown, E. H., ed., 1980, *Geography Yesterday and Tomorrow,* Oxford University Press, New York, 302p.

Burger, W. C., 1981, Why are There so Many Kinds of Flowering Plants? *BioScience* **31**(8):572, 577-581.

Carlquist, S., 1965, *Island Life,* Natural History Press, Garden City, N.Y., 451p.

Carlquist, S., 1973, *Island Biology,* Columbia University Press, New York, 656p.

Colbert, E. H., 1969, *Evolution of the Vertebrates,* Wiley, New York, 535p.

Cox, C. B., I. N. Healey, and P. D. Moore, 1972, *Biogeography: An Ecological and Evolutionary Approach,* Blackwell Scientific, Oxford, England, 179p.

Darlington, P. J., 1965, *Biogeography of the Southern End of the World,* McGraw-Hill, New York, 236p.

Dasmann, R. F., 1973, *A System for Defining and Classifying Natural Regions for Purposes of Conservation,* IUCN Occasional Paper No. 7, International Union for the Conservation of Nature and Natural Resources, Morges, Switzerland.

Dobzhansky, Th., 1951, *Genetics and the Origin of Species,* 3rd ed., Columbia University Press, New York, 364p.

Eicher, D. L., 1968, *Geologic Time,* Prentice-Hall, Englewood Cliffs, N.J., 150p.

Fosberg, F. R., 1976, Geography, Ecology and Biogeography, *Assoc. Am. Geogr. Ann.* **66**:117-128.

Good, R., 1974, *The Geography of the Flowering Plants,* 4th ed., Longman, London, 557p.

Grant, V., 1963, *The Origin of Adaptations,* Columbia University Press, New York, 606p.

Hanson, E. D., 1964, *Animal Diversity,* 2nd ed., Prentice-Hall, Englewood Cliffs, N.J., 118p.

Hargraves, R. B., 1976, Precambrian Geologic History, *Science* **193**:363-371.

Kellman, M. C., 1980, *Plant Geography,* 2nd ed., Methuen, New York, 181p.

Laporte, L. F., 1968, *Ancient Environments,* Prentice-Hall, Englewood Cliffs, N.J., 115p.

Levin, D. A., 1979, The Nature of Plant Species, *Science* **204**:381-384.

Levins, R., 1968, *Evolution in Changing Environments,* Princeton University Press, Princeton, N.J., 120p.

Lewontin, R. C., ed., 1968, *Population Biology and Evolution,* Syracuse University Press, Syracuse, N.Y., 205p.

Lowe-McConnell, R. H., ed., 1969, *Speciation in Tropical Environments,* Academic Press, New York. (Published for Linnean Society of London.) 246p.

MacArthur, R. H., 1972, *Geographical Ecology: Patterns in the Distribution of Species,* Harper and Row, New York, 269p.

MacArthur, R. H., and J. H. Connell, 1966, *The Biology of Populations,* Wiley, New York, 200p.

MacArthur, R. H., and E. O. Wilson, 1967, *The Theory of Island Biogeography,* Princeton University Press, Princeton, N.J., 203p.

Mayr, E., 1969, *Animal Species and Evolution,* Harvard University Press, Cambridge, Mass., 797p.
Mayr, E., 1981, Biological Classification: Toward a Synthesis of Opposing Methodologies, *Science* **214**:510-516.
Mettler, L. E., and T. G. Gregg, 1969, *Population Genetics and Evolution,* Prentice-Hall, Englewood Cliffs, N.J., 212p.
Pielou, E. C., 1979, *Biogeography,* Wiley Interscience, New York, 351p.
Simpson, G. G., 1953, *The Major Features of Evolution,* Columbia University Press, New York, 434p.
Simpson, G. G., 1964, *This View of Life,* Harcourt, Brace & World, New York, 308p.
Simpson, G. G., 1965, *The Geography of Evolution,* Capricorn Books, New York, 249p.
Stebbins, G. L., 1974, *Flowering Plants: Evolution Above the Species Level,* Harvard University Press, Cambridge, Mass., 400p.
Stebbins, G. L., 1981, Why Are There so Many Species of Flowering Plants? *Bio-Science* **31**(8):573-577.
Stebbins, G. L., and F. J. Ayala, 1981, Is a New Evolutionary Synthesis Necessary? *Science* **213**:967-971.
Stoddart, D. R., 1977, Biogeography, *Prog. Phys. Geogr.* **1**:537-543.
Stoddart, D. R., 1978, Biogeography, *Prog. Phys. Geogr.* **2**:514-528.
Stott, P., 1981, *Historical Plant Geography,* Allen and Unwin, London, 151p.
Udvardy, M. D. F., 1969, *Dynamic Zoogeography,* Van Nostrand, New York, 445p.
Volpe, E. P., 1967, *Understanding Evolution,* W. C. Brown, Dubuque, Iowa, 160p.
Voronov, A. G., 1970, Some Problems in the Biogeography of Land Areas, *Soviet Geography* **11**(3):180-188.
Watts, D., 1971, *Principles of Biogeography,* McGraw-Hill, New York, 402p.
Wegener, A., 1966, *The Origin of Continents and Oceans,* translated by John Biram, Dover, New York, 246p.
Wilson, E. O., and W. H. Bossert, 1971, *A Primer on Population Biology,* Sinauer Associates, Stamford, Conn., 192p.
Wulff, E. V., 1943, *An Introduction to Historical Plant Geography,* Chronica Botanica, Waltham, Mass., 223p.

Chapter 2

Variation and Isolation Initiation of Change

Among the more critical elements that set the stage for diversification in organisms is the creation and isolation of genetic variation. Variation's root cause is partly cosmic in that it can be induced by incoming short-wavelength radiation, and partly accidental, in that it can be induced by chemical mistakes. All variation is random and can be traced to fundamental structural and chemical discrepancies in the reproduction of DNA molecules. Once it has formed, a variant may undergo one of several fates: the individual in which it occurs may be purged from the population by premature death to prevent further spread of the trait; the individual may carry the first in a series of imperceptable progressive changes that will ultimately spread throughout a population; or the individual may carry the genetic basis of a crucial attribute that will promote the rise of a new organism. The difference between the second and third of these fates hinges on the degree of present or future isolation that will be experienced by the population containing the variant individual(s). To begin the quest for understanding speciation, therefore, it is appropriate to look at some of the more important aspects of variation and isolation.

ORIGINS OF VARIATION

In nature, all observed diversity arises through variation in individuals and progresses by genetic recombination into populations. By gradual, minute steps, such changes in form and substance weave their way through generations. In any given life form, a comparison of early and recent generations usually reveals gradual differences in their physical appearance. These phenotypic differences represent a process of evolution, or more precisely, phyletic evolution, but they do not necessarily lead to the creation of new organisms. Only at major junctures where differences in form, appearance, or reproductive incompatibility are observed does science recognize the rise of a new species.

Only genetic variation is involved in the process of speciation, but it is not the only form of variation recognized in nature (see Gould and Johnston, 1972). In other words, for any phenotype there are nongenetic (external) and genetic (internal) forces at work. It has been argued by Mayr (1969) and others that in the animal kingdom geographic variation probably does not have a genetic basis but is nonetheless important in explaining differences in the behaviors and homeostatic sensitivities of individuals. It is further possible that such attitudinal or behavioral differences could lead to a condition of reproductive isolation that might eventually result in prominent genetic differences and ultimately in new species.

Genetic Origins

Genetic variation is responsible for the adaptation of populations to their environments. Since no two individuals are genetically identical, the combination of reproductive material to form new offspring ensures that there will always be some measure of variation in a local population. At the same time, since the probability of intrapopulation breeding is greater than that of interpopulation breeding, there is a high degree of genetic stability and similarity in the individuals of a population. Phenotypes display much of this genetic similarity and variation. That is to say, morphological appearance is the outward manifestation of genetic similarities and differences at the population level. These attributes may take the form of spot patterns, numbers of fingers and toes, color of hair and eyes, and similar features. In addition, each attribute can be observed as continuous variation, as in the range of body weights or plant heights within a given taxon, or as discontinuous variation, as in the presence or absence of certain plant or animal markings. In general, the fewer the number of genes involved, the more discontinuous the variation. The array of lady beetles (*Harmonia axyridis*) in Figure 2.1 shows discontinuous color patterns ranging from a few black markings on an orange field (a and b), to large yellow areas on a black field (c and d), and finally, to red markings on a black field (e and f). These variants are all the result of differences in a single gene.

Successful speciation is accompanied by adaptive radiation into a new environment. In animals, radiation is associated with feeding habits, as was amply demonstrated by Charles Darwin in his description of the finches of the Galapagos Islands (Figure 2.2). Some forms of Galapagos finches are ground feeders, whereas others specialize in cactus gardens, and still others are tree feeders. In contrast, adaptive radiation in plants is usually associated with methods of pollination, such as by wind or bees.

Some of the more important aspects of genetic variation can be summarized by introducing the concepts of polymorphism, pleiotropy, mutation, gene flow, and recombination. As shown in Figure 2.1, *polymorphism* involves discontinuous variation within a population. It results

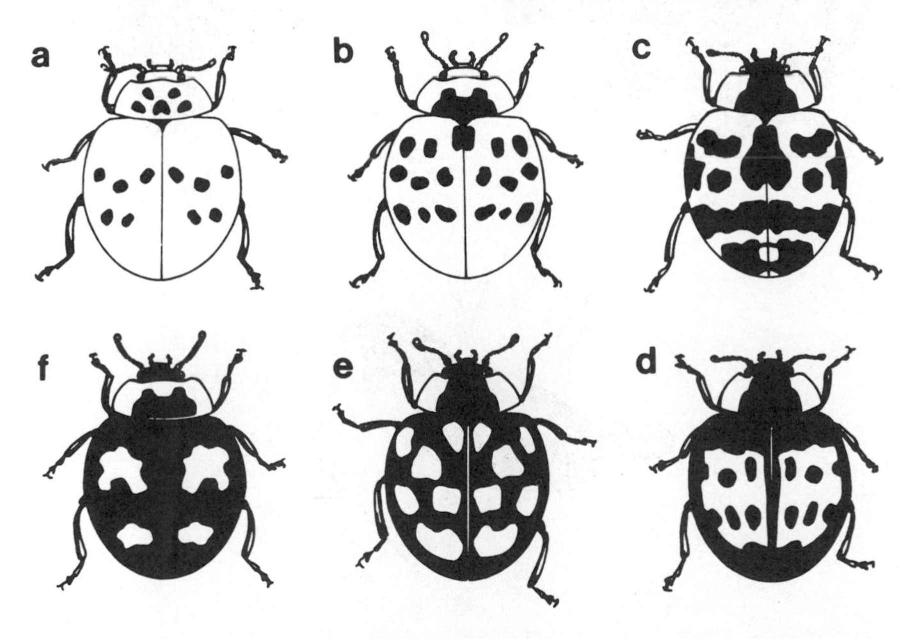

Figure 2.1. Polymorphs of the lady beetle (*Harmonia axyridis*) indigenous to Siberia, Japan, Korea, and China. In natural life, (*a*) is brownish-yellow with black markings; (*b*) yellow-orange with black markings; (*c*) brownish-orange with black; (*d*) black with bright yellow; (*e*) black with red-orange markings; and (*f*) black with red markings. *(From F. J. Ayala, 1978, The Mechanisms of Evolution, Scientific American* **239***(3):57; copyright © 1978 by Scientific American, Inc. All rights reserved.)*

from the simultaneous occurrence of gene arrangements with discontinuous phenotypic effects (e.g., blue versus brown eyes). These effects are not restricted to structural characters but may also include physiological and behavioral characters, provided they are genetically controlled and discontinuous. Male and female are examples of two polymorphs of the same population. Polymorphism should not be confused with *polytypic.* This latter term refers to the recognition of several different units within a given taxonomic level. A genus having several species is said to be

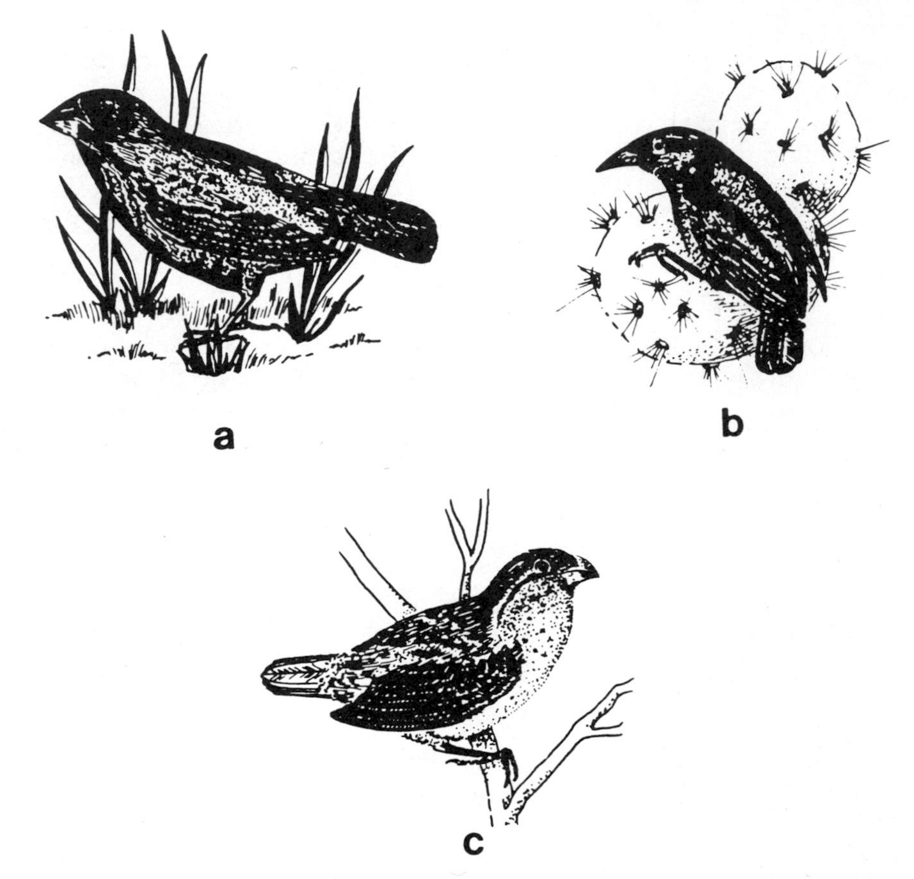

Figure 2.2. Variation in three Galapagos finches that has led to their adaptive radiation into different feeding habits: (*a*) *Geospiza magni rostris,* seed-eating ground finch; (*b*) *Geospiza scandens,* cactus-feeding ground finch; (c) *Camarhynchus psittacula,* insectivorous tree finch. *(Modified from V. Grant, 1963, The Origin of Adaptations, Columbia University Press, New York, p. 520; reproduced with permission of Columbia University Press.)*

polytypic. A local population of any of these species may exhibit polymorphism.

Pleiotropy is a genetic process behind the apparent explanation for why some plants and animals within the same population have different markings. For example, some flower petals may be pink with three red spots at the base, whereas others may be red with five pink spots; some tropical birds may be iridescent green with golden throats, whereas others may be green with greenish-gold throats. One can hardly help but wonder if these kinds of variations have any significance. A past view was that every mark had some survival value associated with it, but the modern synthesis views such variations as important, but not necessarily because they have direct survival value. Rather, the consensus is that they are no more than visual manifestations of variations involving a series of genes whose combinations have multiple and intricately interwoven effects. Genes are not simply molds that stamp out new genes like an assembly line. In other words, genes that control animal size may also effect color (say, of the eyes or the hair), and these in turn may have some selective significance. Those genes that are most important in evolution or selection (such as the ones governing fertility, longevity, or tolerance) may influence a whole series of other attributes that can be seen and measured but are really only consequential. Such pleiotropic characters are often referred to as neutral polymorphs.

Mutation is the basis for the process of evolution. Most botanists agree that it does not occur in giant discontinuous steps, as was once thought by Lamarkian evolutionary theory, but rather in small, imperceptable, random ways. Some zoologists, on the other hand, believe that major evolutionary advances can take place in an individual, and it is for this reason that the fossil record contains gaps. In either case, mutations cannot be predicted at any given gene locus. However, since there are thousands of gene loci at which potential mutations can occur, the chances are very good that mutations provide an enormous and steady source of genetic variation.

Mutation is most often studied in closed laboratory populations, so that observed differences can be attributed to genetic variation within those populations. In nature, however, populations are seldom completely isolated. Most often there is a gene flow between neighboring populations, the strength of which declines with increasing distance or effectiveness of barriers. Gene flow is the means by which genetic variation spreads among neighboring populations. It is in this light that one can imagine the immense biological impact of fifteenth century European contact in the New World. North American populations that were relatively free to intermingle, but which until 1492 were isolated from European populations, provided the basis for a dramatic increase in genetic variation. All of the polymorphic, pleiotropic, and mutant content of both genetic stocks were mingled to produce new offspring (the mestizo) having noticeably different attributes.

Gene flow has been largely neglected as a line of scientific inquiry because of the difficulty in measuring it. It is estimated that between 90 and 99 percent of all "new" genes in a population are contributed by migrants. Gene flow is one of the most essential elements of evolution, affecting, as it does, geographic variation, ecotypic adaptation, speciation, and long-range evolution. The flow of genes across populations is essentially a retarding influence in speciation because of the synthesizing and homogenizing effect on the process. Although there may be observable differences between individuals and populations, there is a tight genetic bond that characterizes and stabilizes the life form.

Recombination is the mechanism that allows genetic variation to occur in sexual reproduction. During meiosis, maternal and paternal chromosomes pair, break in several places, and exchange pieces (called *crossing over*). Through this recombination process a population can undergo genetic variation for many generations without any new input from mutation or gene flow.

In sexually reproducing organisms having a long time span per generation, recombination is the source of new genotypes. In humans, a mutation could be tested for selective significance only once in approximately 25 years (one generation); in mature forests, only once in 100 to 200 years. If mutations were the only way for the population to adapt, the process would be exceedingly slow—slower, probably, than most changes in the environment with which organisms are trying to keep pace. Under these conditions, therefore, the mutation process is considered to be negligible as a determinant of evolutionary change. In asexual reproduction, on the other hand, mutation provides the ability for microorganisms that produce a new generation every few minutes to adapt rapidly to changing environments.

Variation within a population is maintained by the presence of several genotypes at different levels of adaptation. During periods of environmental stress, the less well-adapted types lose population, whereas the better-adapted types survive and perhaps increase. Genotypic variation also permits greater utilization of the environment, because such variety allows organisms to colonize marginal habitats, or altogether new habitats, that become available through geologic or geomorphic events. In other words, genotypic variation within a species counteracts a tendency toward the creation of new species and contributes elasticity to a population. The history of *Homo sapiens,* from approximately the last three million years, is an excellent example of a highly variable, elastic form that has been able to radiate into virtually all terrestrial environments. It is even believed that *Homo sapiens* was intimately involved in the extinction of other close relatives in the genus, namely *Homo erectus* (Leakey, 1977).

That adaptation is maintained through natural selection, which always involves an elimination of genotypes, raises a most important question: How can a population retain an adequate supply of genetic variability when such variability is continuously depleted by natural selection? One answer is that it does so through an equilibrium established between those mechanisms that produce variation (mutation, recombination, and gene flow), and those mechanisms that reduce variation (selection, chance elimination, and extinction), and those mechanisms that protect existing variation (e.g., ecology).

Geographic Origins

Geographic variation is an important element in the total picture of differences observed in organisms. Since no two individuals are identical in all genetic respects, it follows that there must be variation within populations. Similarly, there must be variation between populations. *Geographic variation* is defined as the set of observable differences between spatially segregated populations. Sometimes the terms *spatial variation* and *microgeographic variation* are substituted. Basically, differences in appearance between populations are more readily observed, the greater the distance between them.

Geographic variation was first recognized by taxonomists who, in the course of time and from observations over large areas, began to notice differences between individuals of a given species. At first the differences were ignored or were believed to have no significance. In many cases the individuals were so different that new species were described. Earlier views held that most variations, like those shown in Figure 2.3 for *Achillea lanulosa,* were irrelevant, and that in any case, variation below the generic level was nonadaptive. A growing body of knowledge, however, leads to the inescapable conclusion that geographic variation is not only important, but is also the rule rather than the exception at the species level. Certainly no one would have difficulty separating European, African, and Asian populations of *Homo sapiens.* Similar variations exist in other plant and animal life forms, despite our fallible ability to record them.

Several explanations have been given for those cases where spatial variation appears to be absent. One is that the total geographic range is too small, and the environment too uniform, to display any significant variation. In these cases one might question whether the species is expanding its range and will eventually develop variation between distant populations, or whether the range is diminishing and variation is being eliminated.

A second explanation suggests that the means of dispersal is so great that in practical terms the species can be said to be "cosmopolitan." The common dandelion (*Taraxacum officinale*) is often cited as a cosmopolitan species, yet close examination of widely separated populations shows that they are not alike in every respect.

Finally, there is some basis for arguing that the absence of variation indicates either a stable phenotype or a stable genotype. A high degree of homeostasis may buffer the individuals and populations from change that in distant populations, is sufficient enough to be recognized (see Langley, 1965). In ancient forms like the lung fish, some observers explain the stability as a loss of mutability. One should regard genetic stability as a system that is highly buffered against change, but is capable of change if necessary. In the plant world, species that propagate vegetatively represent genetically stable systems. They tend to have more restricted ranges and to be more susceptible to environmental change. If this mode of reproduction is obligate rather than faculative,

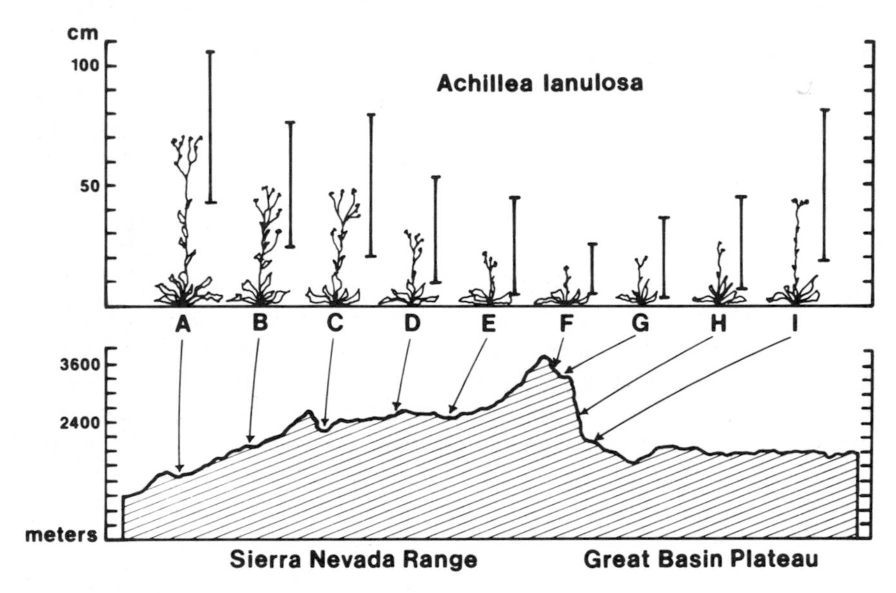

Figure 2.3. Geographic variations in height between populations of *Achillea lanulosa* collected along a transect of the Sierra Nevada Range in California. The vertical bars alongside each plant indicate the range of height variability within a local population of approximately 60 plants. (*A*) Mather, California; (*B*) Aspen Valley; (C)Yosemite Creek; (*D*) Tenaya Lake; (*E*) Tuolumme Meadows; (*F*) Big Horn Lake; (*G*) Timberline; (*H*) Conway Summit; (*I*) Leevining. *(Modified from A. S. Boughey, 1968, Ecology of Populations, Macmillan, London, p. 77.)*

there is little or no genetic variation, and correspondingly little survival potential should change become necessary.

MODES OF ISOLATION

Whenever a population is cut off geographically, genetically, or mechanically, it is said to be *isolated.* Of course, isolation is rarely complete in any one of these senses, so scientists have learned to study the imperfections of nature and deal in reduced rather than zero probabilities. Thus, any type of isolation may reduce but not completely eliminate gene flow between populations. In most cases, it is a combination of isolating mechanisms that effectively cuts off one population from another. Isolating mechanisms evolve over long periods as a consequence of two populations accumulating sufficient genetic differences to make them reproductively discrete. There usually are spatial and temporal components to the separation; otherwise, normal mixing in the gene pool would promote interfertility.

In general, three forms of isolation are recognized: *geographic, ecologic,* and *reproductive.* The last is generally regarded as essential for defining a new species. Reproductive isolation is further divided into external and internal factors, as highlighted in Table 2.1. External factors keep the would-be parents apart; internal factors involve those mechanisms that prevent successful progeny.

In the following examples one should note the complexity of isolating mechanisms and the thoroughness of nature in providing cascading mechanisms that guarantee isolation and thereby ensure for the individual species involved the relative purity of its genetic composition.

Populations of *Achillea lanulosa* (see Figure 2.3) are interfertile but are geographically and ecologically separated. In time, their gene pools may become so distinct that individuals representing eastern and western populations may cease to be interfertile. When this occurs, either a separate species or a subspecies can be recognized.

Red oak *(Quercus rubra)* and white oak *(Q. alba)* maintain themselves as more or less distinct populations because, even though they are interfertile, their F_1 (first generation) progeny are limited in habitat breadth. This condition is a combination of ecological isolation and, perhaps, hybrid breakdown, an internal form of reproductive isolation.

Mallard *(Anas platyrhynchos)* and pintail *(A. acuta)* ducks share a common geographic range and habitat but, for the most part, are reproductively isolated from each other because they do not recognize

Table 2.1. External and Internal Factors in Reproductive Isolation

External factors

Mechanical isolation: some structural difference makes reproduction physically impossible between individuals of interfertile populations.

Ethological isolation: some behavioral characteristics such as courtship dances, and the like, keep would-be parents apart.

Temporal isolation: potential parents are kept apart by having different phenological cycles.

Gametic isolation: chemical differences make reproduction impossible.

Internal factors

Incompatibility: parents are interfertile, but fail to produce offspring.

Hybrid inviability: the offspring do not survive.

Hybrid sterility: the offspring survive, but cannot reproduce.

Hybrid breakdown: the first generation survives, but fewer and fewer offspring survive from succeeding generations.

each others' "overtures." There is some evidence that these two "species" interbreed successfully, and that the F_1 hybrids survive. The F_2 generation, however, is not viable. This situation combines ethological isolation and hybrid breakdown.

Salvia apiana (outwash sage) and *S. mellifera* (foothill sage) are sympatric over Southern California, but they flower in different seasons and are normally pollinated by different insects. In addition, all naturally produced F_1 hybrids are infertile. Laboratory hybrids are semifertile, but F_2 and F_3 generations are less so. This example demonstrates a combination of mechanical and temporal isolation together with hybrid breakdown.

Some tropical orchids (e.g., *Dendrobium*) bloom for only one day in response to a particular sequence of temperatures. Some have a 9-day lag between this temperature sequence and the blossom; others 10 days; others 11 days. The whole population blooms simultaneously. At the other extreme there are species that bloom erratically, or only once in several years. Reproduction with the "wrong" individual, therefore, becomes a purely chance occurrence.

Geographic isolation is said to exist when populations of the same species are separated by a distance great enough to seriously reduce, or effectively eliminate, gene flow by migration or dispersal. Solitary mountain tops, like Mount Kilimanjaro in East Africa, and islands, like Pitcairn in the South Pacific, are easily understandable examples of geographic isolation. Intuitively, one can agree that few species could arrive there safely, except by chance or through human introduction. The gene pools of species already there are developing in the virtual absence of outside genetic material. Less obvious cases of geographic isolation result from fluctuating climates, rising and falling sea levels, glacial transgressions, drifting continents, and the like. Yet each of these processes aids in the creation and breakdown of barriers and migration routes; thus each influences tremendously the ebb and flow of gene pools.

The distances required to achieve isolation depend entirely on the dispersal mechanisms of individual plants and animals (discussed in Chapter 5). Much has been written regarding the ability of species to propagate after disseminules have crossed great distances.

For this reason some students reject the importance of geographic isolation as a prerequisite for speciation. They argue that distance alone cannot effectively reduce or eliminate gene flow. Others, for example Ehrlich and Raven (1969), argue that effective pollen dispersal beyond 1,000 meters is rare, even though such pollen may be carried for thousands of kilometers in upper air currents. If this argument is true, then geographic isolation should be relatively easy to achieve, even in the higher plants. In either case, the question at issue involves the probability of successful migration, multiplied by the probability of subsequent interbreeding between migrant and existing stock. These probabilities may be negligible among animal populations characterized by territorial behavior, but considerably higher among wind-pollinated plants.

Ecological isolation refers to the separation of species within geographically confined communities. For all intents and purposes, the local populations of the two species intermingle, and it is difficult to explain how the two forms initiated reproductive isolation, let alone how they maintain it. One would expect gene flow to counteract any tendency in that direction. In geographically coincident forms, gene flow may not have been reduced by distance so much as by the fact that the two species occupy different niches. It is possible for two closely related species to coexist in nearly identical ranges yet never interbreed because their ecological requirements for survival are different. The finches of the Galapagos Islands shown in Figure 2.2 are a good illustration of ecological segregation that has led to reproductive isolation. Usually ecological isolation is observed as a *fait accompli* at the species level. Populations of the same species are difficult to describe as ecologically segregated since, by definition, their ability to interbreed still exists.

In general terms, ecological isolation can be split into its component concepts of *niche* and *habitat*. *Niche* refers to the mode of food acquisition, or preferred food sources, and *habitat* is the site at which such activity takes place. The wider the niche and habitat, the greater the ability of a species to cope with change, and the greater its potential for maintaining contact with sister populations. Narrower limits generally signal increased ecological specialization which, over many generations, may lead to segregation of populations. The koala bear (*Phascolarctos cinereus*) of Australia, for example, has its rather specialized steady diet of eucalyptus leaves of a specific kind, and it would have a slim chance for survival in an environment where they were absent. Its niche is very narrow. Furthermore, despite the more or less continuous distribution of eucalyptus species in eastern Australia, the local absence of the right species serves as an effective means to separate bear populations. The Koala's habitat is also very narrow.

Horses *(Equus)*, on the other hand, are capable of depending on a far wider range of resources. Their ability to survive on a variety of grass and herbaceous species means thcy can penetrate new environments wherever those resources exist. In times of environmental stress as, for example, during Pleistocene glacial episodes, populations were able to migrate to regions within their ecological tolerance. Populations comprising the species have a greater mobility and are therefore less ecologically segregated.

It must be remembered that species with narrow ecological requirements are no less well adapted than species with wider tolerances, given the availability of resources. As will be shown in Chapter 3, specialization may be regarded as a result of finely tuned adaptation which, after all, is one of the goals of natural selection. It is only in circumstances of sudden environmental change, which may include the introduction of competition, that niche and habitat breadth become important tools of survival.

SUMMARY

The diversity in nature is a complex mixture of nongenetic, genetic, and geographic factors. Some observed variation has recognized survival and adaptive value, but some of it is also a consequence of gene complexes that control other important life functions. All genetic variation originates in mutations and is passed into the gene complex of the population, where it is acted upon through natural selection. In general, phenotypes are the visible outward appearance of genotypes interacting with the environment. These recognizable genotypes arise by gene recombination within a population, although a significant measure of variation may also result from gene flow from immigrants.

The following generalizations can be made based upon current understanding of population genetics. It is apparent from them that both selection and adaptation become crucial topics for consideration in the overall process of speciation.

Phenotypes are produced by genotypes interacting with the environment.
Genotypes are the result of gene recombination within a population.

Most genotypic variation in a population is due to gene flow and recombination, but all of it originates through mutation.

Geographic isolation is purely extrinsic and completely reversible. It does not, by itself, lead to speciation. Its role is to permit undisturbed genetic reconstruction of populations.

There is a tendency in the integrated gene complex to establish ever finer cohesion—in other words, to be finely tuned to survive short-term changes in the environment. Some species are so successful at accomplishing this fine tuning that they reach a point of "evolutionary inertia."

Speciation may therefore be regarded as a process of rejuvenation—an escape from being overadapted.

REFERENCES

Ayala, F. J., 1978, The Mechanisms of Evolution, *Scientific American* **239**(3):56-69.

Boughey, A. S., 1968, *Ecology of Populations,* Macmillan, London, 135p.

Ehrlich, P. R., and P. H. Raven, 1969, Differentiation of Populations, *Science,* **165**:1228-1232.

Gould, S. T., and R. F. Johnston, 1972, Geographic Variation, *Annu. Rev. of Ecol. and Syst.* **3**:457-498.

Grant, V., 1963, *The Origin of Adaptations*, Columbia University Press, New York, 606p.

Langley, L. L., 1965, *Homeostasis,* Van Nostrand, New York, 114p.

Leakey, R. E., 1977, *Origins,* E. P. Dutton, New York, 264p.

Mayr, E., 1969, *Animal Species and Evolution,* Harvard University Press, Cambridge, Mass., 797p.

FURTHER READING

Ayala, F. J., 1968, Genotype, Environment, and Population Numbers, *Science* **162**:1453-1459.

Endler, J. A., 1973, Gene Flow and Population Differentiation, *Science* **179**:243-250.

Kolata, G. B., 1974, Population Genetics: Reevaluation of Genetic Variation, *Science,* **184**:452-454.

Levin, D. A., 1979, The Nature of Plant Species, *Science,* **204**:381-384.

Mayr, E., 1947, Ecological Factors in Speciation, *Evolution* **1**:263-288.

Raup, D. M., 1977, Probabilistic Models in Evolutionary Paleobiology, *American Scientist* **65**(1):50-57.

Schall, J. J., and E. R. Pianka, 1978, Geographical Trends in Numbers of Species, *Science* **201**:679-686.

Wills, C., 1981, *Genetic Variability,* Oxford University Press, New York, 312p.

Yablokov, A. V., 1974, *Variability of Mammals,* Amerind Press, New Delhi, 350p.

Chapter 3

Selection, Adaptation, and Evolutionary Rate
The Fixing of Change

The creation of variation in gene complexes, and its spread throughout a population, sets the stage for differentiation of forms and, ultimately, for increasing diversity. Variation is required to ensure each life form's ability to adapt to continuously changing environments. Although a perfectly adapted state is never achieved, it is evident, judging by population numbers and their geographic ranges, that some species are more successful than others. A species is adapted to its place in nature only by degree, and that degree is determined by the intensity and direction of selection, balanced by the rate of environmental change.

Through the process of selection, new gene complexes can increase their frequency in a population. If a given population then becomes partially or completely isolated, selection pressure can drive sets of favorable complexes toward eventual reproductive isolation from sister populations. This chapter describes some of the finer points surrounding the processes of selection, adaptation, and evolutionary rate, particularly the way these processes influence peripheral populations.

SELECTION

Two types of selection are recognized: *genetic selection,* which refers to the internal factors influencing genotypes within populations of a species; and *environmental selection,* which describes those external influences of nature that promote or restrict the geographic spread of genotypes.

Genetic Selection

Since the earliest years of the Darwinian age, several alternate "theories" of selection have arisen. Not all have enjoyed equal popularity, and some have appeared with little or no foundation whatever. Three of the better-known, but now discarded views were expressed in the Lamarkian, preadaptation, and mutationist theories. The Lamarkian theory held that adaptive characters are caused, rather than selected, by environmental influences, as a result of individuals coping with life's adversities. A classic example is given in the case of giraffe (*Giraffe camelopardalis*), where the neck was said to have lengthened through repeated generations of individuals stretching for food. No mechanism was ever suggested for this process. Nor was a mechanism suggested for the preadaptation theory. This view held that adaptive characters arise in anticipation of coming environmental changes, instead of in response to them. Finally, the mutationist theory held that changes arise spontaneously from within the organism, and that their adaptive value is incidental. The problem was that this theory could not reconcile the random nature of mutation with the obviously nonrandom process of adaptation.

Darwinian selection and its more recent refinement, genetic selection, contains the substance of a viable theory, and the latter, in particular, provides a mechanism for progressive change in organisms. Darwin's concept of the survival of the fittest is really a process of differential survival. It states that among all individuals produced in nature, some die sooner and others survive longer. Critics have objected to this restricted view because it is not "creative;" that is, selection only eliminates the disadvantageous. It fails to produce something new, and under these circumstances it only maintains a status quo. Nevertheless, even in his ignorance of genes and genetic complexes, Darwin recognized that adaptation results from material interaction of organisms and environment. The mechanism of that interaction is selection, and the role of selection is the production of adaptation.

Genetic selection is a process whereby systematic changes in gene frequencies occur from one generation to the next. Simpson (1953) defines selection as a process tending to produce systematic heritable change in populations between one generation and the next. Defined in this manner, selection includes a strong non-Darwinian component that is creative and permits progressive evolution in a population. It would be possible by this definition for all individuals in a population to have exactly the same life span (no differential mortality), but for certain genotypes to have twice the fertility rate of the others.

This broader view of selection further implies that the process favors reproductive adaptation, but not that it favors all advantageous characters. It would be quite possible for certain advantageous attributes to be selected against, if they were linked to a gene complex that resulted in a shorter life span. In sum, genetic selection is a process that has both intensity and direction.

Intensity: Given sufficient variation in a population that is not already optimally adapted to its environment, one can imagine a selection pressure that drives toward better adaptation. The intensity of that pressure should be greater the farther the group is from optimum. As the state of being perfectly adapted to an environment is reached, the intensity should decline. In other words, selection intensity varies as a function of the population's distance from optimum adaptation to a given set of environmental factors. There is evolutionary evidence that most of the time, adaptation maintains a steady pace with changes in the environment. Consequently, it has been said that environmental changes control the rate of evolution and that the mechanism of that control is selection.

If placed on a scale, the intensity of selection for a given character may be viewed as ranging from positive (+1) for perfect favorable selection, through neutral (0), to negative (−1) for complete adverse selection. In natural populations, few characters can be shown to have such strong selective pressures as +1 or −1. Most intensities are so close to neutrality (.001, .0001, or less) that one wonders how the process can account for adaptation or evolutionary change. In cases of mimicry where the resemblance between two insects is nearly perfect, one could argue that the final stages of that trend toward adaptedness had no selective advantage. However, Simpson (1965) points out that over many generations, there is a significant difference between a zero advantage and a very slight one.

All genetic variation existing in a population occurs in the mutations for specific traits. Favorable traits are maintained because their frequency of occurrence is greater than the frequencies for more neutral traits. Blond hair and blue eyes are typical in nordic populations, but certainly other traits are possible. If more neutral genes carry traits that do not adversely influence longevity, fertility, or other characters that greatly affect survival, they may persist in a population in an almost "preadaptive" sense; although the variation did not arise in anticipation of a particular environment but, rather, proved to be adaptive after the environment changed. The intensity of selection for the once more neutral trait is directly related to how adaptive it is to the new environment.

Direction: Selection intensity can be viewed as a rate that effects how frequently genetic combinations occur in a population. As the population continues to become finely tuned to its habitat, however, the direction of change must also be considered. Three types of direction can be described: centripetal, centrifugal, and linear. These are illustrated in Figure 3.1.

In *centripetal selection* there is a tendency for greater reproduction of the "typical" characters in a population than for any newly arising variants. Selection favors the status quo, presumably because all other variation is less than optimum for that environment. This situation generally leads to specialization and a gradual reduction in the number of less well-adapted individuals. It is typical of environments that are comparatively stable.

Centrifugal selection, on the other hand, considers the typical character to be terribly ill adapted. Selection favors a continual divergence from the average individual. It is illogical, however, for a typical individual to be highly ill adapted, except in the case of an abrupt environmental change where the typical character is suddenly highly disfavored.

Finally, in the case of *linear selection,* there are gradual shifts away from a modal character toward one that is slightly more adapted. When this happens there is a corresponding shift in gene frequencies to reflect the finer adjustment. In nature, linear selection is the most frequently observed situation, and one that arises in response to gradual, but continual, changes in the external environment. Progressive shifts occur at about the same rate as changes in the environment. Any considerable lag might leave a population behind in its adaptedness, resulting in either extinction or a new direction in evolution.

By these lines of reasoning, one can begin to understand the complex interrelations between selection and evolution, adaptation, and the environment. Evolution is faster in populations that are least adapted, provided the population size and genetic variability are sufficient to overcome extinction. The least well-adapted individuals in a population will be eliminated rapidly by environmental factors, thereby increasing the gene frequencies in the surviving, better-adapted members. Through this process environment controls the rate of evolution, and selection is the mechanism. Environmental changes are continually taking place, and as a consequence, adaptation is also continuous through selection of the most competitive individuals. Without a process for eliminating ill-adapted individuals, there would be no means for eliminating their contribution to the gene pool. In successive generations such adversities could accumulate and reduce the viability of the entire species. Selection, like mutation, occurs in small stages in individuals or groups; but whereas mutation is random, selection exhibits both rate and direction.

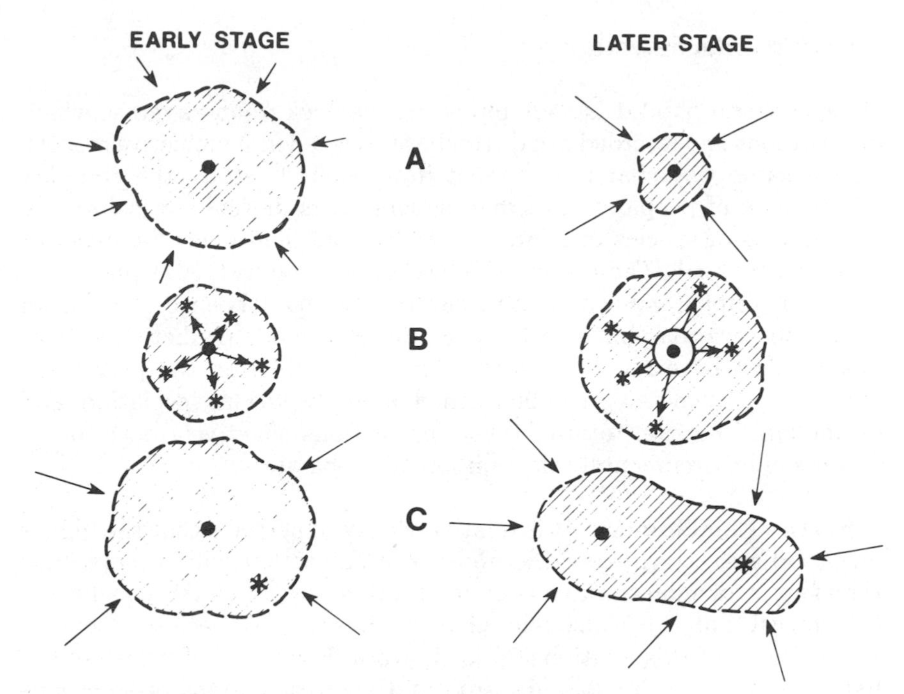

Figure 3.1. Models of direction in genetic selection.
(*A*) *Centripetal selection* (comparatively stable environments). Left: Initial stage showing a "typical" character (black dot) that is always favorably selected in preference to the variants of that character (diagonal lines). Right: In time the variability of that character will be reduced and the selection pressure for the "typical" character will get stronger. The length of the arrows is proportional to the intensity of selection.
(*B*) *Centrifugal selection* (abrupt environmental change). Left: Initial stage showing a number of individuals (crosses) having characters more adaptive than the typical character (black dot at center). Right: through time there is a trend toward selection of the more adaptive characters and against the typical character. Eventually this could lead toward extinction of the typical form.
(*C*) *Linear selection* (gradual environmental change). Left: a case similar to centripetal selection, but where "typical" character (black dot) is replaced by a gradual shift toward a character within the range of variability (diagonal lines) that is slightly better-adapted. Right: in time there is a selection pressure toward the better-adapted state, as shown by the length of the arrows. In this model one gets the feeling for "plasticity" in the genetic make-up of the population and species. Adaptation can be viewed as a continuous pulsation in response to selection.

Environmental Selection

Biological and habitat islands represent the background against which observations and recorded trends can be abstracted into predictive models. In the same way that it is almost impossible to study the complex interactions of people in an urban setting, it is difficult to isolate the interactions of species in a large uninterrupted land mass. MacArthur and Wilson's book, *The Theory of Island Biogeography* (1967), presents a series of graphs, equations, and curves that have been derived from island observations but can be generalized to make predictions about islands that have not been studied as thoroughly. In the process of generalizing, something can be learned about evolution, speciation, and extinction. The sum total of these interactions effectively describes a process of environmental selection for filling habitats.

Species Equilibrium: It is relatively easy to perceive that the bigger an island area, the greater the number of habitats it will possess. Size therefore, is an indirect measure of habitat diversity. From this conclusion, it is logical that size is also related to the diversity of species present. A simple curve of this relationship is shown in Figure 3.2. Comparison of lists of species from islands around the world suggests that this relationship is true, but that islands generally only support a finite number of organisms in small populations.

As with island size, it is also intuitive that in the process of occupying

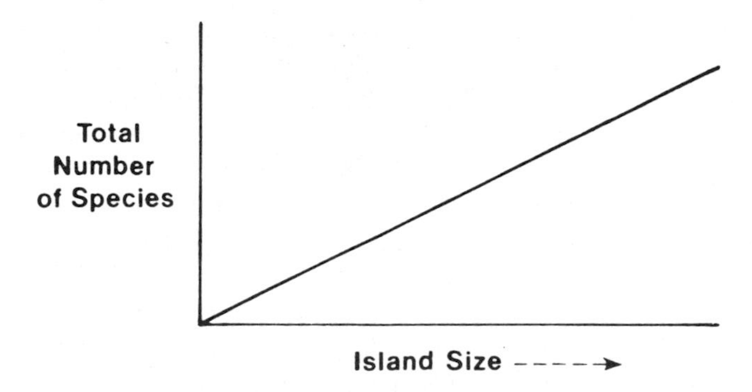

Figure 3.2. Relationship of geographic area to number of species. It is generally true that as area increases, the number of habitats and niches also increases to provide opportunities for more species. *(From R. H. MacArthur and E. O. Wilson, 1967, The Theory of Island Biogeography, Princeton University Press, Princeton, N. J., p. 8; copyright © 1967 by Princeton University Press. Reprinted by permission.)*

any given island, some species will survive and adapt, and others will not. There are habitat and niche factors, competition and tolerance levels, predator/prey relationships, and other considerations. In addition, the rate of adaptation of the colonizers to their new environment will be highly variable. It is reasonable to expect, therefore, that the immigration and extinction rates of island occupants will approach an equilibrium. There will also be a tendency toward speciation within the confines of the island. Basically, the model for this idea is given in Figure 3.3. Initially, immigration is high and extinction is low; but as the number of species increases, extinction increases and immigration decreases.

If the idea of the distance from a source of colonizing species is inserted into the model, Figure 3.4 can be generated from Figure 3.3. The size of the island can also be incorporated insofar as it is a function of the rates of immigration and extinction. Note that the rate of immigration decreases as distance from the source increases, and that the rate of extinction decreases as island size increases. Note also that in this construct, a far, large island may support as many species as a near, small island (their curves intersect respectively at $\hat{S}_0$). A near, large island ($\hat{S}_1$) supports a larger number of species at equilibrium than a small, far island ($\hat{S}_2$).

Colonization Rate: The major difficulty with the above approach is that neither the immigration nor extinction curves can be constructed, nor their rates calculated. In nature, what is actually observed is the result of these interactions, not the process itself. Consequently, what is needed is an ability to relate observations to some measure that is, itself, a result of immigration and extinction. This relation can be made by reference to a measure called the rate of colonization.

The colonization process is dynamic. Both the immigration and extinction rates vary depending upon the number of species already present. In practice, it is difficult to measure the immigration rate at any single moment in time when a number of species are already present. The same is true of the rate of extinction. These problems result from an inability to record each incoming and outgoing propagule. The rate of colonization can be defined, however, as

$$S_t = \int_0^t (I - E)dt$$

where S_t = total species observed at time t, I = number of immigrant species, and E = number of species becoming extinct.

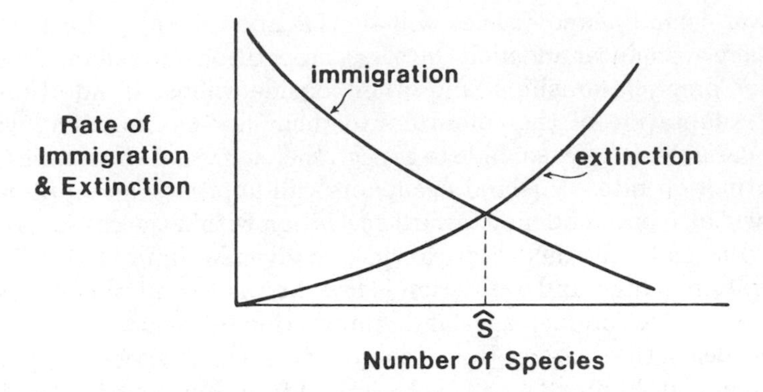

Figure 3.3. Relationship of immigration and extinction rates. When there are few species present, the rate of immigration is high. As greater numbers of species become established, the prospects for extinction of some forms increase. The equilibrium number of species, $\hat{S}$, is defined as that point where the curves cross. *(From R. H. MacArthur and E. O. Wilson, 1967, The Theory of Island Biogeography, Princeton University Press, Princeton, N. J., p. 21; copyright © 1967 by Princeton University Press. Reprinted by permission.)*

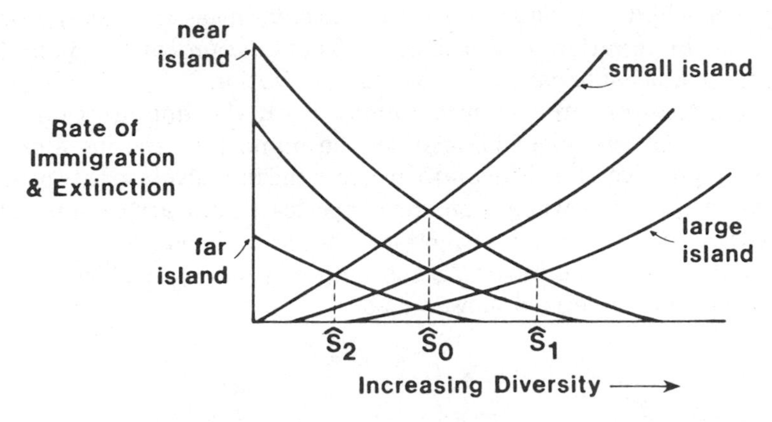

Figure 3.4. Relationship of species equilibrium to source of colonizing species and size of geographic area. *(From R. H. MacArthur and E. O. Wilson, 1967, The Theory of Island Biogeography, Princeton University Press, Princeton, N. J., p. 22; copyright © 1967 by Princeton University Press. Reprinted by permission.)*

The actual shape of the curve for colonization may vary from place to place, or between major taxonomic groups in the same place. In other words, when the equation is plotted using actual data, it is possible to observe either an exponential or a logistic curve.

The theoretical rate of colonization can be compared to actual observed rates where data have been obtained, as for example in places like Krakatau in Java (Figure 3.5). If an actual colonization rate can be

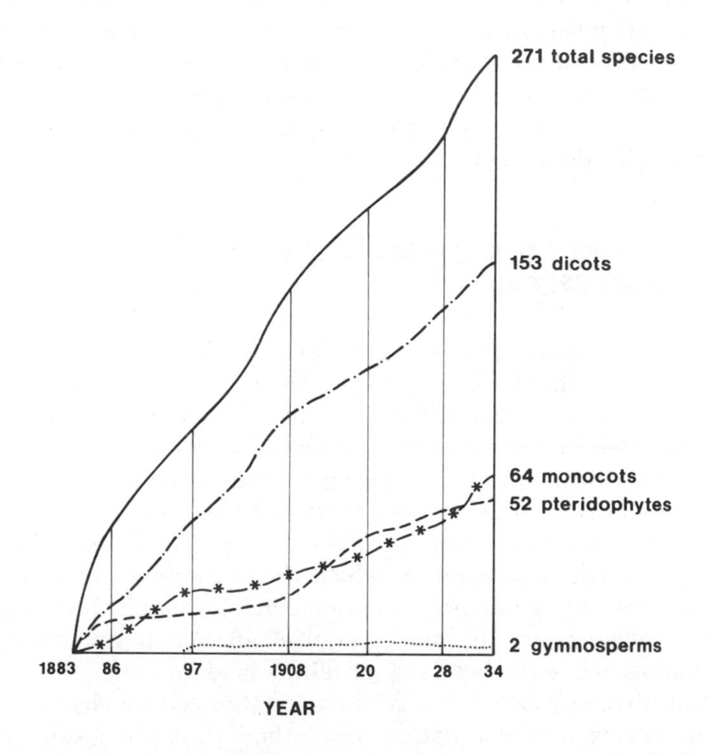

Figure 3.5. Actual colonization curves for major groups in the Krakatau flora. The complete destruction of vegetation on Krakatau following the volcanic eruption around 1880 provided a natural laboratory for measuring the rate of colonization of a barren island. Periodic surveys of the island's vegetation between 1883 and 1934 have provided the data for these curves. Note that the curve for gymnosperms is virtually flat since initial immigration in 1897, but that there has been a rather steady increase in angiosperms (dicots and monocots) since colonization began in 1883. *(From R. H. MacArthur and E. O. Wilson, 1967, The Theory of Island Biogeography, Princeton University Press, Princeton, N. J.; p. 49; copyright © 1967 by Princeton University Press. Reprinted by permission.)*

obtained, one can estimate the immigration and extinction rates through time. However, there are several difficulties with this simplistic model. Definitions for immigration and extinction and for breeding populations need to be formulated. Moreover, it is necessary to assess successional stages and their influence on immigration and extinction. In the case of the flora of Krakatau, for the colonization to continually increase, all forms that became extinct must have been immediately replaced so that at no time was a decrease observed. MacArthur and Wilson (1967) explain this phenomenon by two possible mechanisms: (a) the proximity of Sumatra and Java could have provided an unending reservoir of immigrant material; and/or (b) successional or pioneer species did not become extinct, but merely shifted their habitat preferences as the biotic communities developed.

ADAPTATION OF POPULATIONS AND SPECIES

The process of adaptation involves both the organism and the environment. In this sense, *environment* must include the physical milieu where the population resides, together with that surrounding it, and the internal environment or range of variability of the individuals comprising the population. An adaptive trait, in contrast, is an attribute within an individual that proves to be advantageous in a selective sense. The process of adaptation involves the spread of such an attribute throughout the population in which the individual lives. This process is considered to be universal in organisms, although there are different degrees of success. The surest sign of success is an increase in the population, while the best sign of failure is extinction.

In an altruistic sense, it is conceivable that certain characters may be advantageous to a population as a whole, but not necessarily to its individuals—as in bee and termite societies organized to ensure only a small number of reproductive members. The opposite case may also be true. Groups, as well as individuals, may be favored by narrow variation, as long as conditions are stable; and both may be favored by wide variation when conditions are changing. Since the environment, in the broadest sense, is constantly changing, some degree of variation is advantageous. Selection acts on that variation to produce the best-adapted forms. The general sequence of events and processes can be conceptualized in the trends shown in Table 3.1.

The question of whether all visible characters in a phenotype are adaptive has already been considered. Some students maintain that all characters are adaptive, since none has been proven not to be. By the

Table 3.1 Steps in the General Process of Speciation

Interacting elements	*Through process of*	*produces*
Gene pools with environment →	Variability →	Genotypes
Genotypes with environment →	Selection →	Phenotypes
Phenotypes with environment →	Selection →	Population
Population with environment →	Adaptation →	Taxonomic species

same reasoning, others argue that few characters have been proven to be adaptive, and therefore the concept of universal adaptiveness is not proven.

Measures of Adaptation

Ultimately the survival of a species in its environment depends upon its ability to reproduce; its ability to avoid annihilation by its predators, and its ability to compete for scarce resources. In recent years, considerable model building has gone on in these areas of population biology, with the aim of providing a more holistic approach to traditional concepts in biogeography. In other words, there is a tendency to take a step back from "one-factor" ecology and move toward systems analysis. Wilson and Bossert's book, *A Primer on Population Biology* (1971) provides a good basis for the following introduction to models of population growth, predation theory, and competition theory.

Population Growth: In the absence of other influences, growth in any population can be shown to follow one of two basic patterns. Variations on these patterns are introduced as a consequence of competition for resources and of predator/prey relationships. An exponential mode of increase can be observed in small, isolated populations, if resources are abundant and losses few. More commonly, a logistic growth curve is observed, because in most instances resources are limited.

Basically, the exponential growth model is symbolized in equation 3.1 of Figure 3.6. When plotted, it shows an ever increasing population over time. In the long term, however, it is unrealistic to believe that any population, including that of humankind, can continue to expand in this fashion. When it occurs, exponential growth usually exists for short periods of time in small populations, but sometimes, it also occurs in larger ones that for some reason or another, are expanding their range, niche, or habitat.

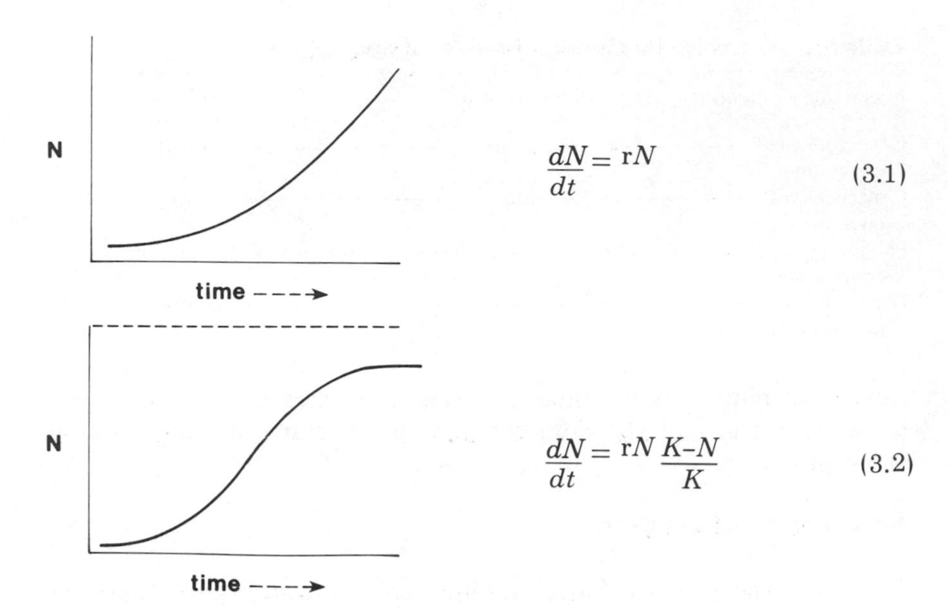

Figure 3.6. Models of population growth. In equation 3.1, describing exponential growth, r = intrinsic rate of increase. It is equal to the birth rate, b_0, minus the death rate, d_0. Both b_0 and d_0 are average numbers of births and deaths per unit of time; r may be negative, if d_0 exceeds b_0. When this happens, the population declines exponentially. There is an optimum called zero growth where $b_0 = d_0$, and $dN/dt = 0$. In equation 3.2 describing logistic growth, K = carrying capacity. When $N = K$, $dN/dt = 0$ (i.e., $b_0 = d_0$). If N is greater than K, dN/dt will be negative and the population will decline to K. The highest growth rate occurs when $N = \frac{1}{2}K$.

Eventually these populations reach a limit called *carrying capacity, K*. It is a simple idea that as a population increases, its growth rate will decrease because the environment can support only a limited number of organisms. The logistic growth model is a simple refinement of the exponential model, but one that takes carrying capacity into account (Figure 3.6, equation 3.2).

There are two difficulties with the logistic curve. First, when the population size, N, is very low (that is, when it is almost extinct), the growth rate is highest. Actually, there appears to be a "threshold" population size, M, below which the population will not regain its stature and, indeed, may well become extinct. This threshold may be expressed as

$$\frac{dN}{dt} = rN\frac{(K-N)}{K} \times \frac{(N-M)}{N} \qquad (3.3)$$

In this equation, if the actual population, N, is less than the threshold population, M, N becomes extinct. Normally N exceeds M by such a large margin that in the logistic curve, N does not reduce to zero unless it is a founder population. Then the chances of extinction are better.

The second problem with equation 3.2 is that it promises more than it delivers. It implies that if the growth rate, $\frac{dN}{dt}$, the population size, and the intrinsic rate of increase, r, are known, then K can be calculated. This is not true in the majority of cases. In general, there is no way of predicting what the carrying capacity is for any given species. Enough is known about K, however, that it can be conceptualized by considering density dependence and independence.

The idea that birth, b_0, and death, d_0, may reach an equilibrium leads to a realization that they are interdependent, as shown in the models in Figure 3.7. In diagram a, when N is low (near the y-axis), b_0 is often high. If b_0 remained high regardless of how large N got, there would be a

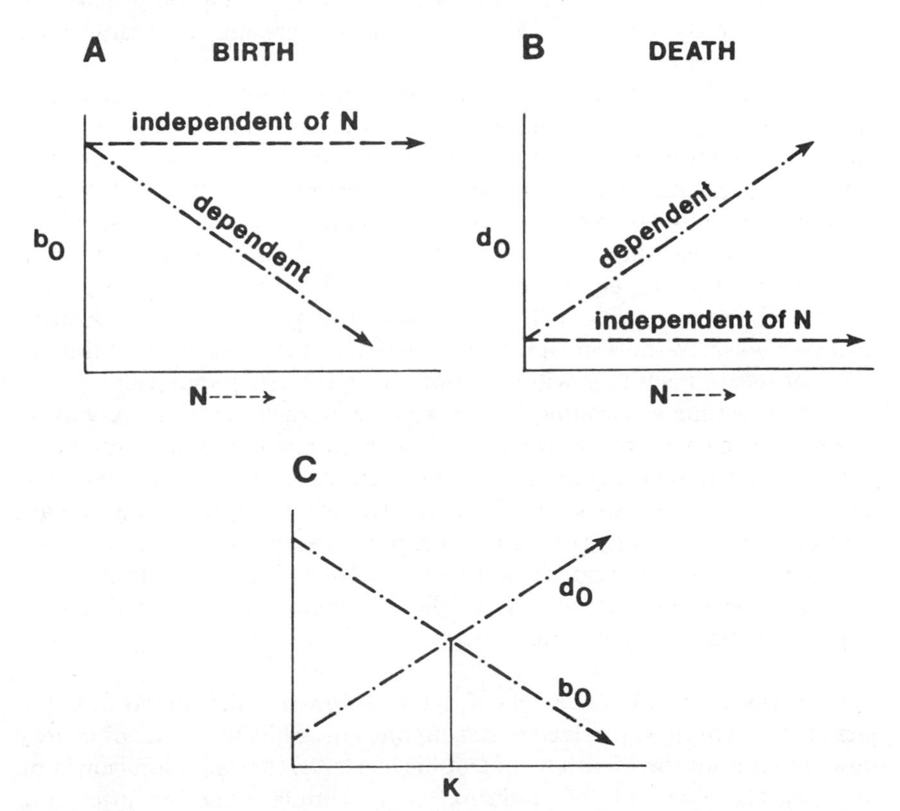

Figure 3.7. Estimating the carrying capacity, K, by considering density-dependent functions of birth and death rates.

straight line parallel to the x-axis (shown as a dashed line). In reality, as N increases, a decline in b_0 is normally observed. This "dependency" is illustrated by the dash-dot line.

In similar fashion, a death rate that was independent of N would plot as a parallel line to the x-axis (Figure 3.7b). However, one normally sees an increase in d_0 as N increases. Where these lines cross (Figure 3.7c), the carrying capacity, K, is defined. It will be shown later that b_0 and d_0 are functions of the environment exemplified by predator/prey relationships; so the implication is that K (the intersect of b_0 and d_0), is a direct outcome of the environment and the availability of resources. It is in this light that competition for resources, and thus competitive ability, assume a central role in speciation.

In density-dependent cases, K can be calculated by observing b_0 and d_0 over long periods and noting their values when they become equal, which is not the same as saying that K can be predicted in advance. In the famous case of lemmings (*Lemmus lemmus*) it may be possible to calculate K rather nicely, but in most cases there are too many intervening variables.

One must also reckon with the idea of positive and negative feedback. An example of positive feedback might be the case of human occupation on the Bolivian Altiplano. Studies have been proposed to determine just what the carrying capacity of the environment is in terms of human populations. It is believed that as the population goes up, there will be more efficient use of the natural resource base. As its resources are used more efficiently, the environment will be able to support more people (that is, its K goes up). To the extent that this is true, there is positive feedback. Clearly, there are limits to such expansion, because an infinitely efficient use of resources will not support an infinite population.

It is true that environmental changes that raise or lower K will be reflected in the population size. In the case of humanity, we have produced our own environmental changes, as for example by conquering diseases. The death rate decreases and N increases, even though b_0 may remain the same. Interestingly, this situation puts man on an almost density-independent status; and theoretically at least, such populations are destined for early extinction. Without density-dependent controls, population size fluctuates randomly.

Predation: *Predation* is one of the most important controls of population size. Through predation, nature has ensured the means of energy flow throughout the hierarchy of trophic levels. In the consideration here, however, the chief role of predation is to stimulate the evolution and development of organisms. Given two species, each characterized by its own birth and death rates, the basic considerations of predation can be reduced to the following equations (called the *Lotka-Volterra equations,* after their authors):

$$\text{Predator} \quad \frac{dN_1}{dt} = B_1N_1N_2 - D_1N_1 \qquad (3.4)$$

$$\text{Prey} \quad \frac{dN_2}{dt} = B_2N_2 - D_2N_1N_2 \qquad (3.5)$$

What these equations say is that the predator growth rate, N_1, is directly related to the number, of prey N_2. (The more abundant food source, the higher the K for the predator.) Conversely, the death rate of prey is directly related to the number of predators.

There are four situations that arise from the Lotka-Volterra models. In the lower left quadrant of Figure 3.8, when both the predator and prey populations are low, there is an opportunity for the prey population to increase. When there are few cats, there are many rats! Under such conditions there will be a trend toward the lower right-hand quadrant—namely, a low predator population, and high prey. When this occurs, the carrying capacity for the predator will be higher, because there are greater resources to support the population. As has already been shown, if K exceeds N, then N will tend to increase—barring any external suppressive agent like human beings.

A high-predator, high-prey situation evolves out of this situation

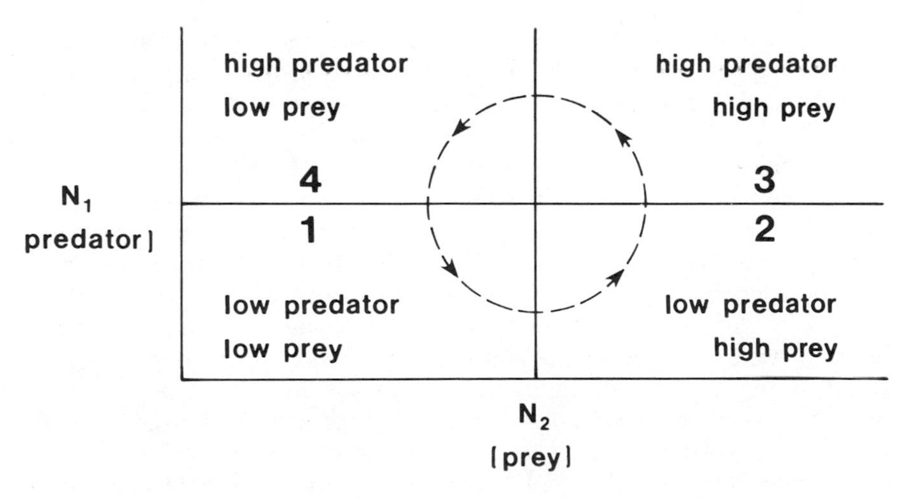

Figure 3.8. The cyclical nature of predator/prey populations.

representing ample food sources. Here the K for prey decreases because the environment has become hostile and can support fewer individuals. As the prey population declines, there is a shift toward the upper left quadrant in Figure 3.8, a case where N exceeds K for the predator population.

If both predator and prey populations (N_1 and N_2) are decreased by equal amounts through an outside agent, there will be a relative increase in N_2 (prey) and decrease in N_1 (predator). This realignment comes about because the product of N_1 and N_2 determines both the birth rate of predator, B_1, and the death rate of prey, D_2. If N_1 and N_2 are reduced equally, the product of those reductions will affect B_1 and D_2. The result of this relationship favors a more rapid recovery of prey than of their predator. Bear in mind also that this relationship migrates up the food chain, because a predator at one level is prey to another. The more predator/prey interactions in an ecosystem, the more stable it will be, because each population will have more alternatives. The stabilizing effect is seen in the ability of predators to switch food sources if one source becomes scarce. It is also seen in the ability of prey (including plants) to switch niches or habitats, as needed, so long as such changes are within the adaptive breadth of the organisms.

Competition: Like predation, *competition* is vital in regulating populations; that is, it is a factor in density dependence. When resources are finite, organisms display comparative advantages and disadvantages in their ability to utilize those resources. Competition, therefore, can be either an interspecific or intraspecific phenomenon. While the process takes place at the level of the individual, its effects are best observed in the population.

Competition is inevitable between organisms, because no two species that are ecologically identical will survive together very long. This axiom is often called the competitive exclusion principle. It is useful in a conceptual way, but it falls short of value in practice because it can never be demonstrated whether the principle really works. If two species coexist, but utilize some of the same resources, one concludes that they are not identical in their total environmental needs and are, therefore, nonexclusive. On the other hand, if one species is excluded, it is assumed that the principle is proven without having really supplied evidence.

Broadly defined, competition may take the form of an interference between two species, or of exploitation. *Interference* "refers to any activity which either directly or indirectly limits a competitor's access to a necessary resource or requirement. It usually operates in a spatial context, as in the case of territoriality, and assures possession of some minimum requirement by one individual or species" (Miller 1967, 8). In contrast, *exploitation* refers to resource extraction. Although both forms of competition lead to the same result, zoologists seem to agree that interference is the more effective mechanism for one species to gain power over a resource at the expense of another species.

Exploitation refers to the ability of a species to locate and make use of unused resources. In this regard, one may equate levels of exploitation with survival strategies. MacArthur and Wilson (1967) describe two such strategies: *r-strategy* and *K-strategy.* The *r-strategists* make up that set of organisms equipped for extremely high growth rates. They are best able to take advantage of short-term environments or ephemeral resources. Examples might include agricultural pests and desert annuals. The *K-strategists* are those organisms that rely on slower growth rates, but higher survivorship. *Homo sapiens* is a good example.

There have been suggestions that organisms have a tendency to exhibit linear selection from r to K as they become better adapted. For example, a geographically isolated population having a low N and undergoing genotype selection to become better adapted to its local environment may exhibit r-selection to rapidly increase its numbers. This usually happens in cases where there is an abundance of resources and little competition, as was the case with the introduction of rabbits into Australia. As N increases, and the chances for extinction decrease, there may be more of a tendency for K-selection. N begins to level off and individuals survive longer. One must view the tenor of this idea with caution. What is really being described is the logistic growth model. Although there may be some evidence for the progression of organisms from r- to K-selection, such changes in adaptedness probably take place over very long periods of time.

Adaptive Zones

If the process of adaptation begins with the creation of a mutant and spreads throughout a population by selection, then it seems reasonable that the primary attributes that distinguish birds from mammals represent different lines of adaptation to substantially different environments. In a broader sense, these substantially different environments are referred to as *adaptive zones.* Adaptations to some of these environmental opportunities have been so successful that they have progressed upwards through higher taxonomic levels. Actually, it is more appropriate to say that over a very long period of geologic time, such adaptations have differentiated into what are now recognized and defined as the higher taxonomic levels—families, classes, phyla, and so on.

The major adaptive zones of air, water, and land are self-evident. In each of these, a wide diversity of life forms has evolved and radiated. However, there are numerous, less obvious adaptive zones that have arisen through the gradual differentiation of life forms and will continue to arise as the history of life unfolds. The arboreal habitat of birds and primates, for example, could not have evolved prior to the successful adaptation of plant species to terrestrial environments. Trees, themselves, can be viewed as the opening of a vast new adaptive territory for life forms to invade. Before the evolution of animal blood, there was no environment for blood viruses; before feathers, there was no home for bird lice; and before the advent of warm-bloodedness, there was little potential for animals to invade the arctic.

Not all adaptive zones are equally prosperous. Some, like the interior of solid rock, appear to have little scope for the evolution of life forms. In fact, according to reports by Panagakos and Friedmann (1978), only the interiors of certain antarctic rocks have been found, so far, to support bacterial species. Decomposed rock in the form of mineral soils, on the other hand, has become the home of thousands of plant and animal species whose interactions represent vastly complex ecosystems.

Moreover, the range of opportunities within adaptive zones is constantly changing. The evolution of earth's oxygen-rich atmosphere clearly sets the stage for air-breathing life forms. Human alteration of this atmosphere by the addition of hydrocarbons, fluorocarbons, ash, and other particulate matter must ultimately stimulate selection pressure in the direction of forms that can tolerate these new conditions (see McLean, 1978). This pressure does not necessarily require the differentiation of new forms, but it most assuredly will require the early elimination of those individuals least able to adapt.

The net effect of constantly changing environments is the creation of new adaptive zones into which organisms infiltrate. Successful forms multiply and differentiate; unsuccessful ones become extinct. To radiate successfully, therefore, organisms must have physical, evolutionary, and ecological access to the new zone. That is, it must be geographically possible for the organism to migrate; it must have an initial tolerance to survive the new habitat while it is adapting (sometimes called *prospective adaptation*); and it must be able to compete with forms already occupying the niche to which it strives to adapt.

Hominid Adaptive Zones: In 1980 there were 4.47 billion living specimens of *Homo sapiens* (Hardin, 1980). Worldwide, the growth rate was estimated at 1.8 percent—equivalent to a "doubling period" of about 40 years. Such reproducing capability places humans in the company of insects and other "lesser" forms of life that can double their population every generation. It has taken less than 20,000 years, less than 700 generations, for our species to virtually subdue the earth through our vast adaptability, versatility, and fitness. Except for watery and perennially frozen locales, humankind can be found everywhere in numbers that tend to exceed the carrying capacities of the ecosystems. Because of our explosive geographic spread and our capacity to extinguish other life forms and pollute or alter environments, it is worthwhile to seek a better understanding of our adaptive stages and the reasons for our success.

Human biogeography has been defined by Terrill as "... the study of the size, distribution and population structure of, and the interactions among, populations; and of the conditions and events leading to similarities and differences among human communities" (1976, 2). Although this may be a useful construct for cultural anthropology, it fails to agree with the essential mood of inquiry being pursued here. The primary thrust of this section is to review the main lines of speciation among the Anthropoidea, particularly the family Hominidae, and to comment on the evolutionary adaptations that have catapulted *Homo sapiens* ever more successfully into new adaptive zones. As such, this section perhaps gives a more biological definition to human biogeography than that used by Terrill.

Modern humankind is the sole surviving species of the genus *Homo,* which in turn is the sole representative of the family Hominidae. It is generally agreed that our family originated some 5 to 6 million years ago in a number of localities reminiscent of today's African savanna. Figure 3.9 shows only the main hominid line descendent from *Ramapithecus.* Other members of the Anthropoidea are shown in Figure 3.10, in which man is shown as a dweller of the forest margin.

According to Weiner, *Homo's* ascendance progressed through four epochs: (1) the emergence from a restrictive arboreal and forested environment; (2) the progressive development of a terrestrial, hunting mode of life, which was made possible through several anatomical adaptations to widen niche breadth (e.g., bipedal stature and the opposable thumb); (3) ecological expansion and population differentiation; and (4) the control and manipulation of environment through the rise of culture and technology. Although they are essentially valid, these few phrases tend to reduce hominid origins to an instant in time and to diminish the drama of 200,000 generations of our ancestors. A partial parade of these ancestors is given in Figure 3.11, and one interesting aspect of the assemblage is that the fossil record shows concurrent existence for the australopithecines and early species of *Homo.* Apparently these forms arose together in several, perhaps widely separated localities.

Leakey (1977) postulates that speciation in the genus *Homo* was complete around 4 million years ago and that there was reproductive isolation from *Ramapithecus, Australopithecus,* and *Pithocanthropus.* Through competition, however, these other forms were quietly overshadowed. As recently as 1.5 million years ago *Homo erectus* was still living alongside the australopithecines. A number of mutually beneficial adaptive zones was shared by all of the early hominids, but only the genus *Homo* continued to enter the zones that have led to modern man.

Figure 3.12 is a representation of the more important adaptive zones occupied by the hominid evolutionary stream. The shift from four- to

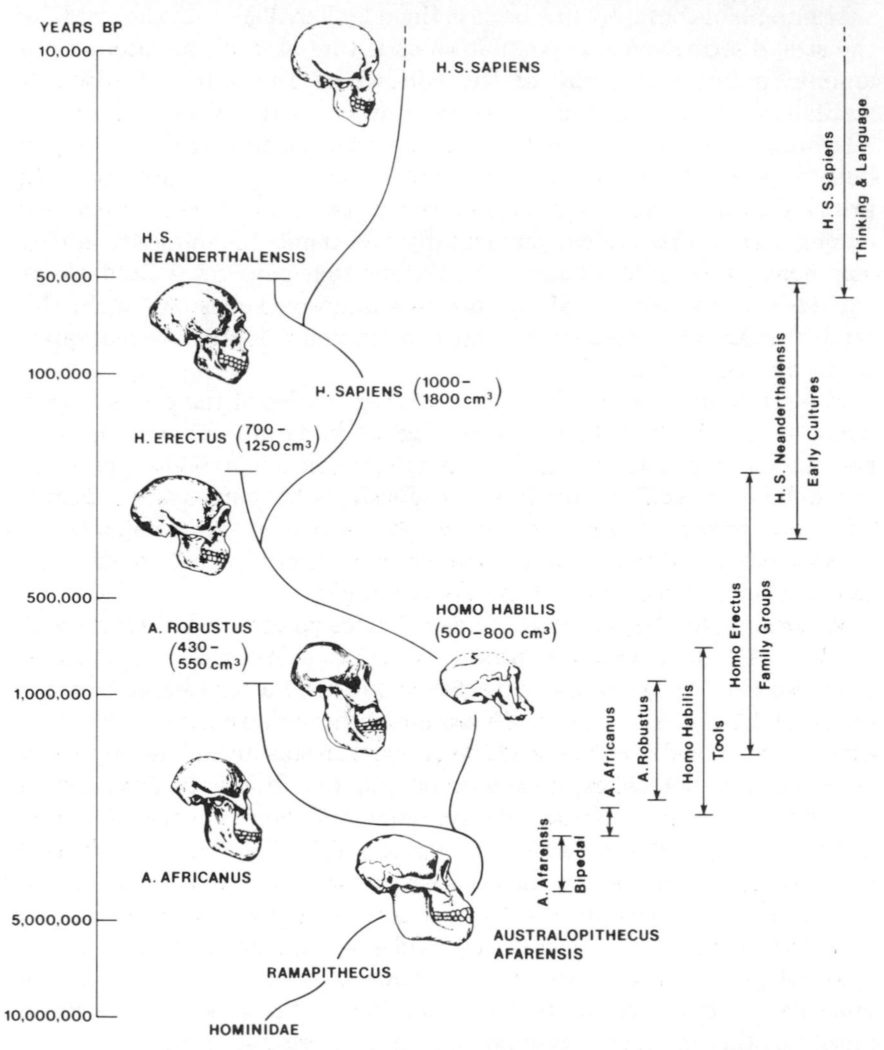

Figure 3.9. Today the family Hominidae consists of a single genus and species, *Homo sapiens*. Throughout the last 5 to 6 million years, our ancestors have included a much wider variety of forms. Some of these are recognized as representing other genera and species within the family, whereas others are still contestable.

two-footed locomotion took place early in the descent of the family, more than 36 million years ago. This ability, of course, freed the forelimbs for grasping. Later, development of the opposable thumb improved, still further, our ancestors' ability to utilize and manipulate resources. Among the opportunities opened by the grasping hand was their improved ability to hunt and gather food; hence arose their eventual transition from herbivory to omnivory.

As *Homo* progressed into group living for food gathering and for protection, the need evolved for cultural practices and institutions, which

Figure 3.10. Selected Anthropoidea in their habitats. Top layer: spider monkey; middle layer: lemur, tarsius, tree shrew, cologus, cercopithecus; bottom layer: gorilla, baboon, man.

arose in concert with expanded cranial capacity and the beginnings of logic and reason. In other words, as our ancestors' ability to think grew, so did their need for incipient cultural attributes. Language is a stupendous adaptive zone that has grown out of these early group-living requirements. While use of language may not be unique to humans, its extension, that of writing and record keeping, is unique. Through this means, our knowledge of how to cope with changing environments and how to manage resources can be passed along as it has been from the last 200 generations of our ancestors (the last tenth of our hominid existence).

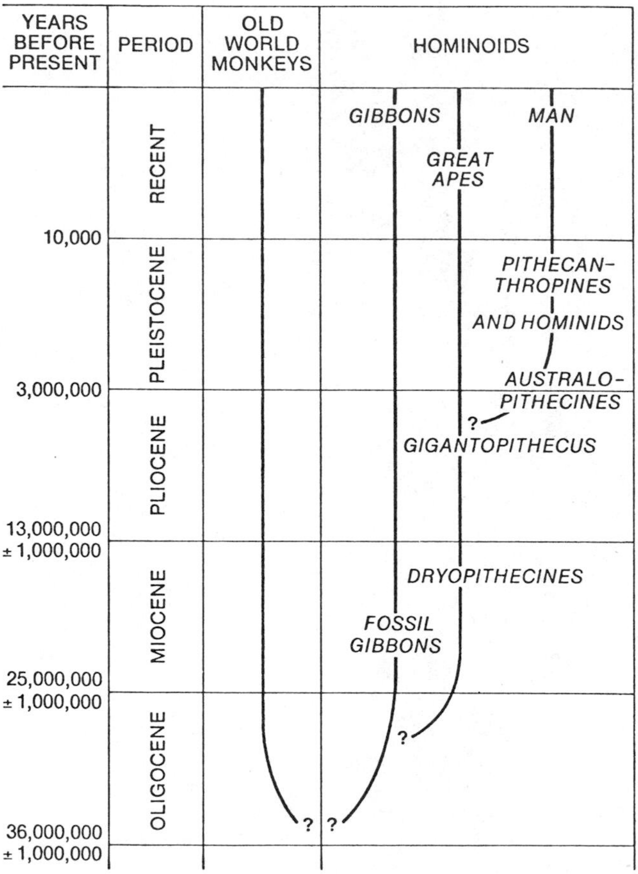

YEARS BEFORE PRESENT	PERIOD	OLD WORLD MONKEYS	HOMINOIDS
	RECENT		*GIBBONS* *MAN* *GREAT APES*
10,000	PLEISTOCENE		*PITHECAN-THROPINES AND HOMINIDS* *AUSTRALO-PITHECINES*
3,000,000	PLIOCENE		? *GIGANTOPITHECUS*
13,000,000 ± 1,000,000	MIOCENE		*DRYOPITHECINES* *FOSSIL GIBBONS*
25,000,000 ± 1,000,000	OLIGOCENE	?	? ?
36,000,000 ± 1,000,000			

Figure 3.11. The hominoids probably split from Old World monkeys more than 36 million years ago. It is uncertain when the hominid line, which includes the australopithecines and pithecanthropines, split from the great apes in terms of behavior and ecological position.

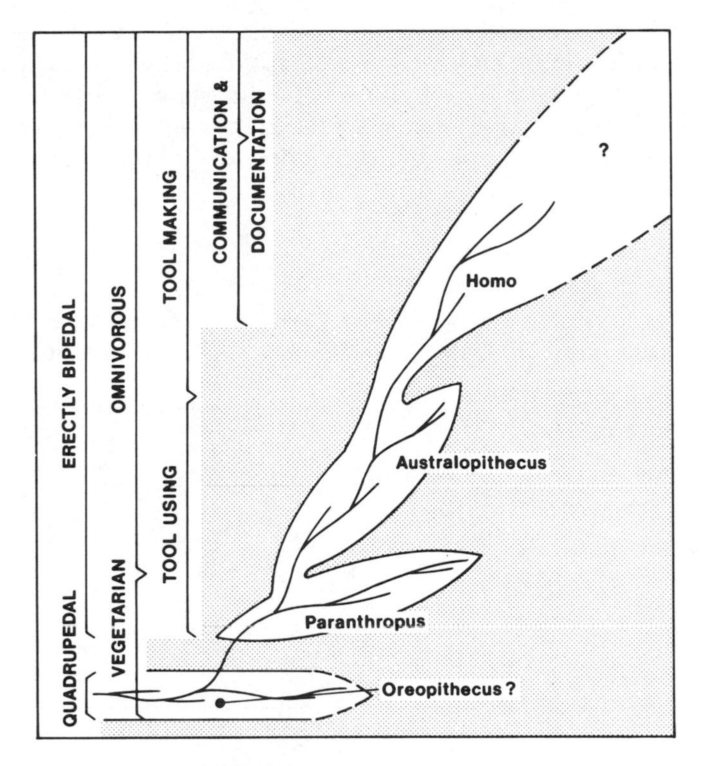

Figure 3.12. Representation of the more important adaptive zones occupied by the hominid evolutionary stream.

If one looks at the entire stream of adaptive zones into which the genus *Homo* has progressed, it comes as no shock that this species could alter the rest of nature so profoundly. Human kind is exceptionally fit to embrace virtually all terrestrial environments. Ironically, it may be fitness (the ability to reproduce in excessive numbers), combined with cultural practices (the inability of populations to live in geographic isolation), that is leading the species into an evolutionary blind alley (see also the section on blind alleys in Chapter 4).

EVOLUTIONARY RATES

Of all the species that have ever roamed this earth, it is estimated that less than 1 percent are now alive. The processes of evolution and speciation have ensured a continuous diversification of forms and the perpetuation of life. From the beginning of the fossil record, organic change has kept pace with environmental change and the availability of new habitats. The branching evolution of reptiles, birds, and mammals in Figure 3.13 illustrates the diversification of life throughout the last 350 million years. The rate of diversification, and the method of measuring it, are important considerations in deciphering the development of world biogeographic patterns.

Total Organic Diversity

Many interesting questions surround the study of evolutionary rates. Will an equilibrium number of taxa ever be reached, or should diversifi-

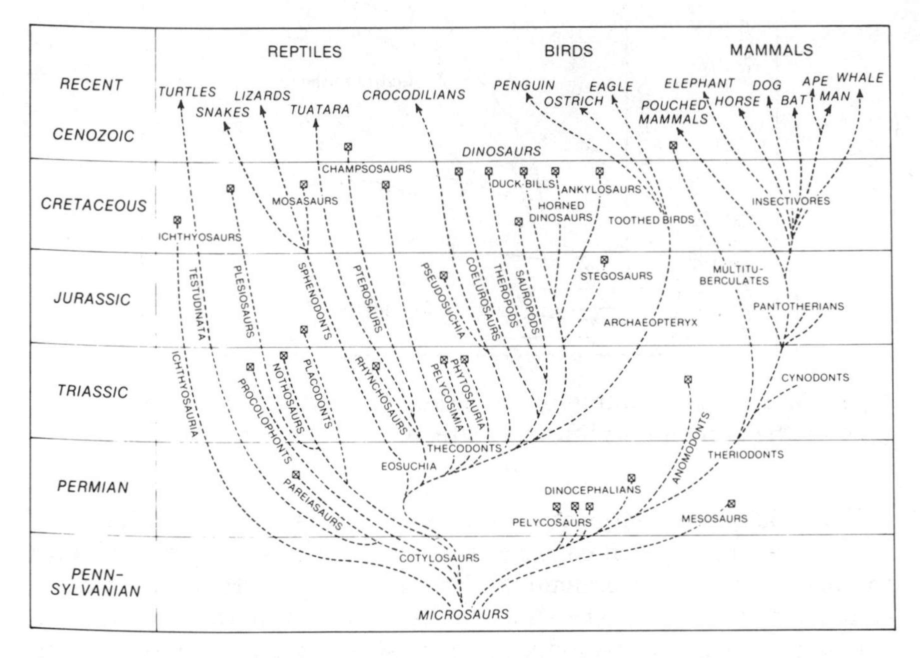

Figure 3.13. Diversification among reptiles, birds and mammals from Carboniferous to recent geologic time. Amphibian turtles arose early and have not undergone extensive diversification since the Mesozoic. Crocodiles also arose early, and are one of the few surviving thecodonts. Most other members of the thecodonts became extinct at the end of the Cretaceous. Progenitors of modern mammals have an early history, but unlike reptiles have achieved extensive adaptive radiation since the Cretaceous period. Insectivory apparently represented a highly successful adaptive zone for early mammals in the latter millenia of the dinosaurian age. The avian world likewise experienced diversification during the Cretaceous, having derived from Jurassic stock known as *Archaeopteryx*.

cation be expected to continue indefinitely? If mass extinctions have taken place in the past, what is the rate of biological recovery? Over the long term, if the rate of speciation exceeds the rate of extinction, is there a geographic pattern to the birth place of new species? The search for valid, quantitative answers to these and other questions continues. For the present, knowledge is based on incomplete data, empirical observations, and conjecture.

The increase in taxonomic diversity since the end of the Precambrian is believed to have followed a trend similar to that shown in Figure 3.14. In this diagram, which represents marine invertebrates only, the major classes and orders achieved their greatest diversity during the Late Ordovician period about 400 million years ago and have continued to decline in numbers ever since. Major extinctions took place during the Triassic and Jurassic periods. The differentiation of families, genera, and species reached their initial peak in Devonian, Carboniferous, and Early Permian times, respectively, and after the Triassic mass extinction, have

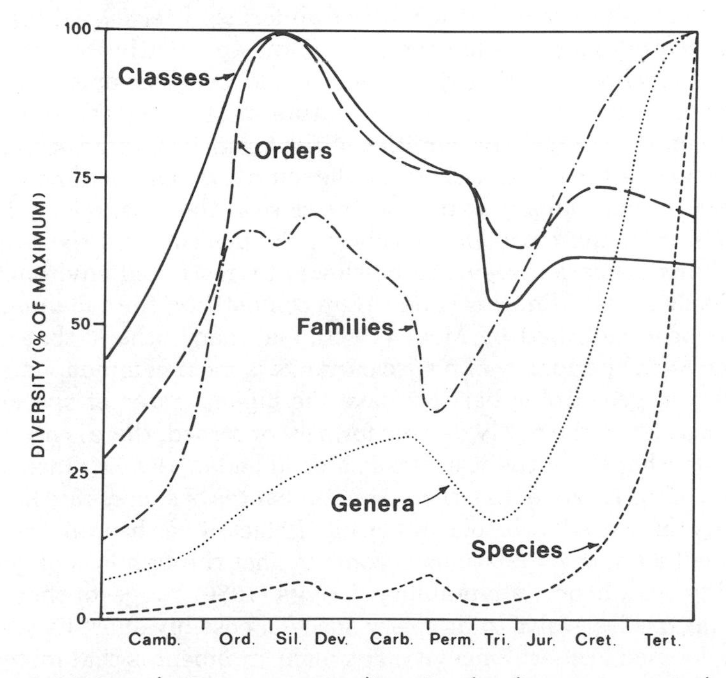

Figure 3.14. Trends in taxonomic diversity for five taxonomic levels of marine invertebrates from Cambrian through Tertiary time. *(From D. M. Raup, 1972, Taxonomic Diversity During the Phanerozoic, Science* **177:***1070; copyright © 1972 by the American Association for the Advancement of Science. Reprinted with permission.)*

achieved their modern measures of diversity. These trends suggest that early adaptive zones, to which some 40 percent of the classes and orders had adaptated, were eliminated. Those classes and orders surviving today, however, are sufficiently plastic and adaptable to undergo rapid diversification at the familial, generic, and specific levels.

For flowering plants, of course, the time scale for most diversification must be compressed, so that the major families evolved during the Middle to Late Cretaceous period (70 million years ago), the genera during the early Cenozoic era (30-70 million years ago), and the species from the Miocene to the present (last 40 million years). From these trends, and from the evidence in Figure 3.13, it does not appear that diversification is complete.

It is also clear from Figures 3.13 and 3.14 that mass extinctions have occurred in the past. Particularly noticeable in Figure 3.14 is the dip in diversity for marine invertebrates at the Permo-Triassic boundary. Similar mass extinctions have occurred in the past for reptile and mammal forms, and history will undoubtedly record a fundamental extinction at the hand of humanity—one that will affect all forms of terrestrial life. The rates of recovery might be fairly rapid, as was apparently the case with marine invertebrates during the Mesozoic and Cenozoic eras; but such rates must surely depend upon the nature of the original extinction mechanism, the competitive abilities of the forms invading the vacated niches, and the genetic plasticity of the survivors. Of the dinosaurian fauna that became largely extinct at the close of the Mesozoic, only the snakes, lizards, and crocodiles survive to the present, and their rate of renewed diversification seems to be slowed by restricted environments (cold-bloodedness is limiting) and severe competition from mammals.

It has been indicated by Mayr (1969), and many others, that differentiation at the periphery of a species' range is most common, and that areas rich in geographic barriers have the highest rates of speciation. However, as far as the newly derived form is concerned, there is no way to predict whether the niche it enters is a dead end or the entrance into a large new adaptive zone. In either case, its chances of success are liable to be greatly influenced by its place of origin. In fact, it has been argued that most small founder populations become extinct through loss of genetic variability and, hence, adaptability. Briggs (1966) suggests that most peripheral species evolve to fit a very specific, possibly short-lived niche and are doomed to short longevity. His main argument is that instead of comparing all evolutionary rates on the same basis, one should distinguish between that form of speciation that occurs in the main center of dispersal, and that taking place in traps and other peripheral areas. The first case is slow but often successful, whereas the second may be rapid but ill fated.

Briggs's argument clarifies why it is that well established dominant species that have arisen at the center of their ranges may so easily compete with, and displace, newly evolved peripheral species. Most of the birds that have become extinct in the past 200 years are peripheral island species that have been replaced by more competitive forms introduced by people. As to why species that have evolved in centers of ranges should be more competitive,

> the concept of relative ecological stability seems to be important. The extreme tropical regions of the world are the major centers for successful evolution and at the same time, they contain the most species and therefore have the most stable ecosystems. The stable ecosystem with its high level of competition provides the proper environment for the production of dominant species. (Briggs, 1966, 288)

Peripheral species lack the continual test of competition in their development and therefore are competitively inferior to (have narrower niche breadth than) species that have developed in the center of their parents' range. These arguments conform essentially to the stasipatric model of geographic speciation, which is further discussed in Chapter 4.

Having these considerations in mind, three types of evolutionary rates can be described: *genetic, phylogenetic,* and *taxonomic.* Genetic rates of change cannot be documented from the fossil record and are therefore of little or no help in analyses of regional development. The rate and process of genetic change is best recorded in the laboratory by using successive generations of isolated populations.

Phylogenetic Rate

Phylogenetic rates are morphological. They can be documented qualitatively from the fossil record by measuring certain anatomical characters, such as the height and width of teeth, or complexes of characters, such as the dimensions of teeth combined with the numbers of toes. In the phylogeny of the horse family since the Eocene, for example, it is apparent that evolution followed a successful adaptive trend from browsing to grazing types and that there were associated anatomical changes from three toes to one (Figure 3.15). Many forerunners to modern *Equus* became extinct through time and in various regions. Among them were the Miocene browser, *Anchitherium,* in the Old World; the Pliocene grazer, *NeoHipparion,* in North America; and the Hippidion group of Pleistocene grazers in South America.

The hypsodont character of horses, which refers to the vertical and horizontal dimensions of cheek teeth, is an excellent example of a rate-of-character evolution. In the north central region of the United

States, fossil beds containing large deposits of equine teeth can be found that represent spatially and temporally distinct populations. These deposits include materials described as *Hyracotherium borealis* from Eocene deposits in Wyoming; *Mesohippus bairdi* from the mid-Oligocene of South Dakota; *Merychippus paniensis* from the late Miocene of

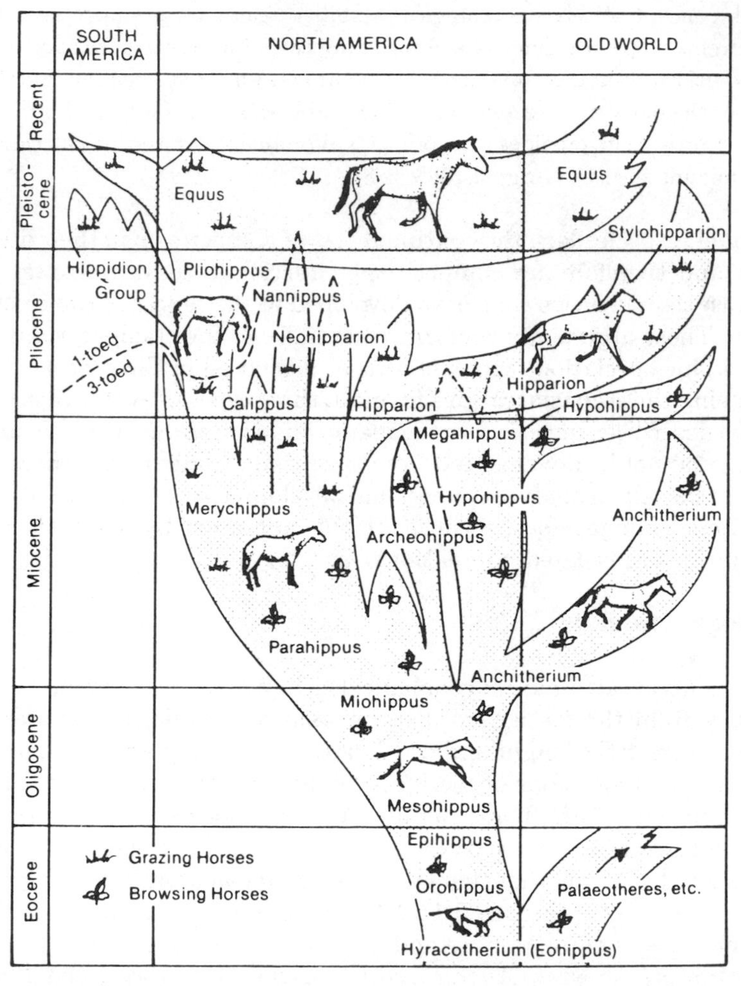

Figure 3.15. Phylogeny of the horse family (Equidae). *(From G. G. Simpson, 1951, Horses: The Story of the Horse Family in the Modern World and Through Sixty Million Years of History, Oxford University Press, New York, p.114; copyright © 1951 by Oxford University Press. Reprinted with permission of Oxford University Press.)*

Nebraska; *Hypohippus osborni* from the late Miocene of Nebraska; and *Neohipparion occidentale* from the early mid-Pliocene of Nebraska. Specific conclusions reached from measurements of many teeth are that there has been an evolutionary tendency toward increased height relative to horizontal dimension, which is coincident with an environmental change toward grasses, and that tooth size is correlated with overall animal size, both within and between populations, and between successive populations.

Taxonomic Rate

Taxonomic rates of evolution can be quantitative or qualitative in scope, but in any case are of questionable value. Both horse and human phylogenies contain examples of taxonomic evolution. One can get a measure of the rate by taking the total number of different genera found in the geologic record over a given period of time, and dividing by the time span. For example, if 10 genera appeared in the record over a 60-million-year span, then the average rate of taxonomic evolution would be .17 genera per million years. What these numbers mean in a true biological sense is unclear, since they are related to rates-of-character evolution. In general, these rates are not constant, but vary considerably depending upon selection pressure. Moreover, the rates of evolution for two characters may change independently, as they might well have done for the height and width of equine teeth, or for teeth and number of toes.

Absolute temporal evolutionary rates are seldom calculable, except in cases of extreme isolation or where events have been carefully recorded. On Mindinao, for example, 18 species and 4 genera of fish have evolved from a single ancestral species in 10,000 years. In the 1,000 years since its introduction by humans, five species of a moth have evolved on the banana in Hawaii. The critical factors in determining these rates are associated with the genetic variability of the original stock, its degree of isolation, and its competitive success. For the vast majority of cases where stratigraphic, cyclical, or sequential data are available, only a comparative evolutionary rate can be estimated, and even these estimates are subject to several sources of error of the type described below.

Range Charts: *Range* in this context refers to the duration of time organisms are believed to have existed. The difficulty is that the first and last occurrences of such organisms in the fossil record are seldom certain. Erosion of the fossil beds and other destructive processes may truncate either end of the record; and since older rocks often have been subject to erosion longer than younger rocks, there may appear to be increased diversification in the younger material.

Incomplete Stratigraphy: Forms that are living today and are also known from the fossil record (say, Cretaceous) are assumed to have lived throughout the intervening period, even if fossils cannot be found for that interval. However, forms that died out quite recently and left no fossil record, except from the Cretaceous, would be described only as Cretaceous forms. Again, since younger rocks contain more fossils of still living forms than older rocks, one could falsely conclude that diversification has been more rapid in the recent geologic past.

Duration of the Stratigraphic Units: Other factors being equal, a rock unit spanning a long period should contain higher fossil diversity than an equally thick unit spanning a short period. The record in this case could indicate greater diversity during the Paleozoic than during the Cenozoic era, because the former was considerably longer.

Intensity of Study: Fossiliferous rocks in western Europe and eastern North America are more likely to be fully studied and to show higher diversity than rocks in other parts of the world. The discovery of especially rich fossil beds like the LaBrea tar pits, the London Clay, or the Baltic Amber may lead to false conclusions about increased diversity of forms at those times, particularly since they are most often found in younger sediments.

Age-Area Relationships: As was seen earlier, diversity is very likely to be area dependent. The larger the area, the greater will be the number of niches and habitats available for occupation. Biologically speaking, most phyla have worldwide distributions. The families comprising a given phylum are usually more restricted to particular climatic zones; and the species are even more restricted to favorable habitats. One can anticipate that several small fossiliferous beds from around the world representing the Miocene would contain a greater diversity than a single large Miocene bed from one locality. Since we do not know the comparative areas of outcrops by age on a worldwide basis, it is difficult to assess errors in the study of evolutionary rates. However, speculation is that Cenozoic rocks cover more surface than Mesozoic or Paleozoic rocks, and that because of this fact, there is a bias toward greater Cenozoic diversity.

Sediment Volume: A thick bed formed by continuous sedimentation may contain more fossils than a thin bed or one formed under discontinuous conditions. Discontinuous sedimentation may preclude the incorporation of many short-lived species, although the bed itself may contain all the relevant families, classes, and orders. The bias introduced by sediment volume, therefore, cannot be separated from age-area relationships. The bias is toward greater apparent diversity at the species level in younger rocks, because portions of the older sedimentary record have been eroded and, with them, their assemblage of species. What is left is a record of higher taxonomic levels but fewer species.

SUMMARY

The processes of selection and adaptation serve to fix genetic variation in a population. The intensity and direction of genetic selection influence the rate of adaptation in a given ecological setting and, ultimately, the rate of evolution at various taxonomic levels. Environmental selection in the form of competition and predator/prey relationships incorporates all the interspecific processes that prevent any species from becoming universally distributed. Adaptive zones represent new environmental opportunities for invasion by peripheral populations having the genetic plasticity to survive the change. The genus *Homo* is an excellent example of a group that entered several highly successful adaptive zones, which permitted the species to radiate throughout the terrestrial world. As geographically or genetically peripheral populations —other than *H. sapiens*—enter new environments, they also enter into a degree of isolation that can finally render them reproductively and geographically distinct. Where such differences become visible through simple observation or through laboratory observation, a new species is named. Our methods for measuring the rates of change are suspect because we have imperfect knowledge of spatial, temporal, and taxonomic relationships.

REFERENCES

Briggs, J. C., 1966, Zoogeography and Evolution, *Evolution* **20**(3):282-289.

Hardin, G., 1980, *1980 World Population Estimates,* The Environmental Fund, Washington, D.C., folio.

Leakey, R., 1977, *Origins.* E. P. Dutton, New York, 264p.

MacArthur, R. H., and E. O. Wilson, 1967, *The Theory of Island Biogeography,* Princeton University Press, Princeton, N.J., 203p.

Mayr, E., 1969, *Animal Species and Evolution,* Harvard University Press, Cambridge, Mass., 797p.

McLean, D. M., 1978, A Terminal Mesozoic "Greenhouse": Lessons from the Past, *Science* **201**:401-406.

Miller, R. S., 1967, Pattern and Process in Competition, in *Advances in Ecological Research,* Academic Press, New York, pp. 1-74.

Panagakos, N., and E. I. Friedman, 1978, *Living Organisms Found Inside Antarctic Rocks,* NASA News Release No. 78-14, National Aeronautics and Space Administration, Washington, D.C., 5p.

Raup, D. M., 1972, Taxonomic Diversity During the Phanerozoic, *Science* **177**:1065-1071.

Simpson, G. G., 1951, *The Story of the Horse Family in the Modern World and Through Sixty Million Years of History,* Oxford University Press, New York, 247p.

Simpson, G. G., 1953, *The Major Features of Evolution,* Columbia University Press, New York, 434p.

Terrill, J., 1976, Island Biogeography and Man in Melanesia, *Archaeol. and Phys. Anthropol. in Oceania* **11**:1-17.

Weiner, J. S., 1971, *Man's Natural History,* Weidenfeld and Nicolson, London, 255p.

Wilson, E. O., and W. H. Bossert, 1971, *A Primer on Population Biology,* Sinauer Associates, Stamford, Conn., 192p.

FURTHER READING

Atchley, W. R., and D. S. Woodruff, 1981, *Evolution and Speciation,* Cambridge University Press, New York, 436p.

Bennett, J. W., 1976, Anticipation, Adaptation, and the Concept of Culture in Anthropology, *Science* **192**:847-853.

Brown, W. L., Jr., 1957, Centrifugal Speciation, *Q. Rev. Biol.* **32**:257-277.

Brown, W. L., Jr., 1958*a,* General Adaptation and Evolution, *Syst. Zool.* **7**(4):157-168.

Brown, W. L., Jr., 1958*b,* Some Zoological Concepts Applied to Problems in Evolution of the Hominid Lineage, *American Scientist* **49**(3):285-299.

Dobzhansky, T., 1961, Man and Natural Selection, *American Scientist* **49**(3):285-299.

Eldridge, N., and J. Cracraft, 1980, *Phylogenetic Patterns and the Evolutionary Process: Method and Theory in Comparative Biology,* Columbia University Press, New York, 350p.

Friedman, E. M., 1982, Endolithic Microorganisms in the Antarctic Cold Desert, *Science* **215**:1045-1053.

Grant, V., 1963, *The Origin of Adaptations,* Columbia University Press, New York, 606p.

Johanson, D., and M. Edey, 1981, *Lucy: The Beginnings of Humankind,* Simon and Schuster, New York, 409p.

Jukes, T. H., and R. Holmquist, 1972, Evolutionary Clock: Nonconstancy of Rate in Different Species, *Science* **177**:530-532.

Kolata, G. B., 1975, Paleobiology: Random Events Over Geological Time, *Science* **189**:625-626, 660.

Le Gros Clark, W. C., 1959, The Crucial Evidence for Human Evolution, *American Scientist* **47**(3):299-313.

Lewontin, R. C., 1978, Adaptation, *Scientific American* **239**(3):212-230.

May, R. M., 1972, Limit Cycles in Predator-Prey Communities, *Science* **177**:900-902.

May, R. M., 1978, The Evolution of Ecological Systems, *Scientific American,* **239**(3):161-175.

McNaughton, S. J., and L. L. Wolf, 1970, Dominance and the Niche in Ecological Systems, *Science* **167**:131-139.

Miller, D. A., 1977, Evolution of Primate Chromosomes, *Science* **198**:1116-1124.

Olsen, P. E., and P. M. Galton, 1977, Triassic-Jurassic Tetrapod Extinctions: Are They Real? *Science* **197**:983-985.

Pilbeam, D., and S. J. Gould, 1974, Size and Scaling in Human Evolution, *Science* **186**:892-900.

Raup, D. M., 1977, Probabilistic Models in Evolutionary Paleobiology, *American Scientist* **65**(1):50-57.

Satinoff, E., 1978, Neural Organization and Evolution of Thermal Regulation in Mammals, *Science* **201**:16-22.

Simpson, G. G., 1965, *Geography of Evolution,* Capricorn Books, New York, 249p.

Valentine, J. W., 1968, Climatic Regulation of Species Diversification and Extinction, *Geol. Soc. Am. Bull.* **79**:273-276.

Van Valen, L., 1969, Climate and Evolutionary Rate, *Science* **166**:1656-1658.

Washburn, S. L., 1978, The Evolution of Man, *Scientific American* **239**(3):194-208.

Watanabe, H., 1972, Periglacial Ecology and the Emergence of Homo sapiens, in *The Origin of Homo sapiens,* F. Bordes, ed., UNESCO, Presses Universitaires de France, Vendôme, France, pp. 271-278.

Wilson, E. O., 1973, Group Selection and Its Significance for Ecology, *BioScience* **23**(11):631-638.

Chapter 4

Speciation and Extinction
The Fate of Change

The preceding discussions have shown that variation and isolation set the stage for organisms to increase in diversity, and that selection, together with adaptation, fashions the ultimate ability of life forms to diverge. This path to differentiation may lead either to speciation or extinction, depending upon the type of genetic variation involved, the adaptability of the population, the size of the initial population, and the external character of the environment.

SPECIATION

Major differences of opinion exist between paleontologists and ecologists, on the one hand, and between botanists and zoologists, on the other, regarding the essential requirements for speciation. Population segregation as a prerequisite for speciation is not accepted in all biological circles, although it appears to be an essential element in all of the models proposed so far. Nor has the apparently simple notion of gradual phyletic evolution leading to the branching off of new forms been eagerly adopted by all. Darwin himself worried that his theory of evolution on the descent of man would be denied through lack of sufficient fossil remains. Indeed, there are still missing links that many paleontologists believe never existed. Instead, they subscribe to a theory of "punctuated equilibrium" —the view that some biological changes occurred so rapidly in geologic time that the gradual morphological trends predicted by Darwin never existed.

There is no question that patterns of speciation are different between the plant and animal kingdoms. They share enough in common, however, that they may be reduced to a manageable number of models. White (1978) and Bush (1975), among others, have described these models in considerable detail and have provided many examples. The brief venture into the field here looks only at the more general models, typically referred to as *allopatric* (or geographic), *sympatric,* and *stasipatric* speciation. These models may be described under the general heading of "gradual" speciation, although, in addition, there are some recognized "instantaneous" mechanisms.

Instantaneous Speciation

There are three recognized possibilities for instant differentiation: *mutation* (genetic "accidents" during recombination), *macrogenesis* (spontaneous generation, or speciation in one gigantic step), and *polyploidy* (chromosome increase or decrease). There is considerable evidence that these processes do occur in nature. In most cases except polyploidy, however, the products are generally not viable in the sense of achieving dominance in a gene pool. A recent example of macrogenesis in apes illustrates the point (Shafer, 1979). In Atlanta, Georgia, the Grant Park Zoo reported that a hybrid ape was produced from the accidental union of two reproductively isolated members of the *Symphalangus,* a male gibbon and a female siamang. Although the offspring survived and was given the pseudoscientific name "siabon," it is highly unlikely that it will ever reproduce. Nevertheless, the incident may be taken as support for the theory that evolution in mammals can proceed by chromosomal rearrangement as well as by mutation of individual genes. In the plant kingdom, such chromosomal alterations have long been regarded as a mechanism for speciation.

Polyploidy: According to Jackson (1976), polyploidy is a spontaneous phenomenon that probably occurs in all eucaryotes, and has been especially important in the evolution of plants. Through this mechanism, an increase or decrease may suddenly occur in the number, size, shape, or arrangement of the thread-like chromatin material along which genes are arranged. Each parent contributes half the total chromosome count, so that when reformed, a diploid individual is produced. In the plant and insect worlds, it often happens that individual chromosomes break or become rearranged in such a way as to result in new offspring characteristics. Figure 4.1 shows a hypothetical pattern for this process.

Because of basic conservation measures for cell space and phosphorus utilization, it is virtually certain that plant species originally had low numbers of chromosomes. Three, for example, seems to be a recurrent haploid number in many plants and insects. Large chromosome numbers are therefore suggestive of increase over time, and are referred to as polyploids. In general, polyploid species ". . . nearly always differ from their diploid progenitors with respect to structure and function. Quanti-

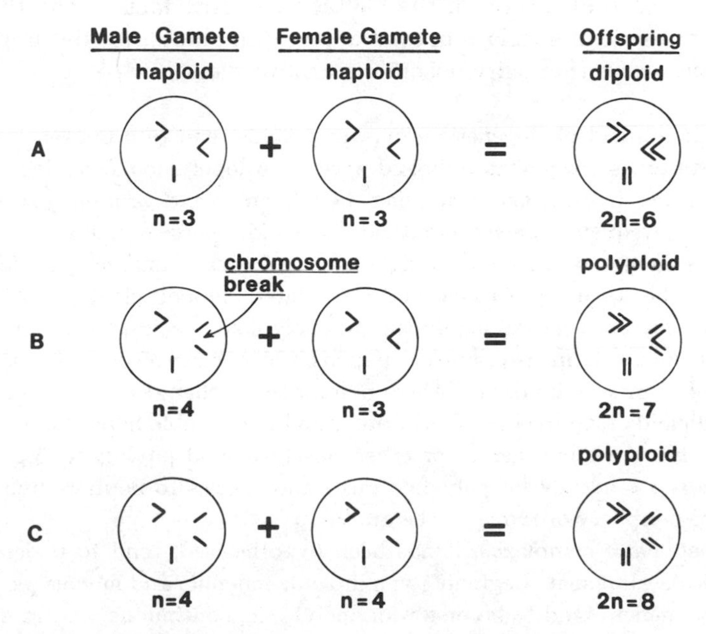

Figure 4.1. Generalized pattern for speciation through polyploidy. The expected reproductive case is given in (*A*) for the union of two haploid gametes to form diploid progeny. It should be indicated that considerable genetic variation may exist in large diploid populations and that in spite of their uniform chromosome number, several genotypes, phenotypes, and ecotypes could exist and be subject to other modes of speciation. In (*B*) one of the chromosomes in the male gamete has broken into two pieces. The polyploid progeny ($2n = 7$) is referred to as an allopolyploid because it originated by the addition of unlike sets of chromosomes. The offspring in (*C*) is referred to as an autopolyploid because its chromosome sets are homologous. It became possible only in the second generation through creation of a female gamete having a haploid count of four ($2n = 8$).

tative and qualitative differences in structure, and new reaction norms cause polyploids to have a tendency toward a different ecology and to occupy habitats and areas different from those of their diploid ancestors" (Cain, 1944, 430).

As recorded by Cain, other principles concerning polyploidy are given in the following statements. He cautions against accepting them as laws and urges that they be regarded as working rules supported by a vast amount of data. Darlington (1968) reiterates this caution by reminding us that even in the plant world, where polyploidy is most frequently observed, over 50 percent of all flowering plant species have remained diploid since their origin in the Cretaceous. This fact means that the prevalent mode of speciation, even among the plants, resides in genetic variation rather than chromosomal rearrangement.

> Within a small phylogenetic group, polyploids tend to occupy more extensive areas than related diploid species. Allopolyploids, having their origin from hybrids and combining the tolerances and variability of both parents, frequently have a foundation for a wider range of tolerance which allows them to occupy wider ranges both ecologically and geographically, than either their diploid progenitors or related autopolyploids. (p. 431)
>
> In complexes containing diploid and polyploid species, the latter tend to predominate in geographical regions which have recently been subjected to great climatic or other environmental changes. In these cases the diploids tend to occupy older areas and more stable habitats. (p. 431)
>
> Within a section, genus, or other closely related phylogenetic group, there is a tendency for polyploid races and species to be more frequent at the periphery or range for the group. (p. 433)
>
> Systematic complexes, it has been hypothesized, tend to undergo a cyclic development. Beginning in a diploid, endemic, and monotypic condition, genera tend to become wide, polytypic, continuous, and in many cases polyploid. In old age, the range of the genus becomes disrupted and the genus may end with only relic polyploid species, endemic in widely disjunct areas. (p. 435)
>
> Certain diploid species, without benefit of chromosomal races, and genera composed only of diploid species, have attained all the ecological adaptations and types of geographical range that other species with polyploid races and polyploid genera have been able to attain. (p. 436)

It is evident from these statements that polyploidy is an instant process for speciation only in the sense that major changes occur within a generation. It may still take several generations for the offspring to distribute themselves into a wider geographic range, or to successfully invade new habitats. It is also clear from Cain's phraseology that although geographic isolation is not a prerequisite to the process, there is a rather quick segregation of polyploids into ecologically noncompetitive habitats.

Gradual Speciation

The concept of gradual speciation dates from the first half of the nineteenth century, and Charles Darwin was its greatest spokesman. Darwin, however, made no distinction between speciation through individuals and speciation through populations. Nor did he clarify the conditions whereby speciation could occur. The idea that geographic isolation might be fundamental to the emergence of new forms was developed by Wagner between 1868 and 1889 but found few supporters. By the turn of the century, sympatric speciation was the favored model, although it was then described as "physiological selection." The theory was so weak by comparison to the mass of evidence in favor of geographic speciation that in the first half of the twentieth century the pendulum reversed toward a requirement for spatial segregation (see Rand, 1948; Huxley, 1938; and Rensch, 1938). In recent years there has been an effort to wed these two basic notions into a more universally acceptable theory.

One of the least refutable aspects of biogeography is the appearance in the fossil record of numerous extinct forms that have modern living relatives. As has been shown in the case of horses, many of these taxonomic units can be arranged in geographic and temporal sequences to form phyletic evolutionary trees. Paleontologists are always on the lookout for new fossil beds to fill in gaps and to complete the tree of life back to its very origins. At the same time, biologists and ecologists are striving to understand the future directions of life based on current genetic and environmental attributes. It is only in the time frame of the present that these endeavors involving geologic and ecological time merge and must be made compatible. Given the vast body of paleontological evidence and our certain knowledge that the selection and adaptation of favorable gene complexes take many generations to achieve, there is little debate that most speciation is a gradual process. What is disputed, however, is the mechanism responsible for the divergence of forms, and the relative importance of its elements.

The Allopatric Model: That geographic separation is central to genetic divergence is so widely accepted in zoological circles that a model of geographic speciation has long been used. A new species is believed to evolve when a population that is geographically distant from other populations of its parent species acquires, during a period of isolation, characters that promote or guarantee reproductive isolation after the external barriers break down. This change is a slow, historical process that single observers can seldom record.

"Macrogeographical" isolation is readily appreciated, since it is clear that populations separated by oceans, deserts, or mountains cannot easily interbreed. The barrier is obviously wider than the mean dispersal distance of the propagules. Controversy arises, however, where suitable habitats are available but disjunct, such as in the case of small forest clearings, or where the habitat seems to be continuous by outward appearances, but where environmental changes such as microclimate or soil fertility may, nevertheless, exert an influence. Distance is involved in either the macro or micro case, and a stretch of unsuitable terrain exists, across which dispersal is inhibited and gene flow reduced. Figure 4.2 illustrates the essence of the model.

In cases of disjunct but close populations, an issue arises over the strength and frequency of gene flow between them. How do populations escape from cohesion in the gene pool long enough to establish their identity? To answer this question requires expanding the conceptual base. Since natural landscapes are rarely uniform over large areas, there is a tendency for species to be distributed into local populations. In highly dissected terrain, for example, there may be ecological segregation of communities according to steepness of slope, aspect, or soil type, such that populations of a given species will be physically separated. Near the margins of the range for that species, peripheral populations can be defined that may also be peripheral isolates in the sense that they are spatially distinct from the main body, but still tied to the main body by gene flow. Gaps are small enough that propagules can cross them. Under these conditions, none of the local populations can drift very far from the gene complexes that are experiencing centripetal selection. Because adaptedness is conditioned by the genetics of the entire species, those populations on the periphery of the range probably are not completely adapted to their local environment. They are striving to become finely tuned, but are prevented from doing so by gene flow.

It is argued by Ehrlich and Raven (1969), and others, that gene flow is not enough to maintain species uniformity by overcoming local mutation and selection. This assertion means that, should it become necessary, peripheral isolates will contain sufficient genetic variation to allow them to adapt to a changing environment. If there is insufficient variation, the population may become extinct when isolated from the main group. It seems to be agreed that the first step in speciation is some sort of ecological change that induces a genetic drift between two adjacent populations. Stated another way, two intraspecific populations develop. That such ecological changes occur over a given area (even if indefinable) preserves the notion of geographic segregation.

Once the peripheral population is geographically isolated, it is referred to as a founder population (Figure 4.3), and the stage is set for genetic drift between the founder and its sister populations. If there is no difference in the environment on either side of the isolating barrier, then there will be a very slow drift in the genetic composition of the separated populations. Given sufficient time, however, the two populations will necessarily drift apart, because there will be no gene flow between them. The mutations that become incorporated into each population will very likely not be the same, and each successful mutation of any gene will have

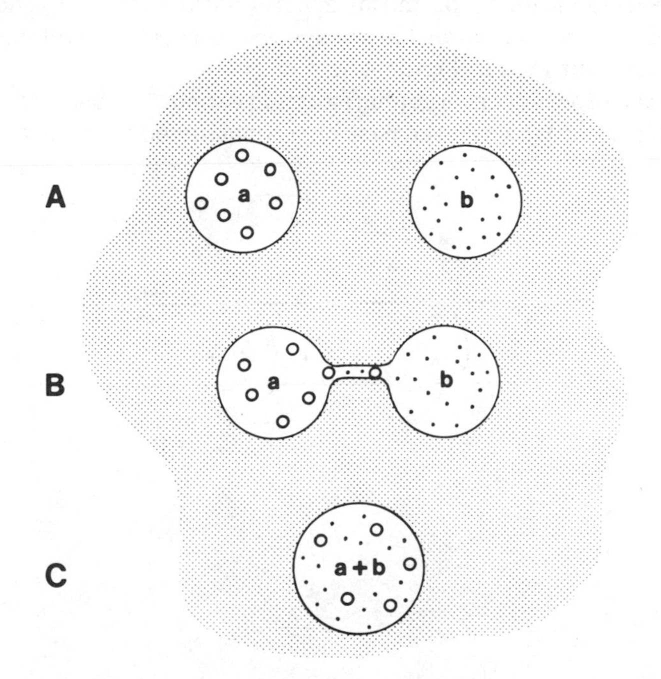

Figure 4.2. Degrees of geographic isolation. In *(A)* two populations of the same species are separated by a distance greater than the mean dispersal distance of individuals or propagules. There is no gene flow, and each population has an opportunity to drift apart genetically. In *(B)* the populations are not completely isolated, so some measure of gene flow between them is still possible. The genetic make-up of each population may be sufficiently different that polymorphs are apparent. Any further reduction in gene flow might result in fairly rapid reproductive isolation. In *(C)* the two populations occupy the same territory, but may be partially or completely segregated into different ecological niches. All three cases confirm the essentiality of either macro- or micro-geographic isolation as an element in speciation.

an influence on future selection. In other words, there will be a progressive evolutionary effect such that the two populations can never again be as genetically similar as they were originally. Consequently, two genotypes will form, and drifting will accelerate.

Isolation of founder populations is essentially a shift from an open to a closed population, as shown in Figure 4.3. The consequent increase in homozygote frequency is enough to set off a genetic chain reaction. In these cases, the smaller the initial population, the more rapid the revolution will be, because there will be less total variability in the gene complexes. In larger populations, the initial genetic variability will tend to slow down the progression toward a new species. Either case will undergo the basic stages outlined in Figure 4.4.

Since the process of geographic isolation cannot be observed directly, a question arises: How can the past be reconstructed to prove the theory?

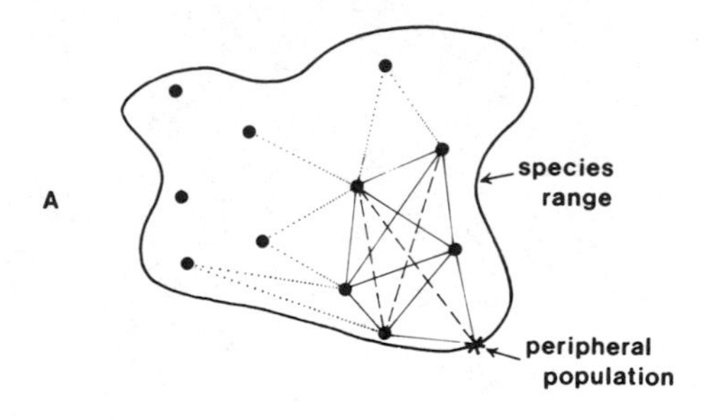

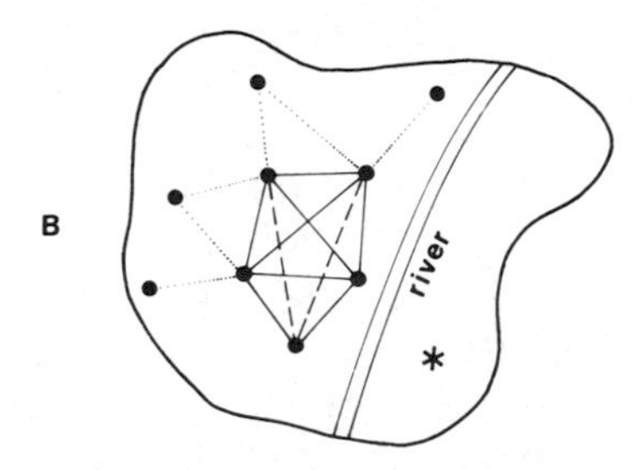

Figure 4.3. *(A)* The location and genetic ties of a peripheral population (designated by an $\star$) within the entire range of a given species at time t_1. *(B)* At time t_2 the peripheral population has been isolated by a barrier and is now a founder population which can experience genetic drift from its sister populations. Solid, dashed, and dotted lines represent varying strengths of gene flow between populations.

The answer, in part, is to arrange populations of an organism into a continuous series and observe the stages through which they have passed. In this regard, it is best to start with the assumption that speciation is a population phenomenon and to look for populations that appear to be incipient species. A related question then arises: Do these incipient species have characteristic patterns of distribution? The answer is that in virtually all cases where an actively evolving genus is found, there are populations that are closely similar in genetics, morphology, and distribution; there are others that are subspecies; there are still others that approach being species; and, finally, there are those that are full species. In many cases, the full species are still allopatric. However, in others, they are sympatric; that is, they appear to the observer to be sympatric because the barriers that once separated them have broken down, and the two species now intermingle but occupy nonoverlapping niches.

That distinct populations remain allopatric until they reach species level is regarded as one of the most convincing proofs of the theory. In other words, through the processes of gene flow and recombination, sympatric populations will tend to remain as a single species and have similar, though certainly not identical, ecological requirements. When these populations become geographically isolated, gene flow essentially ceases. There is almost always a shift in ecological requirements as the

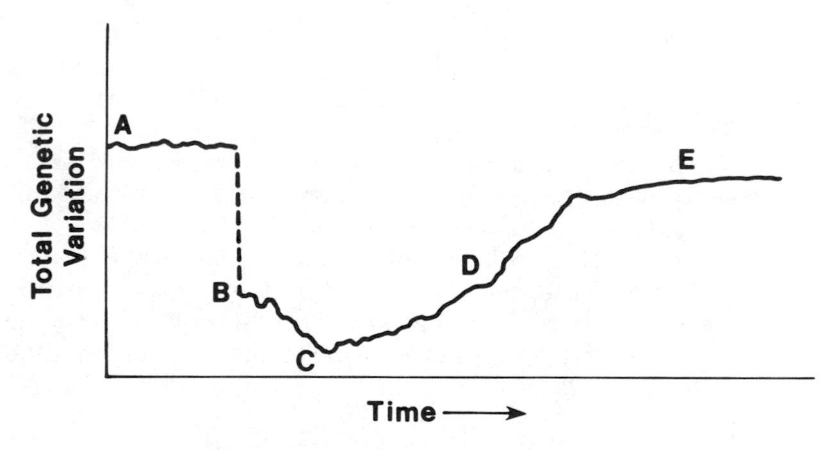

Figure 4.4. Temporal trend in the level of genetic variation experienced in a founder population until it reaches reproductive isolation from the mother species. *(A)* Total level of variation in the mother species; *(B)* residual variation in the founder population after severance from the mother; *(C)*initial loss of variation through extinction of some genes; *(D)* recovery of variability through linear selection; *(E)* total genetic variation in the daughter species.

two populations begin to drift apart genetically. The shift in these niche and habitat requirements is essential; otherwise, should the two forms be reunited, there would be competition leading to selection and possible extinction of one of the forms.

By carefully plotting current geographic ranges of related species, and by evaluating all kinds of borderline cases, it is sometimes possible to reconstruct the probable path of speciation. Overcrowding of niches and habitats ensures that most incipient species become extinct. Thus peripheral populations have no guarantee of becoming new species, but most new species were at one time peripheral populations. These considerations lead to the following situations:

Superspecies: This term refers to a monophyletic group of allopatric forms that are too morphologically different to be a single species, but have never been shown definitely to be reproductively isolated. One assumes that if the polytypic populations were reunited, they could not interbreed. Many of the island birds of Micronesia appear to be superspecies.

Semispecies: These groups are forms that are not yet sufficiently different morphologically to be considered distinct species. The chances are high that reunited populations could interbreed. As these forms drift farther apart genetically, reproductive isolation would probably take place.

Secondary Contact Zone: If the geographic barrier breaks down before reproductive isolation is complete and two semispecies are reunited, their offspring may be rather different from either of the parents and would thus be called a hybrid.

Partial Overlap: In this case, mother and daughter species have achieved reproductive isolation and are in the early stages of being reunited after the collapse of physical barriers.

Multiple Invasion: Early colonists of a species adapt to a new environment and become sufficiently distinct that they remain identifiable despite a second, third, or fourth wave of arrivals of their own kind. Multiple invasion often happens on island groups that are periodically reunited by lower sea levels.

The Sympatric Model: In contrast to an apparent requirement for geographic separation is the belief of many botanists and ecologists that genetic divergence of forms can take place even while those forms occupy the same geographic locale and are in physical contact. The process is referred to as sympatric speciation. Initially, proponents argued that the origin of the isolating mechanisms within the dispersal area of the population differed from that recognized in geographic speciation in that the cause for splitting in gene pools was niche related, rather than habitat related. More recently, for example in White (1978), the sympatric model considers speciation to be adaptation to different habitats or niches within the same area, bearing in mind that populations residing in different habitats are, by definition, microgeographically disjunct. Reproductive isolation is achieved because populations have different ecological requirements and therefore seldom come into reproductive contact with one another.

At least two questions favor the model for sympatric speciation. First, how did so many species of the same genus get to be in the same area if geographic isolation was essential to the origin of those species? Is it not more likely that they arose in the same area? Fedorov (1966) points out that tropical rain forests have the highest species diversity and lowest population numbers of any of the world's terrestrial biomes, and that tropical rain forest environments are also the most uniform. In other words, these environments have the highest niche diversity and lowest habitat diversity of all the biomes. Consequently, if they also have the greatest species diversity, the species most probably arose sympatrically to fill niches. There is further evidence from Ehrlich and Raven (1969) that gene flow is not as cohesive a force as once thought, but that selection regimes and niche adaptation can lead to speciation.

Second, how did all the species of the same area get to be so well adapted to their niches and habitats if they arose in geographically isolated areas? An example of suspected sympatric speciation from entomology is provided by Tauber and Tauber (1977) for *Chrysopa carnea* and *C. downesi* (Table 4.1). They propose that *C. downesi* arose as a new species from *C. carnea* through disruptive selection. They believe the process took place in two steps. First, a stable phenotype arose in *C. carnea* changing the body color of some individuals from light green to dark green. These individuals had higher survival rates in coniferous stands than they had in their original habitat of grassland and meadows. Individuals on the forest-grassland margins, who could migrate to safety, soon began to generate a viable population of dark-green individuals residing primarily in the forest. Apparently *C. carnea* and *C. downesi* became so adapted to their respective habitats that little reproductive contact took place between them. *C. carnea* continued to reproduce several new generations each summer, but *C. downesi* (now virtually isolated from the parental gene pool and subject to further selection within the coniferous habitat), became dominated at two gene loci by alleles affecting their response to photoperiod. As a result, *C. downesi* experienced a genetic drift toward one reproductive cycle in the early spring. With reproductive isolation assured in nature, the two forms continued to diverge genetically.

Table 4.1. Example of Sympatric Speciation in *Chrysopa*

	Chrysopa carnea	*Founder population*	*Chrysopa downesi*
Step 1—Creation of founder population			
Habitat:	Meadows and forest openings		Coniferous forest
Color:	Light green (summer)		Dark green
Color genotypes:	Heterozygote (Gg) selected for; homozygote (GG) selected against	GG individuals are dark green. Those located near the forest edge may survive if they escape into the forest.	Creation of a population of stable genotype (GG) occupying a forest habitat.
Step 2—Initiation of reproductive isolation			
Reproductive cycle:	Long period of reproduction throughout summer governed by dominant alleles D_1D_2.	Recessive alleles d_1d_2 in GG individuals. Creation of one concentrated period of reproduction in early spring.	
Step 3—Recognition of new species			
	Mother species		Reproduction and habitat isolation of GG d_1d_2 individuals leads to recognition of daughter species.

Source: Based on C. A. Tauber and M. J. Tauber, 1977, Sympatric Speciation Based on Allelic Changes at Three Loci: Evidence from Natural Populations in Two Habitats, *Science* **197:**1298–1299.

The above example is interesting for several reasons. First, it appears to be related to habitat and not niche. Second, geographers would argue that habitat isolation is the same as geographic isolation, but on a micro scale. If habitat isolation was the first step in the origin of *C. downesi,* it is not an unambiguous case for sympatric speciation, but rather seems to confirm the essential role for spatial segregation as a precursor of genetic drift. Mayr (1947) and White (1978), with numerous other authors, have stressed that habitats are distinct in space, and surrounded by other habitats that serve as geographic barriers. Whenever it can be shown that two populations require the same habitat but are separated by intervening territory, they are indeed isolated. In the case of *Chrysopa,* that the first step in the process involved a selection pressure toward homozygosity in the gene affecting coloration does not negate the essential fact that separation of the populations was required for further genetic drift to occur.

The Stasipatric Model: A compromise model between allopatry and sympatry is found in the description of stasipatric speciation (Figure 4.5). It argues that a new species can develop through a chromosomal rearrangement in a single individual that initiates genetic isolation from other members in that population. Since gross changes in chromosome number and structure occur in individuals, it may be possible for an individual to act as a point of origin for speciation. The previously cited case of the "siabon" (page 68) may be an example of this process, although the conditions surrounding the siabon's creation were anything

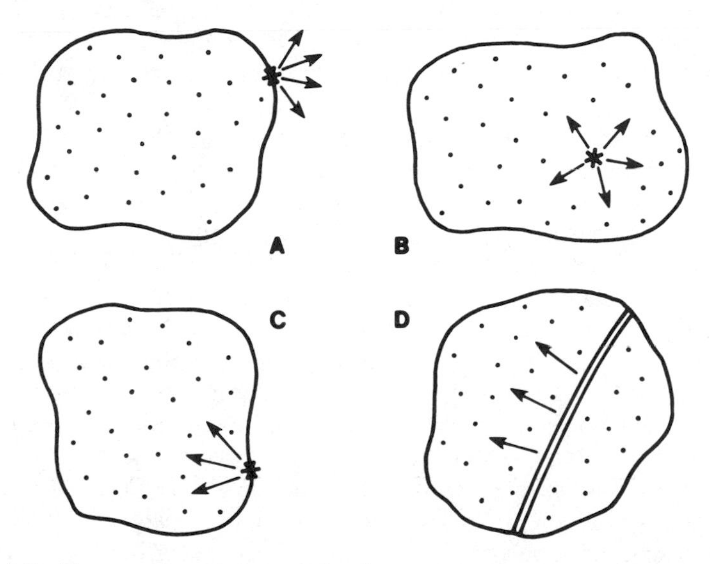

Figure 4.5. The stasipatric model of speciation. In *(A)*, the chromosomal rearrangement becomes established in a population on the periphery of the species' range, and serves as a pioneer by spreading into previously unoccupied territory. In *(B)*, the chromosomal rearrangement becomes established somewhere inside the territory of the species' range and forms a local colony advancing outward. The case in *(C)* is similar to *(B)*, but shows the chromosomal colony originating on the periphery and advancing only in the direction of the existing range. *(D)* shows the result of *(B)* or *(C)* as a slow movement across the territory with a hybrid zone separating the two types. *(From M. J. D. White, 1968, Models of Speciation, Science* **159:***1068; copyright © 1968, by the American Association for the Advancement of Science. Reprinted with permission.)*

but natural. Such individuals may occur anywhere within the species' range, not necessarily only in a peripheral population. Consequently, the stasipatric process hinges on individuals rather than populations for speciation.

Mayr (1970), in contrast, concludes that speciation must be a populational rather than an individual process. His arguments are based on a conviction that the genetic composition of a species is determined very slowly by selection for, and against, mutations within individuals. It would take generations for a successful gene mutation in an individual to become fixed in a population, and in any case, it is highly unlikely that changes in a single gene would so dramatically influence reproduction. That process is controlled by a gene complex, not by a single gene. Chromosome arrangements, however, involve many genes. Even though rearrangements occur infrequently and most are selectively unsuccessful, some have apparently spread over quite large territories. The process is reminiscent of macrogenesis, and in that sense may be regarded as spontaneous.

The stasipatric model is intriguing because it could serve as a meeting ground for hitherto irreconcilable differences between sympatric and allopatric speciation. In botany, for example, it has been impossible to explain on the basis of geographic isolation the high rates of hybridization and species diversity in large genera like *Quercus* (oak) and *Eucalyptus.* The stasipatric model now provides a mechanism that can explain zones of overlap and hybridization without calling for the premature "breakdown" of isolating barriers. It can be seen also that in Figure 4.5a, a chromosomal change taking place at the periphery of a species' range and spreading outward could be described as allopatric; spreading inward (Figure 4.5c), it would be described as sympatric.

An example of speciation following the pattern in Figure 4.5 has been described by Carson (1970) for the evolution of *Drosophila* (fruit flies) in the Hawaiian archipelago. The essence of his case study is diagrammed in Figure 4.6. In the six major Hawaiian Islands there are some 250 species of *Drosophila,* all but a dozen of which are endemic to one or more of the islands. The oldest of the islands, Kauai, is late Pliocene in age (about 5.6 million years), while the youngest, Hawaii, is late Pleistocene (about 700,000 years). Maui is about 1.5 million years old.

Through chromosome analysis, Carson as been able to reconstruct the path of speciation for some eleven species of so-called picture-winged drosophila. Seven of these are presently endemic to Hawaii and have apparently radiated from single founder females representing three ancestral types residing on Maui. Carson postulates that these three fertile females each carried a different chromosomal arrangement from Maui across the Alenuihaha Channel to Hawaii, where further differ-

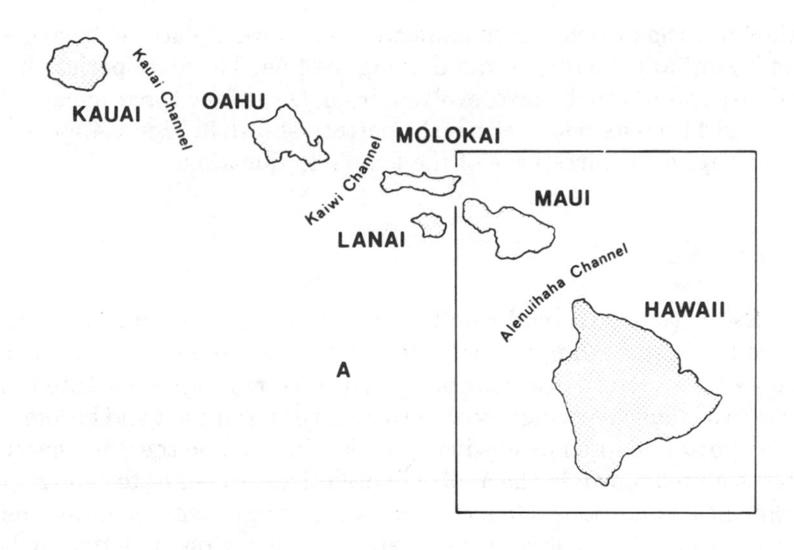

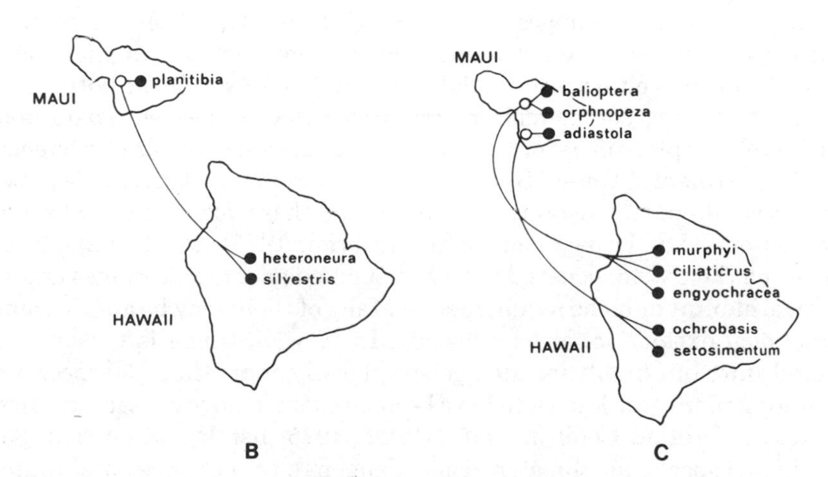

Figure 4.6. Example of stasipatric speciation among *Drosophila* in the Hawaiian Islands. *(A)* The six major islands of the chain. *(B)* postulated route of speciation for *D. planitibia* , *D. heteroneura* , and *D. silvestris,* shown as solid black dots, from an ancestral form, shown as an open circle. It appears that the two Hawaiian Island forms are derived from single female immigrants from Maui. Similar situations are postulated for the eight species shown in *(C)*. *(From H. L. Carson, 1970, Chromosome Tracers of the Origin of Species, Science* **168:***1415, 1417; Copyright © 1970 by the American Association for the Advancement of Science. Reprinted with permission.)*

entiation into the present seven endemic species took place. If true, the pattern is similar to that illustrated in Figure 4.5a. The four species now endemic to Maui which have evolved from the same three ancestral stocks, would then represent either the pattern shown in Figure 4.5b or c, depending upon the birthplace of the female in question.

EXTINCTION

Extinction is the end result of evolution. All life forms on earth today are but a sprinkling of past forms. Simpson viewed extinction as common and survival, the exception. Evolution progresses through selection in a genotype, which exists in a given environment and becomes gradually better adapted to survive there. Extinction, on the other hand, represents a breakdown in the ability of organisms to adapt to changing environmental conditions. Genetic factors that might lead to extinction in a population are generally eliminated by selection of individuals possessing those traits. This process ensures that survivors are adapted. If the environment changes slowly enough that adaptation occurs, then extinction can be avoided; if adaptation does not occur, the implication is that the form is either overspecialized, genetically stable, or both.

The aesthetic, commercial, or research value of species, apart from their ecologic place in nature, can be used to establish a scale of reference. In the continental United States, a list of "rare and endangered" species has been prepared in hopes of protecting these forms from absolute extinction (U.S. Department of the Interior, 1975). Most of the 2,800 forms on the list are believed to have restricted geographic ranges and/or critical habitat or niche requirements. Many of them may be approaching functional extinction, at least locally. In Hawaii, which is not only an island state but highly isolated geographically, more than 250 species of flowering plants are known to have become extinct since contact was first made by Captain Cook in 1770 (USDI, 1975, list B). More than 830 additional species are threatened or endangered. In the continental United States, only 100 species (Figure 4.7) are known to be or presumed to be extinct (USDI, 1975, list C). The pattern on Hawaii is repeated worldwide as humanity continues to upset the balance of plant and animal life on remote islands (Stoddart, 1968). A sampling of the islands most affected is given in Figure 4.8.

The desire to preserve existing species stems not only from the aesthetic and ecological view, but also from the economic view that they may have as yet undiscovered commercial or medical value (see Giesel, 1976). One fact is certain: the elimination of a species reduces the competition for resources and may open an adaptive zone for peripheral popula-

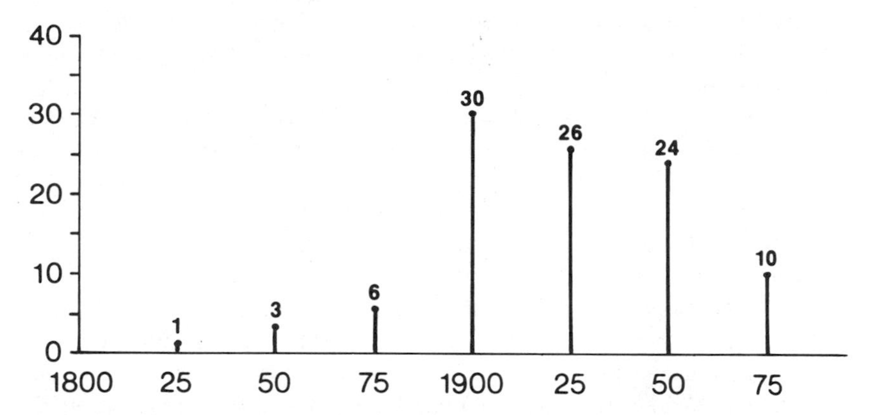

Figure 4.7. History of presumed, or possible, plant extinctions in the continental United States. For plotting purposes, extinction date is assigned to the quarter century following the date of last collection. That is, 30 plants were last collected between 1875 and 1900. [*Source: USDI, 1975, Threatened or Endangered Fauna or Flora, Federal Register* **40**(*127 part V*)*: list C.*]

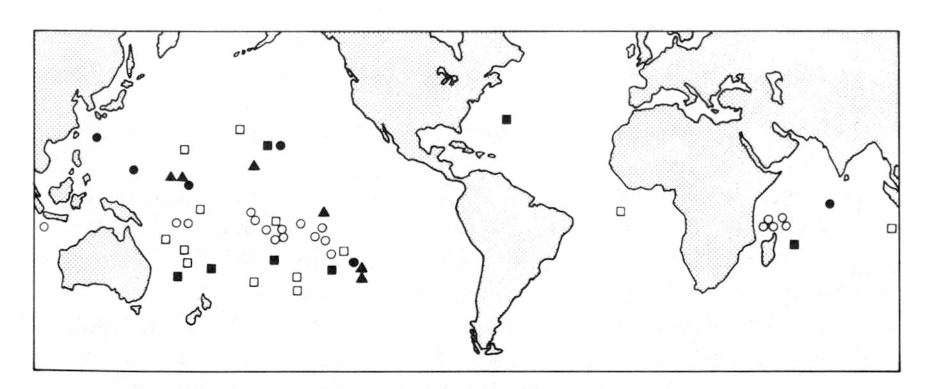

Figure 4.8. The nature and geographic extent of human impact on island biotas. Since many of the species present on each island are endemic, the rate of extinction is increasing dramatically. Black squares are international airports; open squares, airstrips; black dots, military bases; open circles, mining operations; and triangles, weapons-testing sites. *(From D. R. Stoddart, 1968, Isolated Island Communities, Science Journal, April, p. 34; copyright © 1968 by Science Journal. Reprinted with permission.)*

tions of another species to invade. Nature is a master at filling vacated habitats and niches, even though man may not approve of the replacement organisms.

Absolute extinction (which means that the probability of finding living specimens is virtually zero), is most confidently expressed by reference to temporal range charts. These charts indicate the times of first and last appearance of fossil remains in the geologic record. Such charts are questionable at best, because knowledge of world stratigraphy and erosion histories is incomplete (see Olsen and Galton, 1977). Nevertheless, such charts indicate times of wholesale speciation and extinction and, when combined with such other data as paleoclimatic trends, can suggest the extent of humanity's impact on world biota. McLean (1978), for example, correlates the observed increase in atmospheric CO_2 with human activity and suggests that such a fundamental worldwide change of environment may be analogous to those changes that caused the mass extinction of late Mesozoic marine life.

Clear distinctions must be made in the terminology of extinction because of the emotional attitudes generated by some environmental literature. There is a difference between functional extinction and absolute extinction; the former refers to a subviable population that has lost its ability to reproduce and is decreasing in size. One must consider, also, the differences between local and general extinction (see Simberloff, 1976). There are numerous cases of local, even regional and continental, extinctions of organisms that have reappeared safe and thriving in other localities. Dawn redwood *(Metasequoia),* the presumed ancestor of the California redwood *(Sequoia)*, is a well-known example of a species thought to be extinct until rediscovered in China in 1946. Another instance is given by Wetzel, et al, (1975) who have reported living examples of an "extinct" peccary in Paraguay.

What is defined as being extinct, therefore, should be placed on a scale of probability to acknowledge the possibility of rediscovery. On such a scale, the probability that there are living members of the Pleistocene mastodon somewhere in the Arctic approaches very close to zero. On the other hand, the prospect that there may be a few remaining individuals of another species of *Homo* (e.g., *Homo erectus*), that carry such folk names as "Bigfoot," "sasquatch," and the "abominable snowman" might be given slightly more credence. Can it be stated without argument that there are no Lock Ness monsters or that there is no chance they represent the few surviving members of an "extinct" species of marine reptile (see Lehn, 1979)?

There are three general theories for absolute extinction: (a) that organisms "evolve" themselves into a state of overspecialization; (b) that senility leads to their failure to adapt; and (c) that environment is the cause of extinction. The idea that organisms overevolve presumes that evolutionary momentum, in the form of selection intensity and direction, carries certain adaptive trends to inadaptive lengths. *Hypertely* is a good example. It argues that the size of organisms, or of some of their anatomical structures, leads them to a state of being extremely inadapted. The problem is that one cannot know when size is inadaptive, except by subjective means. If whales were to become extinct today, could it be stated that size was a direct factor in their passing? Not likely.

A frequently cited case of evolutionary momentum is given for *Gryphaea*, an extinct oyster-like organism. In this case, one valve actually grew and coiled around the other valve so that the two halves could not be opened, even for feeding. Further study, however, reveals that the development involved several genetic trends, which together aided survival for a considerable length of time. One trend was for increased size, which seems to be common in evolution. A second trend was for increased coiling, which enabled the organism to detach itself from the sea floor. Both trends were apparently positive adaptations, to which the trait of valve closure happened to be linked. But only the largest and oldest individuals ever developed that trait; hence, some have argued that it was actually an altruistic mechanism for promoting youth and perpetuating the population.

Another case of hypertely is that of *Megalocerus*, a Plio-Pleistocene elk with oversized antlers. In the early stages of development, selection may have been for larger body size and larger antlers. Eventually, antler size ceased to be advantageous, so selection was focused on body size. Whatever the pleiotropic relationship between the two characters, both led to increase in antler size. This trend could have resulted in a less than optimum body size, but there is no evidence that either trait led to extinction of the organism.

The clearest general relationship between the characters of organisms and their extinction concerns their degree of specialization. A narrowly adapted organism ranges over a relatively small variety of habitats. It is therefore more likely to be affected by environmental change than an organism that tolerates or thrives in a wider environmental range. Overspecialization is merely specialization that has become disadvantageous because the environment has changed. Precisely the same characters could be highly adaptive at another time or in a different environment.

The idea that senility leads to extinction is regarded as most unlikely. An analogy for the process might be a firework's explosion. Following its origin (the actual detonation), a firework expands upwards, then outwards. In this analogy, each of the bright spots represents a population of individuals linked by a common origin. Each of these populations exists for a time, then begins to fade. In more sophisticated types, there may even be secondary or tertiary explosions. Eventually, there is a recession of area and a fading of intensity, until extinction is finally observed. The problem with this analogy is that in a continually reproducing population, there is nothing that corresponds to the process of aging. In the fireworks example, the individual intensities of light are doomed to fade. All of the relics will become extinct. But, biological relics are not a forecast of extinction. They may very well form the basis for a secondary expansion.

Arguments against senility as a cause of extinction are offered by the case of ammonites, one of the most often used examples of senility. As usually described, the so-called senile forms of ammonites began to appear 100 million years before their last appearance in the fossil record. Some of these forms actually lasted the longest in geologic time and were extremely abundant. Ironically, the so-called normal, or youthful, forms became just as extinct as the senile forms.

Lastly, there is the idea that environment is the cause of extinction. This view is generalized as a loss of adaptation. To say that "loss of adaptation" causes extinction is like saying "loss of existence" causes extinction. The significance of the statement, however, is that it settles the main arguments over whether extinction results from internal or external factors—it results from both. To explain particular cases of extinction requires knowing what happened to the environment and what prevented the population from adapting. Some environmental changes are so abrupt that no population could adjust, as with volcanic eruptions, hydrogen-bomb tests, human intervention, and the like. In general, small or highly uniform populations are most vulnerable to change. Moderate to large populations, and highly variable ones, are most likely to survive (see Ayala, 1968). Changes that will most likely lead to extinction include the appearance of new competitors or predators, the disappearance of food sources, or the appearance of new pathogenic organisms.

An interesting example of a near extinction brought about by a changing environment is reported by Temple (1977; see also Owadally, 1978). Apparently, a coevolutionary adaptation existed between *Calvaria major*, a sapotaceous tree endemic to the island of Mauritius, and *Raphus cucullatus*, the now extinct dodo bird. The stone of the *C. major's* fruit is believed to have adapted to the exceedingly high crushing capacity of the dodo gizzard. In fact, the endocarp became so thick that it could not split unless first passing through the dodo's digestive system. Human discovery of Mauritius in the early 1500s represented an explosive change in the environment of the dodo, and through our efforts, the species was extinct by 1681. It is possible that no *C. major* has germinated naturally since that time. Here is a case of an environmental chain reaction. It also illustrates a case of "overadaptedness" on the part of *C. major*, or at least a case of overdependence on another life form for survival.

Traps and Blind Alleys

The concept of *traps* and *blind alleys* refers to situations where species, or certain of their attributes, become extremely restricted with regard to further adaptive possibilities. Consequently, they survive either as phylogenetic relics, or they become extinct in the face of environmental change to which they cannot adapt. *Trap* is most often applied to environmental conditions and *blind alley* to internal genetic factors in the organisms and populations, although this distinction is not always clear.

Islands of all types and sizes tend to be evolutionary traps. Even though some large islands, like Australia and Madagascar, have been important centers of evolution, they may become traps if strongly isolated or isolated for long geological periods. Geographically speaking, Australia is a continent, but biologically it is an island. South America was also a biological island until its connection to North America via the Isthmus of Panama. For the most part, islands are colonized by immigrants from continents or larger nearby islands. The colonists become very specifically adapted to a relatively small number of habitats and niches. Once adapted, they are seldom able to return to their original areas and, worse perhaps, if invaded successfully by a second wave of migrants, they are particularly liable to extinction. Humankind, and the organisms we have introduced, have caused more extinction on islands than on continents. This statement applies even to Australia, largest of the world's present islands, even though recent evidence from the Lancefield Swamp in southern New South Wales suggests a less catastrophic human affect than previously thought (Gillespie, et al., 1978).

Traps: Lakes and river systems may become traps if they are truly isolated or if they develop peculiar ecological conditions. Consider, for example, Great Salt Lake in the United States, or Lakes Balkash and Baikal in the Soviet Union. All three are classic examples of isolated lakes exhibiting high degrees of endemism; that is, they contain forms that have originated in these places and are found nowhere else on earth. Some of the forms in Lakes Balkash and Baikal are so unique that the Soviet government is as concerned about environmental pollution in those lakes as the United States government is about pollution of the Great Lakes, particularly Lake Erie. In the same vein, the Columbia and Frazer river systems are examples of delicate environments as far as salmon are concerned. The salmon are trapped to the extent that their reproduction and survival are tied to these river networks. Numerous other examples from the fish world confirm that drainage networks represent effectively isolated environments.

Mountain peaks can become evolutionary traps if they stand alone or if they rise to such heights as to be characterized by isolated temperate, boreal, or arctic climatic regimes. Volcanic peaks such as Mount Kilimanjaro in Africa, Mount Kinabalu in Borneo, and a number of others in New Guinea are classic examples of isolated mountain traps.

Blind Alleys: Blind alleys are described as genetic mechanisms that fix organisms into irreversible modes of development. *Apomixis* and self-fertilization in the plant kingdom reduce the capacity for genetic recombination that is the source for much of a species' genetic variation. When these forms of reproduction are obligatory, and not an alternative to bisexual reproduction, their effect is to greatly reduce variation. Although reduced variation improves the adaptedness of the population at a given point, it stringently limits capacity for change.

Specialization in plants and animals does not necessarily lead to extinction, although there are obvious limits to specialization. A horse, for example, cannot have fewer than one toe; nor can the weight of land animals exceed the bearing strength of bone. The first mammals were really extremely specialized reptiles, and mammals certainly do not represent a blind alley. If carried too far, the idea of blind alleys can become meaningless. Since the end point of any evolutionary line is extinction, all adaptations or specializations could be considered blind alleys. Obviously, such an interpretation would reduce the concept to nonsense.

At best, blind alleys and traps can only be assessed after the fact of extinction, and then only rarely. The influences of humanity aside, it is doubtful whether any living forms can be considered to have entered a trap or blind alley, because no one knows for sure what adaptive zones they may yet evolve into. One point is certain, however: evolutionary traps and blind alleys, among other mechanisms, lead to isolation, and it is this process that is significant in speciation.

Man the Extinctor

It is reported that twentieth century man has accelerated the rate of plant and animal extinctions 1,000 percent over what it was at the close of the reptilian era 70 million years ago. The International Union for Conservation of Nature and Natural Resources (IUCN) put the 1974 rate of extinction at one species per year. As recently as 1950, the rate was one per decade. At the close of the Cretaceous period, the rate is estimated to have been one per 1,000 years. Around the world today, some 20,000 species of flowering plants are thought to be endangered. The cause for this rapid rise in the number of endangered and extinct species is almost universally agreed to be humankind. Our species is not only an excellent competitor for food and natural resources; we are also capable of successfully inhabiting virtually all terrestrial habitats.

Over 200 large animal genera became extinct without phyletic replacement during the late Pleistocene. The pattern varies according to the time that hunting groups arrived but embraces all major land masses to a greater or lesser degree. Figure 4.9 shows the general path of destruction, and Figure 4.10 is a list of North American extinctions, not all of which are attributable to humans. There is no concomitant extinction reported for this period in the plant kingdom, and this fact alone causes one to suspect that climatic change was not the sole, or even a major, stimulus for animal extinction.

In North America, extinction began to accelerate about 11,000 years ago. It started in South America about 10,000 years ago; in Australia about 13,000 B.P.; in New Zealand about 900 B.P.; in Madagascar about 800 B.P.; and in Eurasia 13,000-11,000 B.P. Among the animals lost were giant marsupials in Australia; moas and one species of tortoise in North America; a largely unknown fauna from South America; and three genera from Europe. In Africa and Southeast Asia extinctions predated those in other parts of the world, coinciding with the end of the Acheulean cultural stage (about 40,000-50,000 B.P.).

The weight of world evidence is that megafaunal extinctions during the Quaternary closely follow the ascent of humankind. In North America, the Llano Estacado contains evidence for the extinction of species of

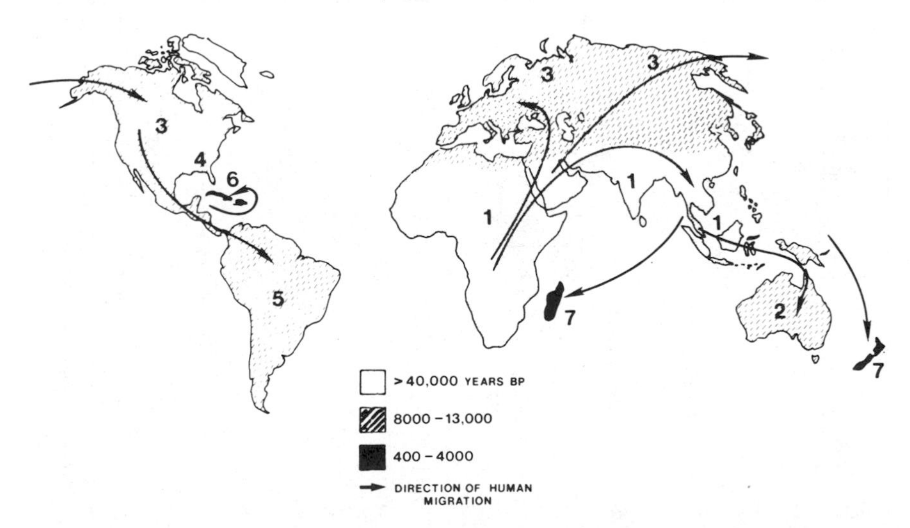

Figure 4.9. Generalized pattern of late Pleistocene and Recent megafaunal extinction around the world. As prehistoric hunters moved out of Africa and Southeast Asia into the Australian region, Eurasia, North America, and finally, South America, they effectively destroyed large numbers of species. The impact continues today and is most readily observed on isolated islands.

mammoth, horse, camel, peccary, and four horned antelope, all during a short, relatively dry phase (the Scharbauer Interval) following the arrival of humans during the preceding pluvial (the Blackwater subpluvial). In Africa, too, there are fewer forms today than there were during the early Pleistocene. In the case of Africa, fire may have been the agent for wholesale elimination of mammals, and human response to these reduced numbers may have been the development of improved hunting techniques and equipment.

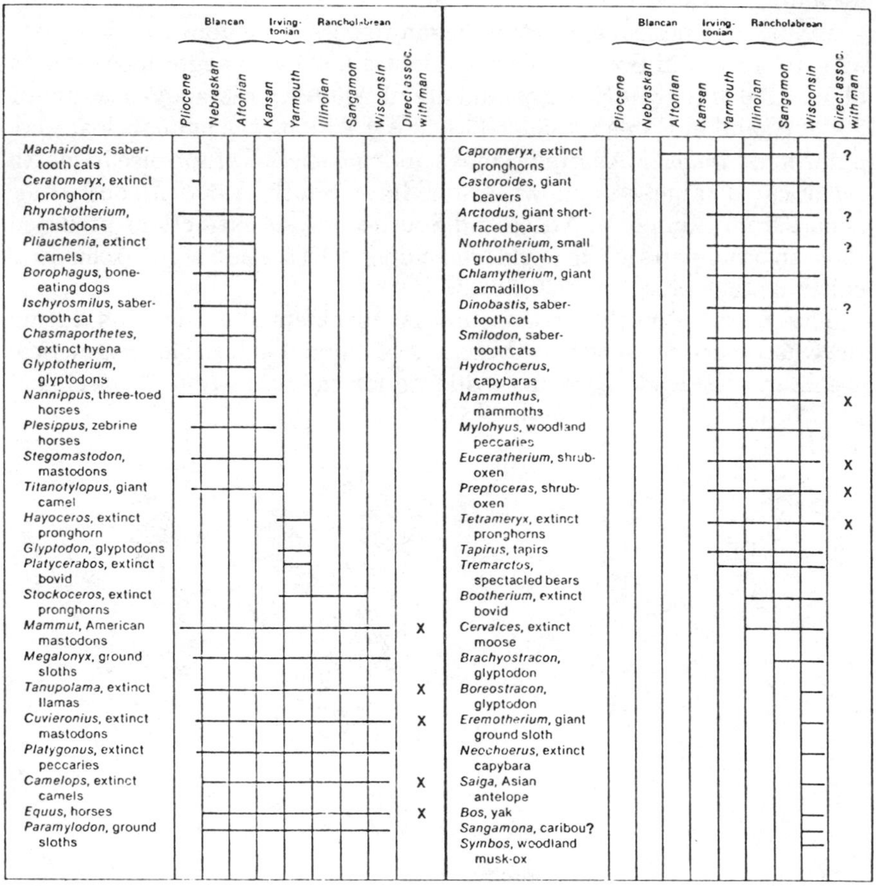

Column headings (both halves): Blancan (*Pliocene*, *Nebraskan*, *Aftonian*); Irving-tonian (*Kansan*, *Yarmouth*); Rancholabrean (*Illinoian*, *Sangamon*, *Wisconsin*); *Direct assoc. with man*

	Direct assoc. with man
Machairodus, saber-tooth cats	
Ceratomeryx, extinct pronghorn	
Rhynchotherium, mastodons	
Pliauchenia, extinct camels	
Borophagus, bone-eating dogs	
Ischyrosmilus, saber-tooth cat	
Chasmaporthetes, extinct hyena	
Glyptotherium, glyptodons	
Nannippus, three-toed horses	
Plesippus, zebrine horses	
Stegomastodon, mastodons	
Titanotylopus, giant camel	
Hayoceros, extinct pronghorn	
Glyptodon, glyptodons	
Platycerabos, extinct bovid	
Stockoceros, extinct pronghorns	
Mammut, American mastodons	X
Megalonyx, ground sloths	
Tanupolama, extinct llamas	X
Cuvieronius, extinct mastodons	X
Platygonus, extinct peccaries	
Camelops, extinct camels	X
Equus, horses	X
Paramylodon, ground sloths	
Capromeryx, extinct pronghorns	?
Castoroides, giant beavers	
Arctodus, giant short-faced bears	?
Nothrotherium, small ground sloths	?
Chlamytherium, giant armadillos	
Dinobastis, saber-tooth cat	?
Smilodon, saber-tooth cats	
Hydrochoerus, capybaras	
Mammuthus, mammoths	X
Mylohyus, woodland peccaries	
Euceratherium, shrub-oxen	X
Preptoceras, shrub-oxen	X
Tetrameryx, extinct pronghorns	X
Tapirus, tapirs	
Tremarctos, spectacled bears	
Bootherium, extinct bovid	
Cervalces, extinct moose	
Brachyostracon, glyptodon	
Boreostracon, glyptodon	
Eremotherium, giant ground sloth	
Neochoerus, extinct capybara	
Saiga, Asian antelope	
Bos, yak	
Sangamona, caribou?	
Symbos, woodland musk-ox	

Figure 4.10. Megafaunal (>50 kg) extinctions in North America from the Pliocene to Recent. *(From C. W. Hibbard, et al., 1965, Quaternary Mammals of North America, in The Quarternary of the United States, H. E. Wright, and D. G. Frey, eds., Princeton University Press, Princeton, N. J., copyright © 1965 by Princeton University Press. Adapted Table 2, p. 513, by permission of Princeton University Press.)*

In southwestern Australia, 17 species of Pleistocene marsupials are no longer present. An additional 34 species managed to survive. It is known that humans first appeared in this region during the late Pleistocene, and it is believed that their impact was to fragment the number and distribution of populations, thus hastening their extinction. Neither man, nor dingo *(Canis dingo)* could have extinguished these species through hunting alone.

Guilday (1967) is one of several authors who argue that humankind was not the "*cause magnifique*" for late Pleistocene extinctions, but rather that differential extinction took place because of the greater needs of larger animals for food, space, and water during the onset of worldwide desiccation. This same desiccation stimulated adaptations in *Homo sapiens* toward animal husbandry and agriculture. Elimination of the large herbivores is believed to have led to the elimination of their carnivorous predators.

In diverse habitats, many herbivores can be supported, as is obviously the case in modern day East Africa. During times of crisis, however, each must retreat to its most efficient habitat and suffer its fortunes. If there are few habitats and if these suffer desiccation or other crises, the forms present will be thrown into severe competition. One can envision this model for the North American Great Plains, with the bison *(Bison)* being the "winner." Smaller animal forms are less severely affected, since they can retreat to microhabitats barred to larger forms. Figures 4.11 and 4.12 illustrate the models for differential extinction as a function of habitat size, niche requirements, and animal size. In Figure 4.11, it is apparent that large animals with overlapping niche requirements enter into competition, with resulting extinctions. Small animals suffer fewer extinctions. In Figure 4.12, the model shows that the Saharan region suffered massive large-animal extinctions compared to the southern and eastern savannas.

SUMMARY

Speciation and extinction represent the conceptual beginning and end of the life stream. Although there is a more or less mature understanding of the elements defining these processes, our knowledge of how they function is still in its infancy. If the idea that speciation is a population phenomenon prevails, it appears that segregation sets the stage for genetic drift and eventual reproductive isolation. If, on the other hand, chromosome rearrangements are important modes for speciation, then individuals may figure prominently in the spontaneous generation of new organisms. The distinctions between allopatric,

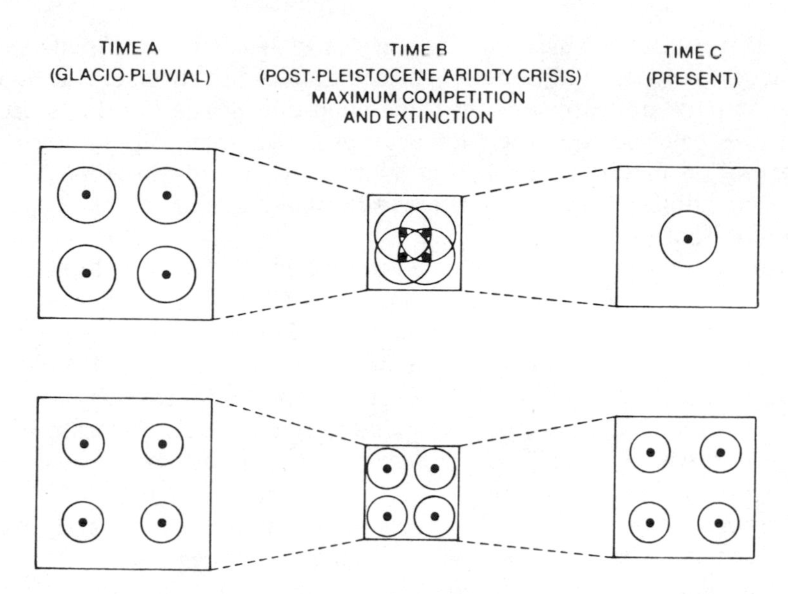

Figure 4.11. Differential extinction of larger and smaller animals as a result of compressed habitats and niches during phases of Pleistocene aridity. The size of squares represents area of major habitat. Circles represent scope of ecological requirements. The size of dots indicates the relative size of animals. *(From J. E. Guilday, 1967, Differential Extinction During Late Pleistocene and Recent Time, in Pleistocene Extinctions: The Search for a Cause, P. S. Martin and H. E. Wright, eds., 1967, p. 124; copyright © 1967 by Yale University Press. Reprinted by permission of Yale University Press.)*

sympatric, and stasipatric speciation influence the details and acceptability of the gradual versus spontaneous, chromosomal versus mutational, and isolation versus congeographic theories. Ample evidence supports the validity of all of the models; and as has been found repeatedly in science, nature probably proceeds along all of these paths to ensure a continual growth in diversity.

Extinction is as difficult to assess as speciation. Pronouncements that a given form no longer exists must be accompanied by caveats relating time and space. Functional extinction, local extinction, absolute extinction, and states of being endangered are all quite separate and, in fact, lead one to favor "probable" rather than "actual" extinction. It is no doubt true that humanity is single-handedly diminishing the world's animal and plant diversity, but there are limits even to our capacity. Humanity should strive to preserve as many existing types as possible, but should we not also use our knowledge to find replacement organisms?

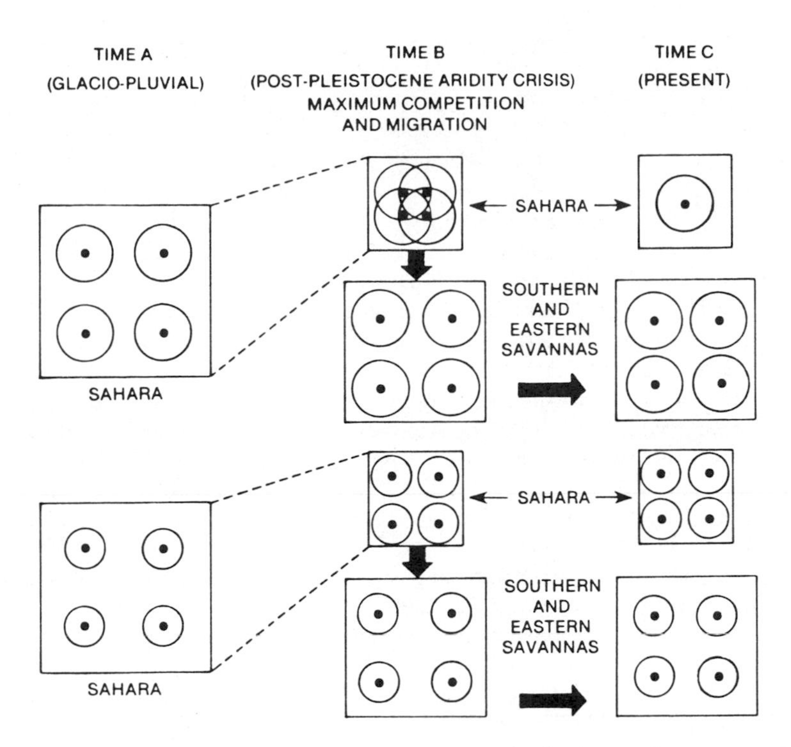

Figure 4.12. Explanatory model for the survival of African fauna. Symbols are as given in Figure 4.11. By this reasoning it is evident that the southern and eastern savannas of Africa would experience fewer extinctions as a result of Pleistocene aridity than would the Sahara. *(From J. E. Guilday, 1967, Differential Extinction During Late Pleistocene and Recent Time, in Pleistocene Extinctions: The Search for a Cause, P. S. Martin and H. E. Wright, eds.,1967, p. 127; copyright © 1967 by Yale University Press. Reprinted by permission of Yale University Press.)*

REFERENCES

Ayala, F. J., 1968, Genotype, Environment, and Population Numbers, *Science* **162**:1453-1459.

Bush, G. L., 1975, Modes of Animal Speciation, *Annu. Rev. Ecol. and Syst.* **6**:339-364.

Cain, S., 1944, *Foundations of Plant Geography,* Harper and Row, New York, 556p.

Carson, H. L., 1970, Chromosome Tracers of the Origin of Species, *Science* **168**:1414-1418.

Darlington, D. C., 1968, *Chromosome Botany and the Origins of Cultivated Plants,* 2nd ed., Allen & Unwin, London, 231p.

Ehrlich, P. R., and P. H. Raven, 1969, Differentiation of Populations, *Science* **165**:1228-1232.

Federov, An. A., 1966, The Structure of the Tropical Rainforest and Speciation in the Humid Tropics, *J. Ecol.* **54**(1):1-11.

Giesel, J. T., 1976, Ecological Genesis of Endangered Species: The Philosophy of Preservation, *Annu. Rev. Ecol. and Syst.* **7**:33-56.

Gillespie, R., D. R. Horton, P. Ladd, P. G. Macumber, T. H. Rick, R. Thorne, and R. V. S. Wright, 1978, Lancefield Swamp and the Extinction of the Australian Megafauna, *Science* **200**:1044-1048.

Guilday, J. E., 1967, Differential Extinction During Late Pleistocene and Recent Times, in *Pleistocene Extinctions: The Search for a Cause,* P. S. Martin, and H. E. Wright, Jr., eds., Yale University Press, New Haven, Conn., pp. 121-140.

Hibbard, C. W., D. E. Ray, D. E. Savage, D. W. Taylor and J. E. Guilday, 1965, Quaternary Mammals of North America, in *The Quaternary of the United States,* H. E. Wright and D. G. Frey, eds., Princeton University Press, Princeton, New Jersey, pp. 509-525.

Huxley, Jr., 1938, Species Formation and Geographical Isolation, *Linn. Soc. London Proc.* **150**:253-264.

Jackson, R. C., 1976, Evolution and Systematic Significance of Polyploidy, *Annu. Rev. Ecol. and Syst.* **7**:209-234.

Lehn, W. H., 1979, Atmospheric Refraction and Lake Monsters, *Science* **205**:183-185.

Martin, P. S., 1967, Prehistoric Overkill, in *Pleistocene Extinctions: The Search for a Cause,* P. S. Martin and H. E. Wright, Jr., eds., Yale University Press, New Haven, Conn., pp. 75-120.

Mayr, E., 1947, Ecological Factors in Speciation, *Evolution* **1**:263-288.

Mayr, E., 1970, *Population, Species, and Evolution,* Harvard University Press, Cambridge, Mass., 453p.

McLean, D. M., 1978, A Terminal Mesozoic 'Greenhouse': Lessons from the Past, *Science* **201**:401-406.

Olsen, P. E., and P. M. Galton, 1977, Triassic-Jurassic Tetrapod Extinctions: Are They Real? *Science* **197**:983-986.

Owadally, A. W., 1978, The Dodo and the Tambalacoque Tree, *Science* **203**:1363-1364.

Rand, A. L., 1948, Glaciation, An Isolating Factor in Speciation, *Evolution* **2**:314-321.

Rensch, B., 1938, Some Problems of Geographical Variation and Species Formation, *Linn. Soc. London Proc.* **150**:275-285.

Shafer, D. A., 1979, Hybrid Ape Cited as "Evolutionary", *Albuquerque Journal,* July 28, p. B-3.

Simberloff, D., 1976, Species Turnover and Equilibrium Island Biogeography, *Science* **194**:572-578.

Simpson, G. G., 1953, *The Major Features of Evolution,* Columbia University Press, New York, 434p.

Stoddart, D. R., 1968, Isolated Island Communities, *Science Journal,* April, pp. 32-38.

Tauber, C. A., and M. J. Tauber, 1977, Sympatric Speciation Based on Allelic Changes at Three Loci: Evidence from Natural Populations in Two Habitats, *Science* **197**:1298-1299.

Temple, S., 1977, Plant-Animal Mutualism: Coevolution with Dodo Leads to Near Extinction of Plant, *Science* **197**:885-886.

U.S. Department of the Interior, 1975, Threatened or Endangered Fauna or Flora, *Federal Register* **40**(127, part V):27824-27924.

Valentine, J. W., 1968, Climatic Regulation of Species Diversification and Extinction, *Geol. Soc. Am. Bull.* **79**:273-276.

White, M. J. D., 1968, Models of Speciation, *Science* **159**:1065-1070.

White, M. J. D., 1978, *Modes of Speciation,* Freeman, San Francisco, 456p.

FURTHER READING

Atchley, W. R., and D. S. Woodruff, eds., 1981, *Evolution and Speciation: Essays in Honor of M. J. D. White,* Cambridge University Press, New York, 436p.

Axelrod, D. I., 1967, *Quaternary Extinctions of Large Mammals,* University of California Publications in Geological Sciences, Vol. 74, University of California Press, Berkeley, Calif., 42p.

Cronquist, A, 1968, *The Evolution and Classification of Flowering Plants,* Houghton-Mifflin, Boston, 396p.

Cuellar, O., 1977, Animal Parthenogenesis, *Science* **197**:837-843.

Ehrlich, P. R., and A. Ehrlich, 1981, *Extinction: The Causes and Consequences of the Disappearance of Species,* Random House, New York, 305p.

Eldredge, N., and S. J. Gould, 1972, Punctuated Equilibria, in *Models in Paleobiology,* T. J. M. Schopf, ed., Freeman Cooper, San Francisco, pp.82-115.

Endler, J. A., 1977, *Geographic Variation, Speciation and Clines,* Princeton University Press, Princeton, N.J., 248p.

Grant, P. R., 1981, Speciation and the Adaptive Radiation of Darwin's Finches, *American Scientist* **69**(6):653-663.

Grant, V., 1981, *Plant Speciation,* 2nd ed., Columbia University Press, New York, 563p.

Hallam, A., ed., 1977, *Patterns of Evolution as Illustrated by the Fossil Record,* Elsevier, New York, 592p.

Harper, C. W., Jr., 1975, Origin of Species in Geologic Time: Alternatives to the Eldredge-Gould Model, *Science* **190**:47-48.

Hendrickson, H. T., 1978, Sympatric Speciation: Evidence? *Science* **200**:345-346.

Johanson, D. C., and T. D. White, 1979, A Systematic Assessment of Early African Hominids, *Science* **203**:321-330.

Levin, D. A., 1979, The Nature of Plant Species, *Science* **204**:381-384.

Mayr, E., 1972, The Nature of the Darwinian Revolution, *Science* **176**:981-989.

Mayr, E., 1976, *Evolution and the Diversity of Life,* Harvard University Press, Cambridge, Mass., 721p.

Merrilees, D., 1967, Man the Destroyer: Late Quaternary Changes in the Australian Marsupial Fauna, *J. R. Soc. West. Aust.* **51**(1):1-24.

Myers, N., 1976, An Expanded Approach to the Problem of Disappearing Species, *Science* **193**:178-180.

Ricklefs, R. E., 1978, Paleontologist Confronting Macroevolution, *Science* **199**:58-60.

Russell, D. A., 1982, The Mass Extinctions of the Late Mesozoic, *Scientific American* **246**(1):58-65.

Schall, J. J., and E. R. Pianka, 1978, Geographical Trends in Numbers of Species, *Science* **201**:679-686.

Stebbins, G. L., 1971, *Processes of Organic Evolution,* 2nd ed., Prentice-Hall, Englewood Cliffs, N. J., 193p.

Taylor, R. W., 1978, Nothomyrmecia macrops: A Living Fossil Ant Rediscovered, *Science* **201**:979-985.

Templeton, A. R., 1981, Mechanisms of Speciation—A Population Genetic Approach, *Annu. Rev. Ecol. and Syst.* **12**:23-48.

Udvardy, M. D. F., 1969, *Dynamic Zoogeography,* Van Nostrand, New York, 445p.

Wetzel, R. M., R. E. Dubos, R. L. Martin, and P. Myers, 1975, Catagonus, an "Extinct" Peccary, Alive in Paraguay, *Science* **189**:379-380.

Chapter 5

Dispersal and Distribution The Spread of Change

Time and space considerations are at the heart of reproductive isolation and speciation. Without them, any given locality would maintain a universally interbreeding panmictic population through normal reproduction. Gene exchange, however, is reduced or prevented by distance and by the intervention of unsuitable habitats. Allopatric distributions represent geographic isolation, whereas sympatric distributions represent ecologically isolated groups. The question is, how can isolation be achieved? At any moment in history every species has a definable geographic distribution. The aim of biogeography is to discover what those patterns are, how they evolved, and under what conditions they change. Within the context of these questions, both active and passive means for dispersal, and different types of distribution patterns must be considered. For the case of active dispersal—that is, where propagules are disseminated or animals moved—several mechanisms, modes, and avenues can be identified. The case of passive dispersal or vicariance biogeography is addressed in Chapter 6.

DISPERSAL MECHANISMS

Basic dispersal mechanisms, which operate at the individual and population levels, are listed below.

Anemochores are organisms dispersed by wind. Locusts in flight are examples of anemochores in the sense that they are influenced by wind speed and direction. The plant kingdom is particularly rich in species whose seeds or spores are disseminated by wind.

Hydrochores are organisms dispersed by water. The coconut *(Cocos nucifera)* is often given as an example.

Anemohydrochores are organisms dispersed by either wind or water. Many island species are adapted to this mechanism as a means of populating nearby islands.

Anthropochores are organisms dispersed by people. Most introductions have limited distributions and little competitive ability. Exceptions to this include such invasive species as the tumbleweed *(Salsola)* and salt cedar *(Tamarix pentandra)* in the United States, and the rabbit (*Lepus* spp.) and prickly pear cactus (*Opuntia* ssp.) in Australia.

Phoresy refers to a symbiotic relationship where one form physically transports another. Examples are mites carried by insects, fleas by dogs, and the like. All anthropochores are phoretics but not all phoretics are anthropochores.

These mechanisms are easy to define but difficult to prove except by example. They imply that successful dispersal relies on internal coevolutionary adaptations as well as on external factors. Some authors (see Eisley, 1970) suggest that there is a "struggle" between internal dispersal, which strives to expand a species' range, and the environment, which strives to prevent expansion by presenting barriers. Sometimes the environment wins and sometimes the organism wins. Although this idea is clearly anthropomorphic, it appears that the environment acts as a filtering device to prevent all species from occupying all environments. If this were not true, dispersal mechanisms would ensure a uniform biosphere. Environment, acting for and against successful dispersal, is the basis of geographic and ecologic isolation, as well as of speciation; but the process is no more a struggle than any other natural process.

Modes

By the mechanisms described above, species expand geographically. Many have adapted to several of these mechanisms. Indeed, it can be said that all species possess the potential to be anthropochores, in addition to their naturally evolved mechanisms. Once a mechanism is set in motion, the modes by which the speed of a species' advance into new territory can be described as slow penetration, migration, and irruption.

Slow Penetration: In the case of *slow penetration,* individuals gradually advance into favorable territory with each succeeding generation. Figure 5.1 illustrates the migration of the nine-banded armadillo *(Dasypus novemcinctus)* into the Gulf Coast region of the United States. It took almost 100 years for the species to spread from southern Texas to southern Arkansas, but fewer than 10 years for it to spread westward from its point of introduction in Florida to Gulf Port, Mississippi. The barrier to more rapid advance appears to be distance relative to animal speed.

Most examples of slow penetration are observed in plants. Since these life forms are not mobile, except as uprooted introductions or as propagules, geographic spread, as such, is usually recognized over several generations. Observations over time, however, can provide a rate and direction of spread. The salt cedar *(Tamarix pentandra)* was introduced into the southwestern United States in 1877 (Harris, 1966) and has subsequently spread by natural means to virtually all of the waterways of the Texas gulf rivers, the Rio Grande, and Colorado river systems. Figure 5.2 shows the extent of the occupation as of 1961.

Good's theory of tolerance (1974) is a useful construct for understanding the essence of plant migrations, including that of the salt cedar in the

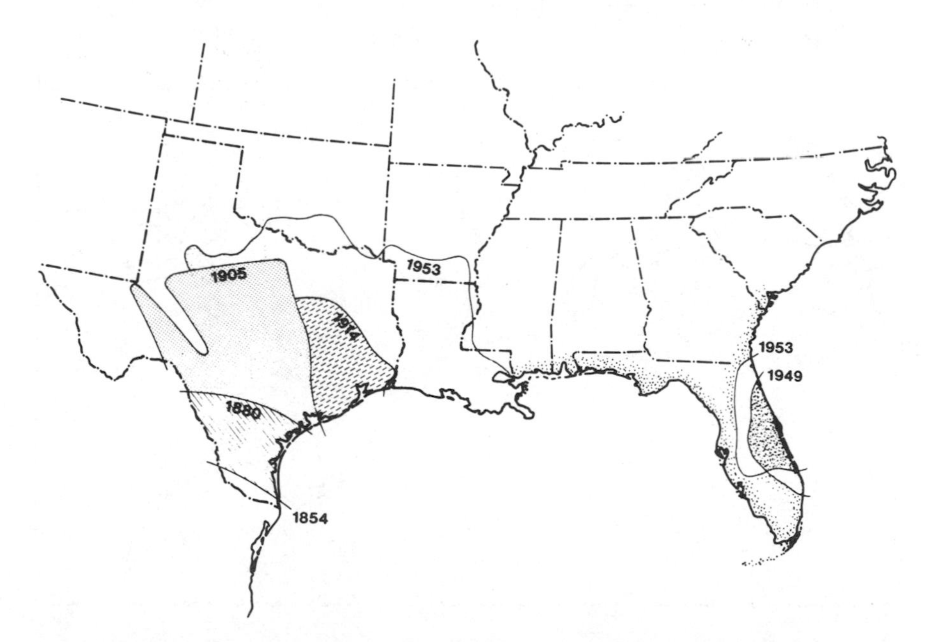

Figure 5.1. Advance of the nine-banded armadillo *(Dasypus novemcinctus)* by slow penetration. *(From G. D. Buchanan and R. V. Talmage, 1954, Geographic Distribution of the Armadillo in the United States, Texas Journal of Science* **6:***142–150; copyright © 1954 by the Texas Academy of Science. Reprinted with permission.)*

Southwest. His theory provides an underlying theme for understanding the geography of all flowering plants, namely, that much of their shifting geography may be explained by the expansion and contraction of areas forced by climatic shifts that have occurred since mid-Tertiary (Miocene) times. The plants have achieved these distributions because they have minimum, optimum, and maximum environmental conditions under which they can survive and reproduce. If the environmental conditions change beyond the limits of their "tolerance range," then survival will depend upon their ability to propagate seed into areas where they can survive. There are significant consequences for the model of geographic speciation with respect to the populations of plants in the van and rear of this type of migration.

The essence of the process is given in a hypothetical example in Figure 5.3. The irregular shape represents the actual and potential

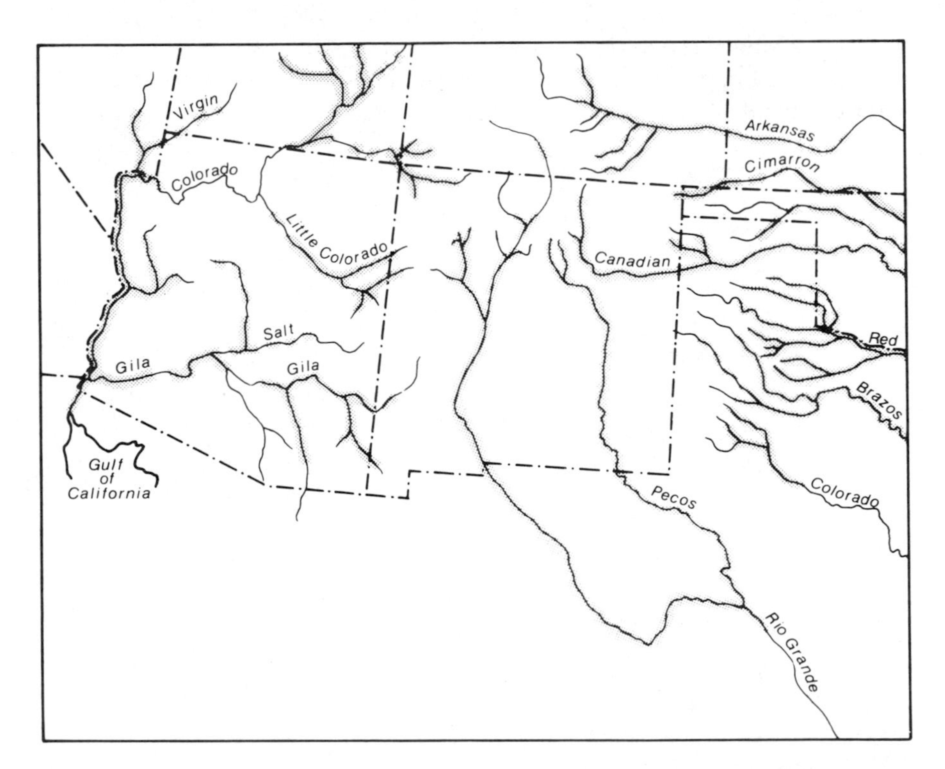

Figure 5.2. Distribution of salt cedar *(Tamarix pentandra)* in the Southwest. The species was introduced into Texas in 1877 and has spread throughout the major drainage basins in the last century. *(From T. W. Robinson, 1965, Introduction, Spread and Areal Extent of Saltcedar (Tamarix) in the Western States, U.S. Geological Survey Professional Paper 491-A, Plate 1.)*

geographic range of a plant species in the taiga region just south of the tundra boundary (10°C July isotherm). At some time, t_1, propagules from individual specimens growing near the northern boundary of the actual range may be spread both inward, into the existing range, and outward

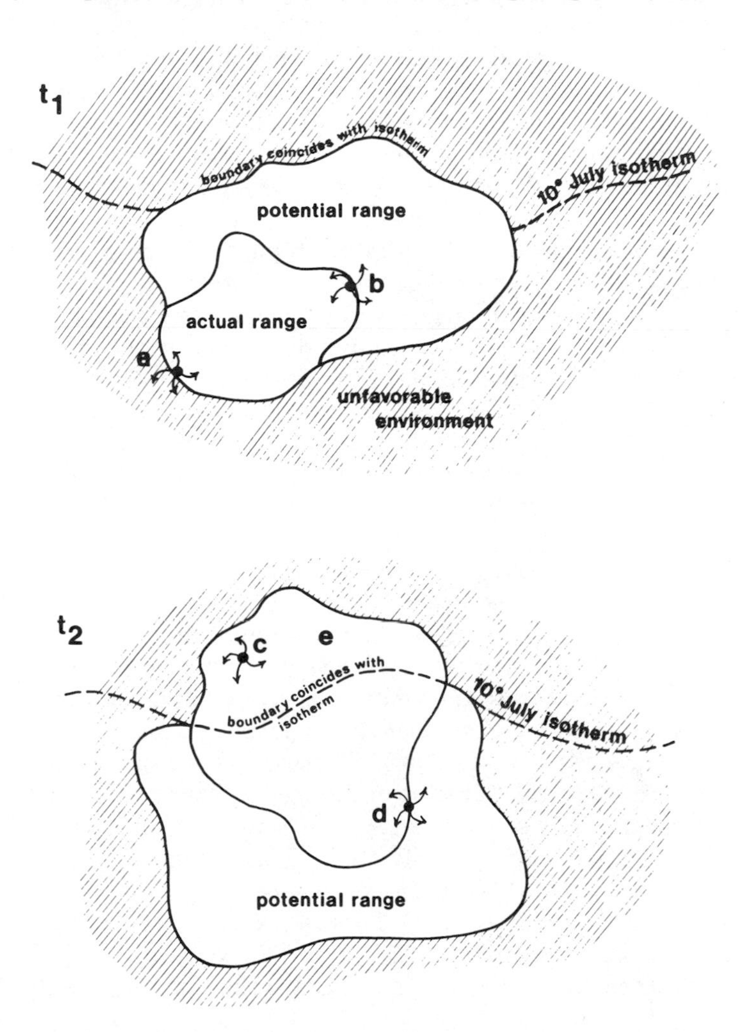

Figure 5.3. Models of the theory of plant migration, as expressed by Good (1974). (above) Initial distribution at time, t_1, of a species near the 10°C July isotherm. The species is expanding northward into its potential range. (below) As a result of climatic shifts the same species at t_2 is advancing southward into a new potential range and becoming relict or extinct in its northern range. *(a)* through *(e)* show directions of propagule dissemination from various points within the actual range.

beyond the species' potential range. If the outward-bound propagules successfully germinate, the range of the species will expand in those directions. The internal physiological tolerances of the species have not been exceeded by the physical conditions of the environment.

Suppose, however, that at some later time, t_2, a glacial advance begins from the area poleward of the tundra. This change would mean that the southern boundary of the tundra (10°C July isotherm) could shift southward. Under these new conditions, propagules from the northern part of the range would have to germinate within the tundra climate to maintain or expand the species' range. Most would not survive in this locality. Outward-bound propagules on the southern margins of the actual range, however, which in t_1 might have been physiologically unadapted to the warmer environment, can now successfully germinate in the cooler conditions imposed by the glacial episode.

Over time, the result is a recordable slow penetration of the species into new territory. The actual range of that species at any given point in time is really a picture of differential survival. More to the point, if one superimposes the stasipatric model of speciation (p. 79), or invokes the idea of chromosomal rearrangement, it becomes immediately clear that speciation processes are intimately linked to processes of geographic dissemination.

Migration: In contrast to imperceptably slow penetrations there are annual migrations to feeding grounds and breeding areas, and *irruptions* due to population pressures. African megafaunal migrations of wildebeest (*Connochaetes* spp.) and zebra (*Equus* spp.), like bird migrations around the world, are usually triggered by seasonal environmental change. They begin as population phenomena, but lead to a reunion of the entire species in the new seasonal range. These kinds of migrations ensure gene flow within the species, at least in those cases where breeding takes place in a geographically smaller range. The Serengeti Plains region contains the most spectacular concentration of mammalian megafauna of any place on earth. In 1974 there were more than 1 million wildebeest (*Connochaetes* spp.); 600,000 Thomson's gazelle *(Gazella thomsoni)*; 200,000 zebra (*Equus* spp.); 65,000 buffalo *(Syncerus caffer)* and more than 20 other species of grazing animals in the area (McNaughton, 1976).

Irruption: Irruptions caused by population pressure lead to a temporary expansion of a species' range and generally subject local populations to environmental stresses beyond their tolerance. The population invades a surrounding territory in search of resources but often cannot reproduce in the expanded area. Individuals who remain behind form the basis for renewed population growth in the core areas. The cases for desert locust *(Schistocerca gregaria)* the African migratory locust *(Locusta migratoria)*, and red locust *(Nomadacris septemfasciata)*, are perhaps among the best known examples. Figures 5.4a, b, and c show their respective invasion and outbreak areas.

The life cycle of locusts has four main stages (Uvarov, 1966). After hatching, young hoppers undergo a series of developmental molts. Throughout these changes they remain essentially nonflying and nongregarious. When concentrated into small areas, however, they often turn gregarious and form marching bands. The length of time in each

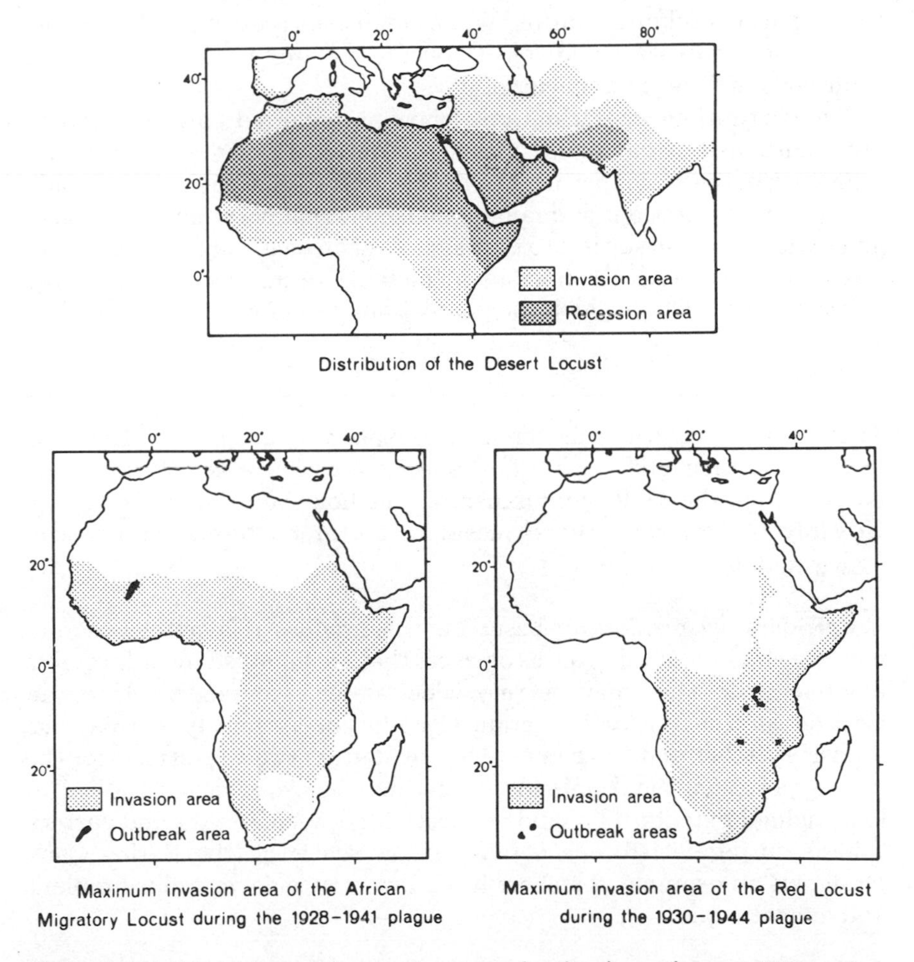

Figure 5.4. *(A)* Invasion and recession areas for the desert locust. *(B)*Invasion and outbreak areas for the African migratory locust, 1928-1941. *(C)* Invasion and outbreak areas for the red locust, 1930-1944. All three species irrupt in plague years into territories where they cannot survive.

hopper stage varies according both to species and geographic location and to seasonal climatic factors. According to Hemming (1968), delays in development may be of survival value to avoid adverse environmental conditions, such as a long dry season. Winged, immature adults emerge from the final hopper molt. Generally, this stage coincides with a recession of suitable habitat during the adverse season, so that crowding takes place. Under these conditions there is often a change in behavior from solitary to gregarious. In most years, decimation by predators, together with adequate supplies of food and space for the young adults, prevent the formation of flying swarms. When conditions are not good, however, swarms of variable size and density result, and are capable of traversing thousands of miles in a matter of days.

These irruptions or infestations follow no preferred path but, rather, are conditioned by day-to-day weather patterns. The individuals cannot reproduce in the areas they invade, so the expansion of their range is only momentary. Migrations and range expansions that occur over a period of generations are also subject to the vagaries of chance, but the dispersal of forms can be described in more predictable terms because there are definable paths along which the populations can move.

Avenues

The avenues along which dispersal takes place are traditionally described in terms of their ability to funnel species into tolerable new locations having much the same environment as the homeland. Simpson (1965), Udvardy (1969), and others, recognize corridors, filters, and sweepstakes routes.

Corridors: *Corridors* are best viewed as avenues of favorable environment, such as mountain passes or rivers that are bounded by unfavorable environments. They may be very wide, as for example, the Eurasian Corridor or the Rocky Mountain Corridor, or extremely narrow, like roadside ditches and fence lines. They operate as two-way streets for the passage of plants and animals, sometimes across wide latitudinal or longitudinal extents. The land connection between Alaska and eastern Siberia during Pleistocene periods of low sea level (the Bering Land Bridge) is an example of a short-term, but very wide, corridor for plant and animal migration.

Filters: *Filters* are viewed as favorable avenues along which the chances of successful migration are high. Generally, however, one observes an attenuation of species diversity from one end to the other, because the environments along the funnel are not equally propitious for all of the forms. The mammalian fauna between Guatemala and Colombia through the Isthmus of Panama, shows a progressive decrease from north to south, leaving mainly the bats and rats as the most successful users of the avenue. Figure 5.5 is another example of a filtering mechanism showing the decline in weevil genera *(Cryptorhychinae)* along a transect from Australia and New Guinea to beyond the Marquesas Islands. In this case, the barriers of water and distance between the islands act together to inhibit dispersal.

The role of islands as *stepping stones* in species dispersal and isolation is so important that it warrants a momentary digression. An assessment of the influence of stepping stones really revolves around the efficiency of species in dispersing as anemochores and hydrochores. The basic question

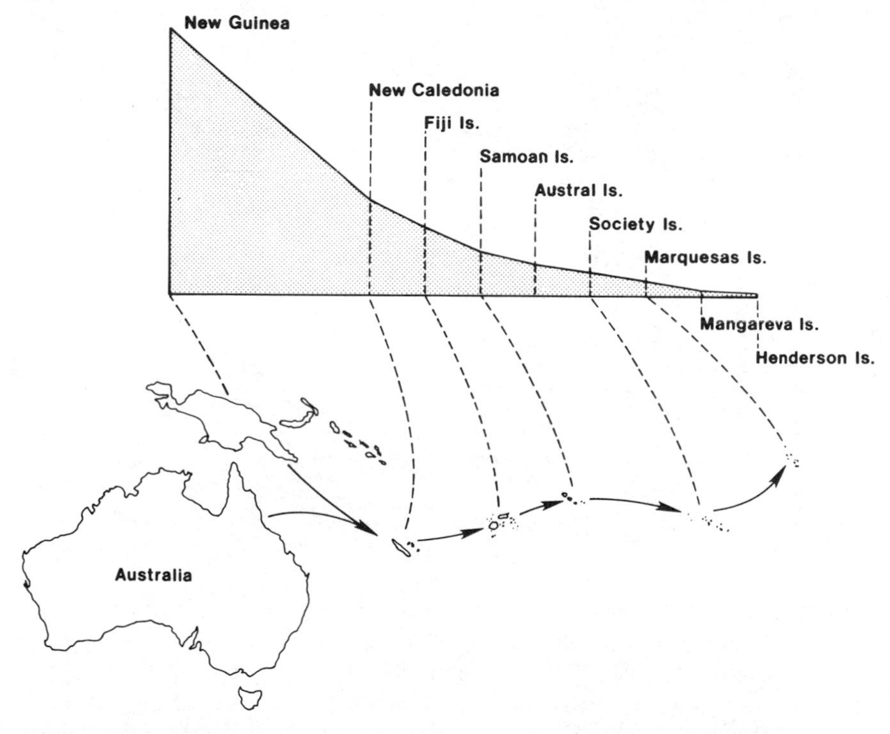

Figure 5.5. Example of a migration filter in the weevil family. The height of the bars forming the funnel are proportional to the number of genera found on each island or island group. Water and distance are the major barriers preventing uniform distribution of all weevil genera. *(From M. D. F. Udvardy, 1969, Dynamic Zoogeography, Van Nostrand, New York, p. 90; coyright © 1969 by Van Nostrand Reinhold Company. Reprinted by permission of Van Nostrand Reinhold Company.)*

can be approached by asking what the probability is that a given propagule arrived on a recipient island from a stepping stone, W_1, rather than from a more distant source region, W_2. The answer depends, in part, on how many propagules are leaving the islands, W_1 and W_2. Figure 5.6a presents

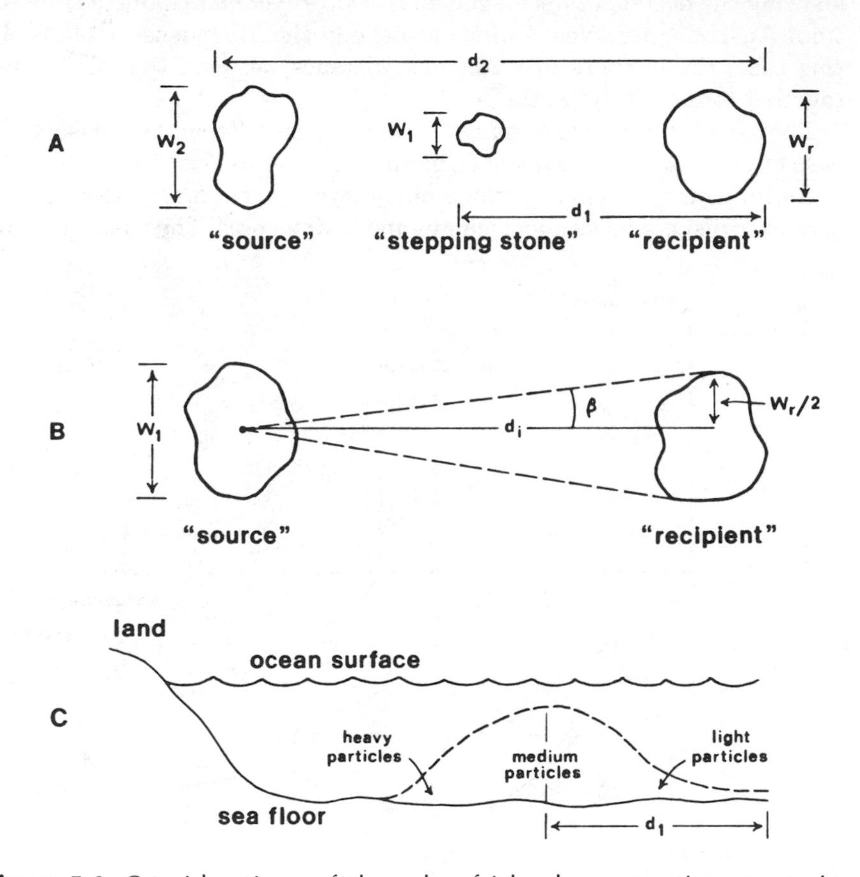

Figure 5.6. Considerations of the role of islands as stepping stones in species dispersal and migration. *(A)* W_1 is a stepping stone between a more distant source, W_2, and a recipient island, W_r, where W is the width of each island respectively. Distances between the islands are given as d_1 and d_2. *(B)* Model taking into account the effect of predominant wind direction as it affects the distribution of anemochores. Successful dispersal is a function of the cotangent of half the width of the recipient island and its distance from a source. *(C)* Model describing deceleration of propagules with increasing distance from a source. In the analogous situation of offshore sediments, the heaviest particles settle nearest to shore and the lightest particles settle farther out. A normal gaussian distribution is anticipated.

the major elements in the argument. If the islands are assumed to be more or less circular, then two probabilities exist:

If the mean dispersal distance, Ω, is long with respect to the area of the islands and the distance between them, then propagules can come from anywhere on either of the source islands, W_1 or W_2, and the stepping stone is immaterial. The probability is thus a function of the area of the islands, Πr^2. The only caveat is that the propagule must settle on land, not water, to successfully migrate.

If Ω is short with respect to island size, only propagules from the island periphery are likely to be successful migrants. This likelihood is a function of the circumference of the island, ΠD, and the stepping stone assumes greater importance as a probable source.

Another aspect of the question is, how many of those propagules that depart ever arrive at a suitable habitat? This issue can also be thought of as the probability of surviving. MacArthur and Wilson (1967) discuss three such probability distributions:

A uniform distribution is achieved if Ω is very long with respect to d_2, the distance from the most remote source to the recipient island. In this case, distance is obviously no longer an important consideration, and the probability of successful dispersal is a function of the ratio of the sizes of the islands to the area of water.

An exponential distribution is observed if the propagule moves in a constant direction and ceases to move farther after a set amount of time. In other words, success declines with increasing distance, as indicated by the expression in

$$aw_i^2 e^{-d_i/\Omega}, \text{ assuming that } \Omega >> w_i \tag{5.1}$$

The term $e^{-d_i/\Omega}$ is a negative exponential that conveys the notion of distance decay. W_i is the source island and a is a constant. This distribution is thought to be a prevalent form of dispersal for plants and insects that disseminate as anemochores. A refinement on the function should take into account dispersion along the path of dispersal, as illustrated in Figure 5.6b. The angular probability for successful dispersal is the ratio of arc subtended by the recipient island to 360 degrees, symbolized as

$$\frac{2\tan^{-1}(w_r/2d_i)}{360^\circ} aw_i^2 e^{-d_i/\Omega} \tag{5.2}$$

In terms of air pollution, this equation might describe the chances of a given particle emitted from a smokestack of landing at any given point along the mean dispersal radius. With both particulate air pollutants and biological forms, a finite distance can be traversed that depends on the size and weight of the propagules, the strength of the wind, and other factors.

A normal distribution is observed when deceleration, according to weight, occurs with increasing distance from the source, as for example, sediment drifting to sea after issuing from the mouth of a river. In Figure 5.6c, the part of the curve labeled as d_1 is the most important for island-hopping propagules. Figure 5.6b is an aerial view of 5.6c. In normal distributions, the number of propagules, n_i, passing by air or water from one island to another, over a distance d_i, can be given as

$$n_i = \tan^{-1}(w_r/2d_i)aw_i^2\, 1 - \frac{1}{d_i} e^{-d_i^2}, \tag{5.3}$$

These models can be much more complicated when applied to island groups and entire archipelagos. Tobler, Mielke, and Detwyler, (1970), for example, attempted to explain similarities in the floristic composition of 11 islands near New Zealand on the basis of their positions and sizes. The authors' assumption was that immigration from outside the 11 islands has been negligible since the Eocene or earlier. An equally important, but unstated, assumption was that the vegetation on each island has reached equilibrium, the $\hat{S}$ of Figure 3.4. The basic model is given as $C_{ij} = C_{ii} \cap C_{jj}$; that is, the number of shared or common species, C_{ij}, between any pair of islands is the intersection of their individual species lists, C_{ii} and C_{jj}. The species present on any given island, C_{ii}, consists of source species, S_i, comprising the initial stock, plus migrants from outside, M_i. The source species are assumed to be equal to the number of endemics.

For any pair of islands, *i* and *j*, the number of shared species equals those migrating from *i* to *j*, and those migrating from *j* to *i*. The total number of species that could migrate depends on the size of the initial stock and the distance between the islands. These numbers can be symbolized as

$$M_{ij} = S_i e^{-b_i d_{ij}} \quad ; \quad \text{and } M_{ji} = S_j e^{-b_j d_{ji}} \tag{5.4}$$

These two equations can be further simplified, if we assume that $d_{ij} = d_{ji}$, which is true in a mathematical sense, but not in a biological sense when ocean currents, wind patterns, temperature gradients, and other factors are taken into account. In the above two equations, *b* is the distance-decay rate expressed in species per kilometer.

The problem staggers the imagination when one considers not just two islands, but the entire mix of 11 and the myriad decay rates connecting them. Additional conceptual problems would include those outlined below:

The floras of the islands are not equally well known. The model assumes a complete list for all islands and that the identification of species is uniform, accurate, and complete.
The islands are of different ages; hence they have had migrants coming and going for different periods of time.
Within the study area the positions of the islands with respect to one another have, in all likelihood, changed.
Environment is not uniform over the entire study area, and some species may not have been able to migrate because their tolerances were exceeded.
No account is taken of speciation rates or population dynamics.

Sweepstakes Routes: The third avenue of migration is called a *sweepstakes route* because the chances of successful migration by any organism are almost nonexistent. Nevertheless, even where physical barriers can be seen to inhibit the dispersal of organisms, there are still those forms that somehow manage to enter new regions. Australia is not completely isolated from the rest of the world, because we can identify more recent arrivals than those of the early Tertiary. Certain forms will make such a migration; others may make it only halfway; and others may not even be able to start. As will be shown in Chapter 8, the Isthmus of Panama represents a corridor for mammalian species moving southward, but a sweepstakes route for southern forms moving northward.

Human Transports

The discussion of the impact of humankind on biotic distributions has been purposely delayed until now because the fates of our introductions are varied as to the modes and avenues of subsequent geographic spread. *Homo sapiens* is a highly vagile form and has, by accident or design, provided a means of long-distance dispersal for an incredible number of plant and animal species. The theoretical constructs presented earlier in this chapter possess limited meaning once humans are introduced into the virgin milieu; and since humanity is ubiquitous on earth, our total impact on the status of community equilibrium, species diversity, competition, mean dispersal distances, and the like, defies complete understanding. The history of plant and animal dispersal is much harder to decipher because of *H. sapiens;* especially that attributable to early human migrations, about which we have no records.

Basically, there are two kinds of introductions: *intentional and accidental.* Throughout history, the accidental types have received greatest notoriety, in large part because they affect human populations. Consider, for example, bubonic plague introduced from the fleas of rats into Renaissance Europe, in about 1347 A.D. (see Pyle, 1979); small pox introduced from Europe into the New World shortly after 1492; and syphilis introduced from the Caribbean into Europe, in about 1500 A.D. These introductions are cases of disease vectors having irruptive modes in a new environment where the local host populations had no resistance. Barriers preventing the natural migration of these vectors were effective prior to the human vehicle.

In the agricultural sector, great effort is expended to prevent the accidental human transport of crop pests. For years California and Arizona have maintained border checks on interstate transport of citrus fruit to prevent the spread of pests from Florida and Texas. On a more general level, the United States and many other countries regulate the importation of soil and biological material of any kind by means of permits, treatment, quarantine, or confiscation. Interestingly enough, such measures are only minimally successful. Unpublished studies in Australia, by Nigel Wace, indicate that the accidental transport of species is phenomenal. It took several greenhouses to accommodate the seeds he recovered from car wash sludge, vacuum cleaner bags off international airlines, and scrapings from the shoes of international conference attendees. Even though only a few of these transported species would likely have survived in their new environment, Dr. Wace's work reveals the incredible efficiency of man as a means of accidental anthropochory.

Intentional transport of useful food and fiber species has been a part of human development since our progenitors became bipedal. In historic time, the greatest impact has been felt since the Age of Discovery. In the last half-millenium, human beings have consciously transported, and retransported, hundreds of species—some useful and some not so useful, (see Sauer, 1952). Throughout the days of exploration, it was common practice for mariners to plant food crops and deposit gravid pigs on islands as sustenance for crews that might be shipwrecked.

The net result of all this accidental and intentional activity, the details of which are mostly unrecorded, can only be appreciated today by reference to the complete species composition list of a particular locale. In a 400-square-mile area of interior Queensland, Australia, for example, a list of 262 vascular plant species contained almost 2 percent introduced forms (Morain, 1970). These introductions included two species of grass *(Rhynchelytrum repens* and *Setaria glauca)* from tropical and subtropical

Africa; two herbaceous species *(Aeschynomene brasiliana* and *Richardia brasiliensis)* from central and tropical South America; and one species of shrub *(Lantana camara)* that has become virtually cosmopolitan in the middle and lower latitudes.

DISTRIBUTION TYPES

Existing knowledge about species distributions comes from sitings and historical documents. From these records, small-scale maps are often produced to show the overall geographic range of a species without regard to its specific habitat. There are several approaches to mapping species distributions, and none of them is entirely satisfactory.

Dot maps that show locations of known occurrence seem to be a common format. Their major shortcoming is that they only show what is known of a distribution, not what is yet to be discovered. No matter how carefully prepared, they cannot show the actual distribution, because no one can be sure what it is.

A refinement of the dot map is the *choropleth map.* It is the easiest to produce and is sometimes more misleading than helpful. Its function is to delineate a known distribution on the assumption that the species occupies all appropriate habitats within the enclosed area. A refinement of Udvardy's map of the dwarf shrew *(Sorex nanus)* is given as an example (Figure 5.7). Usually these maps give little or no information about localized concentrations or voids in the distribution. They merely circumscribe the known locations at a given point in time, in this instance 1959. In the case of the dwarf shrew, one might justifiably redraw the boundary, as shown by the dashed line, thereby eliminating territory from the presumed range. Further inaccuracies are pointed out by the asterisks, which indicate additional, more recent sitings in 1960 and 1961.

Another mapping approach to the choropleth map shows only the peripheral breeding areas connected by a line. The reasoning is that such maps at least show the minimum area of occurrence. They do not show extralimital occurrences of the species; nor can one assume that the species is continuous over the inscribed area. An example for crested tit *(Parus cristatus)* in Figure 5.8 illustrates the case. Each arrow points to a documented breeding site, and an area has been inscribed to fit the data. Although the species is assumed to be present in the area, the internal structure of the range has not been entirely documented.

Lastly, there is what Udvardy calls the *accurate distribution map,* an example of which is given in Figure 5.9 for the wood pigeon *(Columba palumbus).* These maps consist of a combination of lines and symbols to show areas of main occurrence, extralimital localities, and a host of other information. The problem is that they are sometimes so complicated as to be confusing.

Once a distribution map is produced by some logical set of standards and procedures, the distribution it shows must be described. In many instances, the map itself will generate hypotheses as to how the distribution evolved. Many different kinds of distributions can be recognized, of which the following are examples.

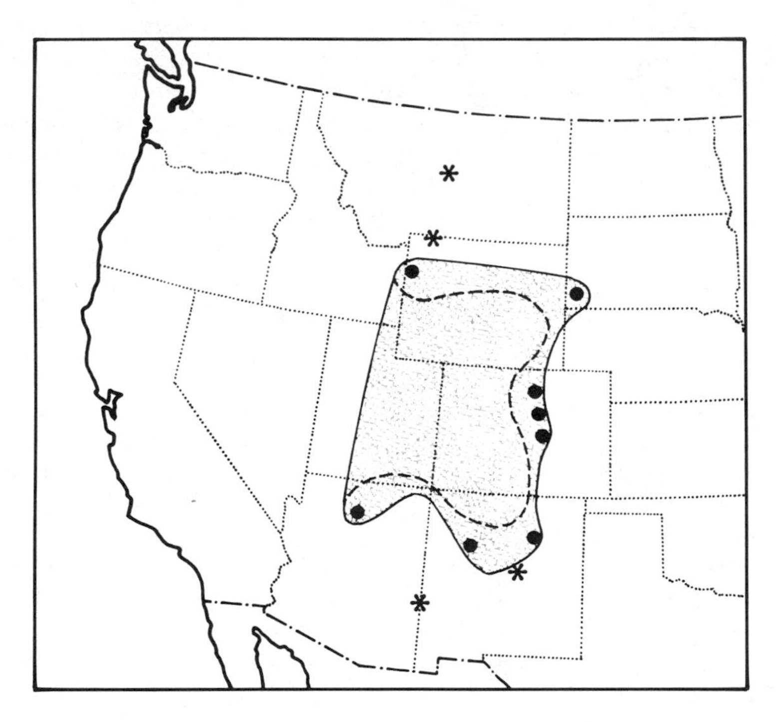

Figure 5.7. Choropleth map showing the distribution of the dwarf shrew *(Sorex nanus).* A line enclosing all known collection sites is used to indicate the species' range. The solid line is that given in Udvardy (1969); the dashed line represents an equally possible interpretation of the data (shown by black dots). Since the map was produced, additional specimens have been collected at the places marked by asterisks. *(From M. D. F. Udvardy, 1969, Dynamic Zoogeography, Van Nostrand, New York, p. 164; copyright © 1969 by Van Nostrand Reinhold Company. Reprinted by permission of Van Nostrand Reinhold Company.)*

Short-Term Occupancy

Many animal species have restricted winter breeding ranges and expanded summer territories. This situation is referred to as seasonal occupancy. The distribution of the red-backed shrike *(Lanius collurio)* in Figure 5.10 shows the species' nesting area in southern Africa. Each year in autumn, the species congregates along a major flyway through Greece, and proceeds southward across the Mediterranean and the Great African Rift. A secondary path crosses Yemen and Somalia. The genetic integrity of the species is maintained through this breeding mechanism. In spring, the

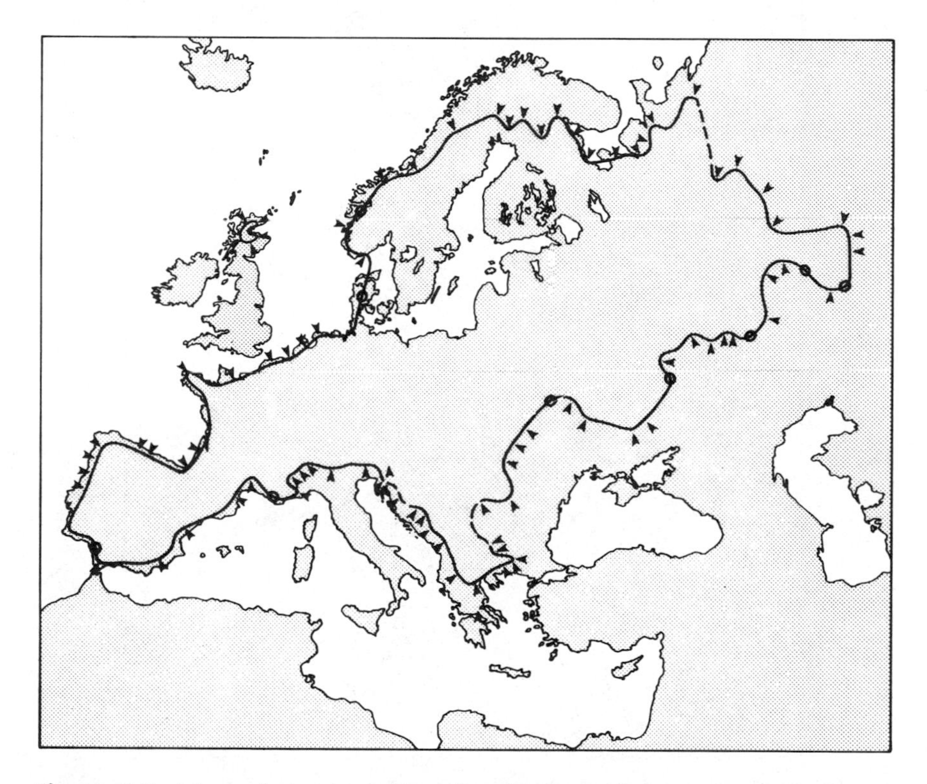

Figure 5.8. Map of the breeding distribution of the crested tit *(Parus cristatus)*. Arrows point to known peripheral breeding sites. The area inscribed by these is assumed to be the range of the species. *(From E. Stresemann, L. A. Portenko, and G. Mauersberger, 1967, Atlas Verbreitung Palaearktischer, Vögel II, Leiferung, Akademie-Verlag. Reproduced by permission of Akademie-Verlag Berlin.)*

flocks migrate northward through Ethiopia and Saudi Arabia, and expand throughout most of Europe and the Caspian.

Other species congregate for summer breeding and disperse in winter when food supplies are scarce. The distribution of the barren-ground caribou *(Rangifer tarandus arctica)*, shown in Figure 5.11, is a typical example. Herds arrive on the tundra in spring, calve there, and remain

Figure 5.9. Example of a so-called accurate distribution map of the wood pigeon *(Columba palumbus)*. Large circles indicate a more or less regular occurrence in the area, while dots show single records of importance. Limits, where known, are indicated by a continuous line. *(Modified from Ekman, 1922, Djurvärldens Utbredningshistoria på den Skandinaviska Halvön, p. 614. Reprinted with permission of Albert Bonnier, Stockholm.)*

throughout the summer. In autumn (August) they migrate southward into the taiga. Their winter extent is larger in area and fluctuates in size depending upon the severity of winter conditions.

Sterile expatriation refers to situations when a species expands its range into territory where it cannot survive or reproduce. The case for locusts has already been mentioned. The Lemming *(Lemmus lemmus)* is

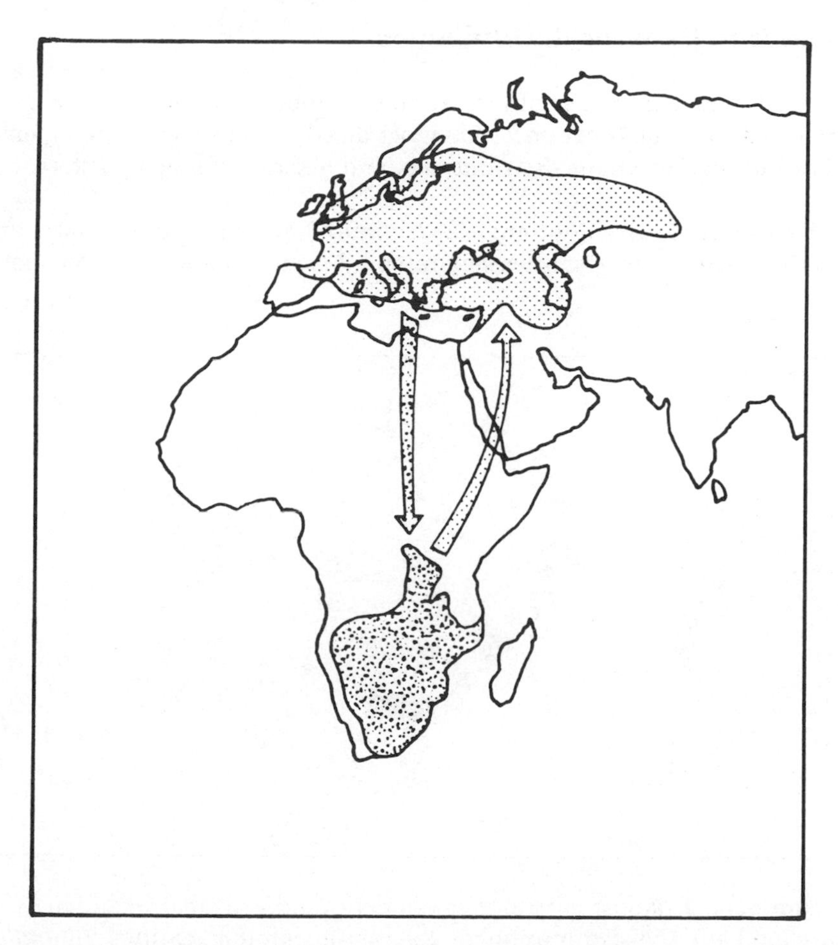

Figure 5.10. Example of winter concentrated seasonal occupancy as observed for the distribution of the red-backed shrike *(Lanius collurio)*. *(From M. D. F. Udvardy, 1969, Dynamic Zoogeography, Van Nostrand, New York, p. 155; copyright © 1969 by Van Nostrand Reinhold Company. Reprinted by permission of Van Nostrand Reinhold Company.)*

another good example; its distribution is shown in Figure 5.12. This small mammal is notorious for its periodic outbreaks due to population pressure. Whole populations have been observed swimming seaward to their death. The permanent breeding area is in the alpine tundra and birch woods of Scandinavia (the black area in Figure 5.12). In peak years, individuals can be found wandering, but not reproducing, throughout most of Scandinavia.

Long-Term (Permanent) Distribution

Under the general rubric of long-term distribution, observers have described extralimital occurrences, disjunct distributions, relict or refugional distributions, vicarious distributions, and a series of global patterns.

Extralimital Distribution: *Extralimital* patterns are shown by species that occupy localities separated by some distance from the main

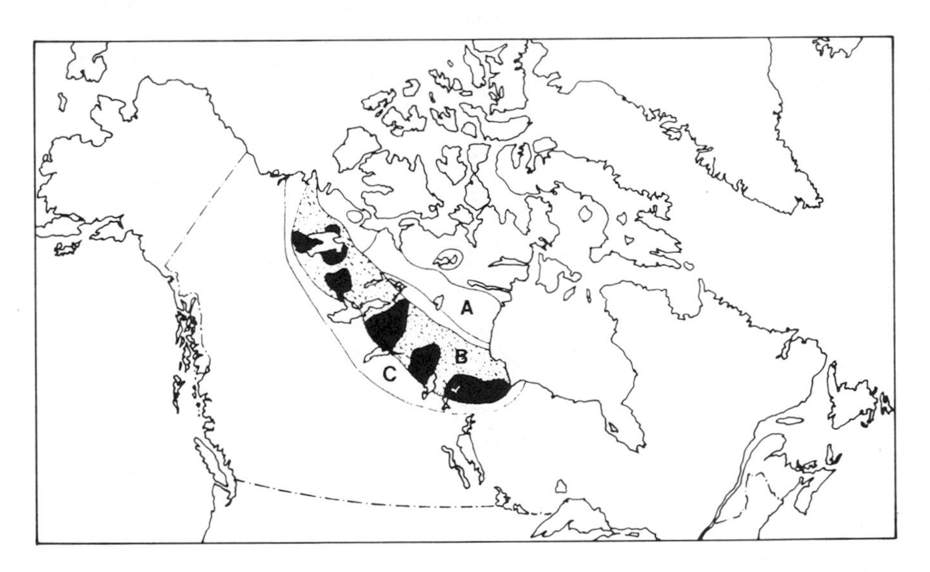

Figure 5.11. Example of summer concentrated seasonal occupancy as observed for the distribution of the barren ground caribou *(Rangifer tarandus arctica)*. *A*. summer range; *B*. late summer and early fall range; *C*. winter range. Patterns associated with the letters indicate areas of highest herd concentration in 1948 and 1949. *(From M. D. F. Udvardy, 1969, Dynamic Zoogeography, Van Nostrand, New York, p. 157; copyright © 1969 by Van Nostrand Reinhold Company. Reprinted by permission of Van Nostrand Reinhold Company.)*

area of occurrence but still within the mean dispersal distance of the individuals. The pattern of the green-tailed towhee *(Chlorura chlorura)* in Figure 5.13 shows numerous extralimital localities along the Missouri and upper Mississippi valleys and along the east coast of the United States. The predominant breeding area, however, is in the Basin and Range and Colorado Plateau provinces. Although the species congregates in northern Mexico in winter, the summer distribution shows essentially extralimital breeding sites. Not all extralimital patterns are of this type, and one wonders whether there may be a great deal of subspecific variation accumulating in the dispersed populations.

Disjunct Distribution: *Disjunct* distributions reveal areas of occurrence that greatly exceed the mean dispersal distance of the populations

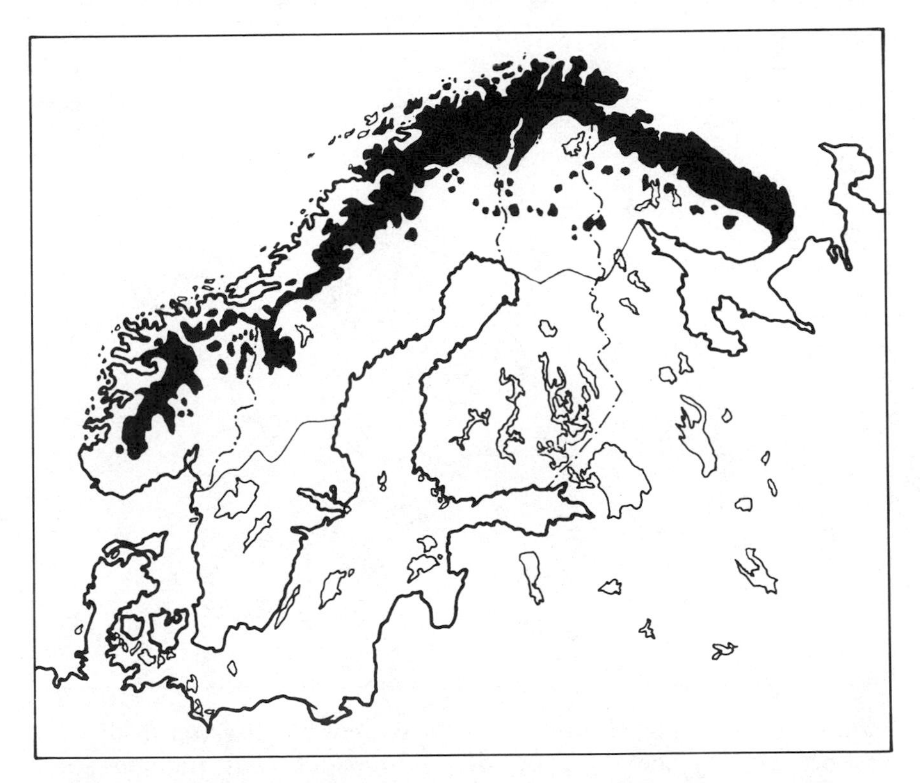

Figure 5.12. Example of irruptive sterile expatriotism, as observed in the distribution for lemmings *(Lemmus lemmus). (From M. D. F. Udvardy, 1969, Dynamic Zoogeography, Van Nostrand, New York, p. 159; copyright © 1969, by Van Nostrand Reinhold Company. Reprinted by permission of Van Nostrand Reinhold Company.)*

comprising the species. In the example of the arctic nuthatch *(Sitta canadensis),* shown in Figure 5.14, there are three primary distribution areas in North America, the Caucasus, and northern China and Korea. The distribution in North America has fossil relatives dating from the late Pleistocene of Southern California. In Europe, there is a small relict population on Corsica. These breeding areas are so disjunct that behavioral studies indicate that individuals from each would not mate (Lohrl, 1960). Some authors consider the populations to be reproductively distinct and argue that they should be elevated to the rank of species (Vanrie, 1959).

Relict Distribution: *Relict,* or *refugional,* distributions are displayed by species whose range is either more restricted than their potential range or confined to specific environments beyond the dispersal distance

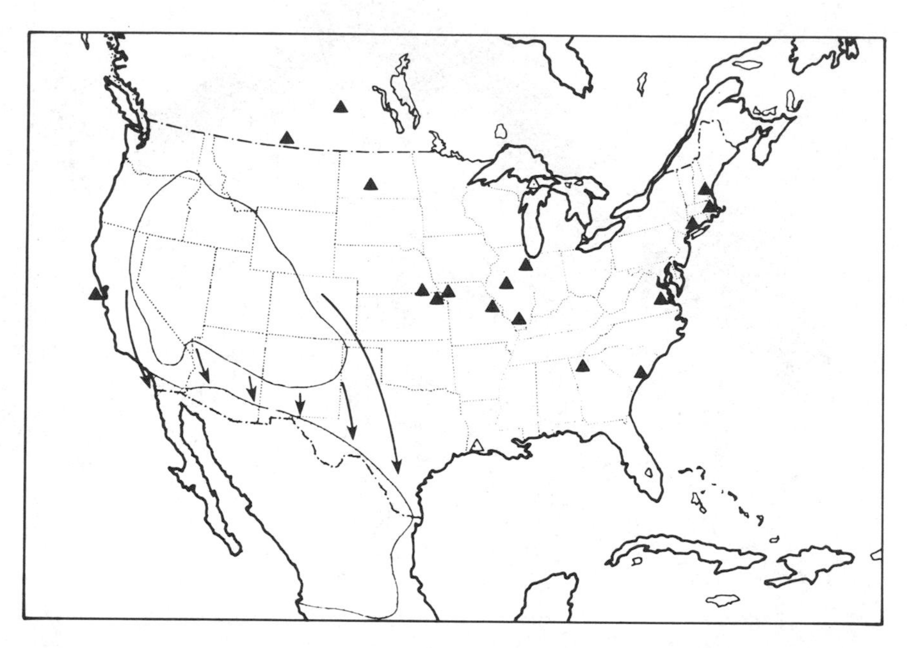

Figure 5.13. Example of an extralimital pattern as observed in the distribution of the green-tailed towhee *(Chlorura chlorura).* Triangles show breeding sites outside the main summer habitat in the Basin and Range Province. Populations move south to northern Mexico in winter. *(From M. D. F. Udvardy, 1969, Dynamic Zoogeography, Van Nostrand, New York, p. 161; copyright © 1969 by Van Nostrand Reinhold Company. Reprinted by permission of Van Nostrand Reinhold Company.)*

of the main area of occurrence. Such situations may be caused by human intervention, as is the case for the giant land tortoise (*Testudo* spp.) in Figure 5.15, or by climatic shifts of the type described by Good's theory of plant migration (see Figure 5.3). Figure 5.16 shows a Pleistocene glacial relict distribution for wood frogs *(Rana sylvatica)*. The main distribution is boreal but refugia exist in the Rocky Mountains of Idaho and Colorado, as well as in southeast Kansas and northwest Arkansas. Areas that serve as relicts of one species often turn out to support relicts of entire communities because of their peculiar environmental inertias acquired from a previous age.

Vicarious Distributions: Lastly, there is a distribution that applies to closely related species, and sometimes entire communities, rather than to individual species. This type is known as vicarious distribution. When two species occupy similar habitats separated by geographic barriers, they are clearly allopatric and are referred to as "true" vicars (Figure 5.17). Two species that occupy similar habitats, but have somewhat different ecological requirements are referred to as "ecological" vicars. They are often ecological substitutes in zones of overlap (Figure 5.18). The distributions of the North Pacific guillemots *Cepphus columba* and

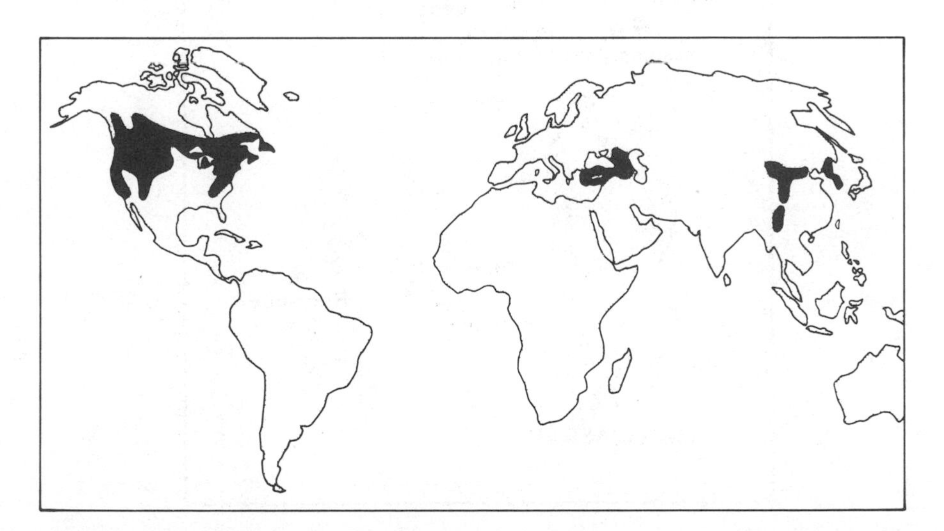

Figure 5.14. Example of a disjunct distribution (shown in black) as observed in the range of the arctic nuthatch *(Sitta canadensis)*. *(From M. D. F. Udvardy, 1969, Dynamic Zoogeography, Van Nostrand, New York, p. 179; copyright © 1969 by Van Nostrand Reinhold Company. Reprinted by permission of Van Nostrand Reinhold Company.)*

C. carbo in Figure 5.17 most likely arose from common stock sometime during the Pleistocene, and are now allopatric except on one of the southernmost Kurile Islands. The case of the swamp and marsh rabbits *Sylvilagus aquaticus* and *S. palustris* in Figure 5.18 shows not only an extensive geographic separation, but also an ecological separation along their zone of overlap in Alabama and Georgia. They are species of the same subgenus but differ in size.

Global Distributions

Over the years the worldwide distributions of hundreds of species, genera, and families have been plotted. Not unexpectedly, recurrent patterns can

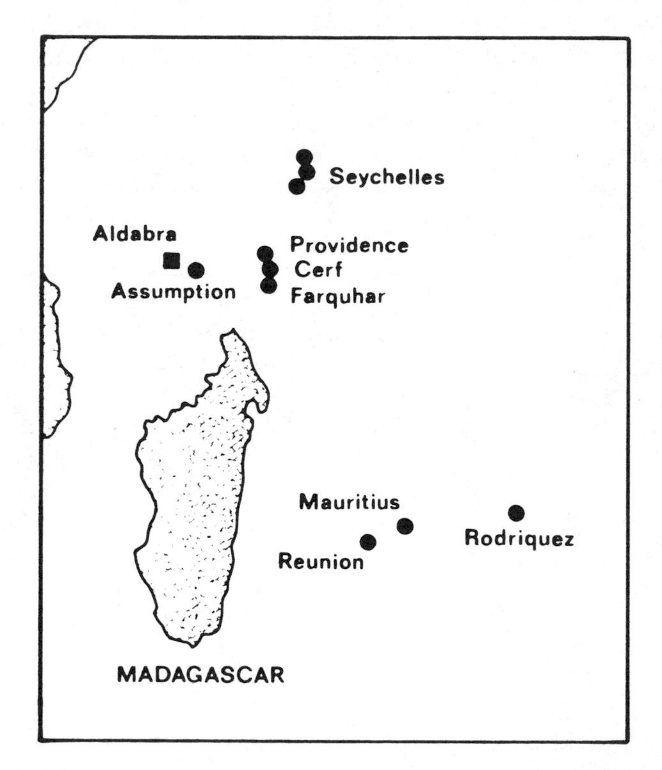

Figure 5.15. Example of a relict distribution as shown by the retreat of the giant land tortoise *(Testudo spp.)* since the 1600's. Dots indicate distribution in 1600's; the squares, the distribution in 1968. *(From D. R. Stoddart, 1968, Isolated Island Communities, Science Journal, April, p. 37; copyright © 1968 by Science Journal. Reprinted with permission.)*

be described in general terms, and these patterns aid biogeographers in understanding the origins and dispersals of major biotic elements. Figure 5.19 shows the most frequently observed patterns. They are described as follows:

Circumpolar: distribution encircles one of the poles.
Panboreal: distribution encircles the high latitudes.
Pantropical: distribution includes Old and New World tropics.
Amphipacific: disjunct on both sides of North Pacific.
Amphiatlantic: disjunct on both sides of the Atlantic.
Amphitropical: disjunct on both sides of the equator.

Figure 5.16. Example of a Pleistocene glacial relict distribution as shown by the wood frog *(Rana sylvatica)*. The main distribution today is shown by the stippled pattern. Outlying localities are shown in black. *(From M. D. F. Udvardy, 1969, Dynamic Zoogeography, Van Nostrand, New York, p. 213; copyright © 1969 by Van Nostrand Reinhold Company. Reprinted by permission of Van Nostrand Reinhold Company.)*

Bipolar: disjunct in Northern and Southern hemispheres, as in temperate Europe and subtropical Africa.

Amphiamerican: disjunct on Atlantic and Pacific sides of tropical North America.

SUMMARY

The genetically and environmentally controlled means by which species disperse are fundamental attributes that result in an observed pattern of distribution. In large measure, these patterns are the first line of evidence regarding the places of origin, rates and directions of advance or contraction, and general adaptability of the organisms involved. Higher taxonomic levels of terrestrial plants and lower taxonomic levels of animals have evolved several mechanisms for their dispersal. Many forms have adapted to more than one mechanism to ensure their wider distribution. The vagility of human beings, together with our modern ability to traverse the world in hours, means that almost all organic life

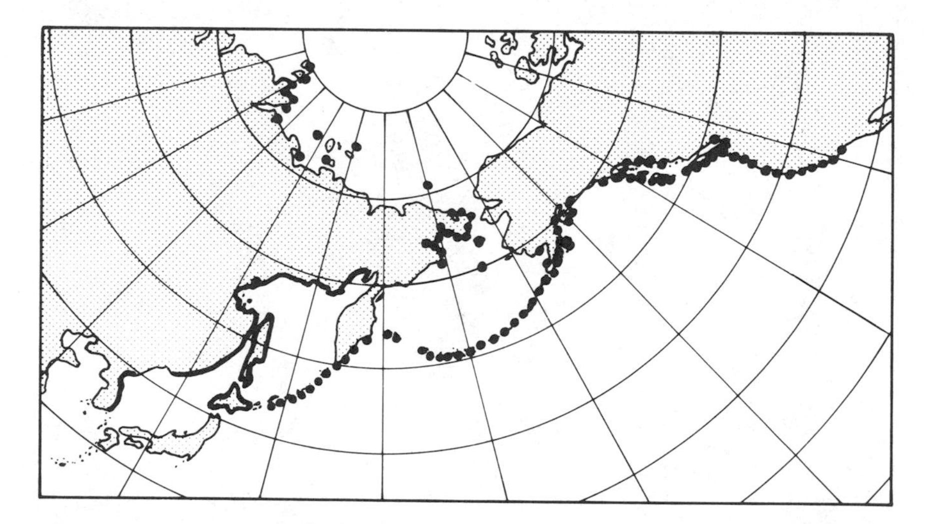

Figure 5.17. Example of a true vicarious distribution, as observed in two North Pacific guillemots: *Cepphus columba* is shown by the large dots and *Cepphus carbo* by the solid line. These two forms probably originated from a common ancestor during Pleistocene glacial episodes that disrupted the ancestral range. *(From M. D. F. Udvardy, 1963, Zoogeographical Study of the Pacific Alcidae, in J. L. Gressitt, ed., Pacific Basin Biogeography, p. 98, 99; reprinted by permission of Bishop Museum Press.)*

has defacto become adapted to anthropochory. The magnitude of accidental and intentional human introductions is rapidly altering predator/prey, competition, and other relationships to the extent that ecological balances and natural processes are becoming disrupted. Even the most remote island refuges and interior isolated localities show the effects of human introductions.

Given the foggy history of humanity's early and accelerating influence, one can understand the difficulty of identifying species whose distributions are natural enough to shed light on how they evolved and dispersed. Present-day patterns can be classified into a relatively few distribution types. Proper interpretation of these patterns can point toward individual as well as collective origins, so long as human influence is accounted for.

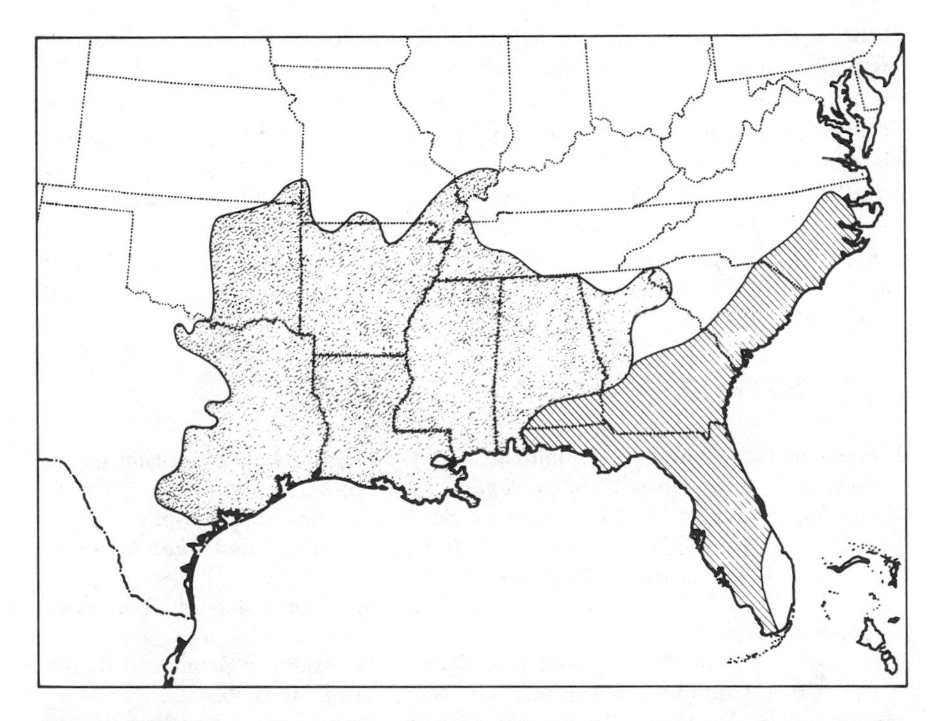

Figure 5.18. Example of the distribution of two ecological vicars, the swamp rabbit *(Sylvilagus aquaticus)*, shown by diagonal lines, and marsh rabbit *(Sylvilagus palustris)*, shown in stipple. In their region of overlap they are ecological substitutes. *(From M. D. F. Udvardy, 1969, Dynamic Zoogeography, Van Nostrand, New York, p. 193; copyright © 1969 by Van Nostrand Reinhold Company. Reprinted by permission of Van Nostrand Reinhold Company.)*

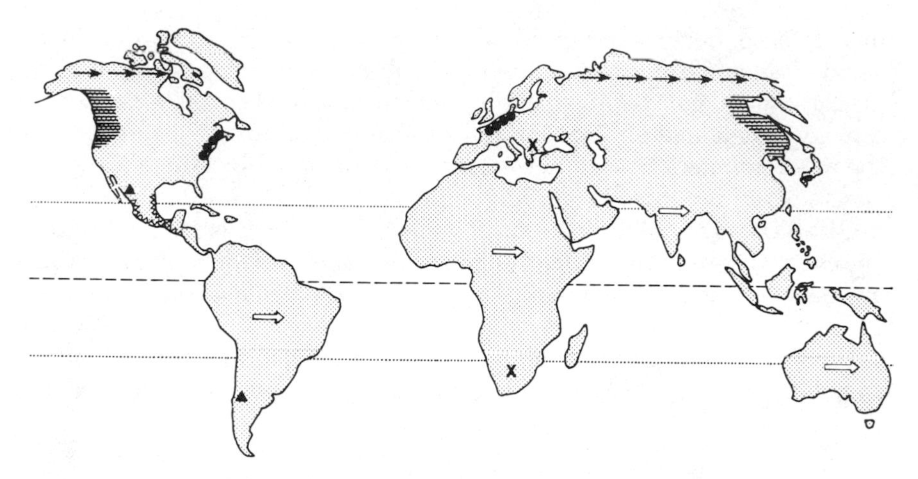

Figure 5.19. Types of global distribution. Circumpolar (holarctic) distributions are shown by arrows in North America and Eurasia; amphipacific distributions by horizontal lines; amphiatlantic distributions by black dots; amphitropical distributions by triangles in Mexico and Chile; bipolar distributions by crosses in temperate Europe and subtropical Africa; amphiamerican distributions by cross hatch in tropical North America; pantropical by open arrows. *(From M. D. F. Udvardy, 1969, Dynamic Zoogeography, Van Nostrand, New York, p. 221; copyright © 1969 by Van Nostrand Reinhold Company. Reprinted by permission of Van Nostrand Reinhold Company.)*

REFERENCES

Buchanan, G. D., and R. V. Talmage, 1954, Geographic Distribution of the Armadillo in the United States, *Texas J. Sci.* **6**:142-150.

Eisley, L., 1970, *The Invisible Pyramid,* Scribner's, New York, 173p.

Ekman, S., 1922, *Djurvärldens Utbredningshistoria på den Skandinaviska Halvön,* Bonnier, Stockholm, 614p.

Good, R., 1974, *The Geography of the Flowering Plants,* 4th ed., Longman, London, 557p.

Harris, D. R., 1966, Recent Plant Invasions in the Arid and Semi-Arid Southwest of the United States, *Assoc. Am. Geogr. Ann.* **56**(3):408-422.

Hemming, C. F., 1968, *The Locust Menace,* Anti-Locust Research Center, London, 31p.

Lohrl, H., 1960-61, Vergleichende Studien uber Brutbiologie und Verhalten der Kleiber Sitta whiteheadi und Sitta canadensis, *Ornithology* **101**:245-264; **102**:111-132.

MacArthur, R. H., and E. O. Wilson, 1967, *The Theory of Island Biogeography,* Princeton University Press, Princeton, N. J., 203p.

McNaughton, S. J., 1976, Serengeti Migratory Wildebeest: Facilitation of Energy Flow by Grazing, *Science* **191**:92-94.

Morain, S. A., 1970, Ecological Segregation of Phytogeographic Elements, Upper Burdekin Valley, North Queensland, Kansas University, Lawrence, Kans., Ph.D. Diss., 203p.

Pyle, G. F., 1979, *Applied Medical Geography,* Scripta Series in Geography, Wiley, New York, 282p.

Sauer, C., 1952, *Agricultural Origins and Dispersals,* American Geographical Society, New York.

Simpson, G. G., 1965, *The Geography of Evolution,* Chilton, New York, 249p.

Stresemann, E., L. A. Portenko, and G. Mauersberger, 1967, *Atlas der Verbreitung Palaearktischer,* Vögel II, Leiferung Akademic-Vg, Berlin, 15 maps.

Tobler, W. R., H. W. Mielke, and T. R. Detwyler, 1970, Geobotanical Distance Between New Zealand and Neighboring Islands, *Bio-Science* **20**(9):537-542.

Udvardy, M. D. F., 1963, Zoogeographical Study of the Pacific Alcidae, in *Pacific Basin Biogeography,* J. L. Gressitt, ed., Bishop Museum Press, Honolulu, pp. 85-111.

Udvardy, M. D. F., 1969, *Dynamic Zoogeography,* Van Nostrand, New York, 435p.

Uvarov, B., 1966, *Grasshoppers and Locusts,* vol. 1, Cambridge University Press, London, 481p.

Vanrie, C., 1959, *The Birds of the Palearctic Fauna. A Systematic Reference. Order Passeriformes,* Witherby, London, 762p.

FURTHER READING

Andrewartha, H. G., and L. C. Birch, 1954, *The Distribution and Abundance of Animals,* University of Chicago Press, Chicago, 782p.

Carlquist, S., 1981, Chance Dispersal, *American Scientist* **69**:509-516.

Hallam, A., ed., 1972, *Atlas of Paleobiogeography,* Elsevier, New York, 500p.,

Krebs, C. J., 1972, *Ecology: The Experimental Analysis of Distribution and Abundance,* Harper and Row, New York, 694p.

Mayr, E., 1976, *Evolution and the Diversity of Life,* Harvard University Press, Cambridge, Mass., 721p.

Nelson, G., and D. E. Rosen, eds., 1981, *Vicariance Biogeography,* Columbia University Press, New York, 594p.

Pijl, L. van der, 1969, *Principles of Dispersal in Higher Plants,* Springer-Verlag, New York, 154p.

Raven, P. H., and D. I. Axelrod, 1974, Angiosperm Biogeography and Past Continental Movements, *Ann. Missouri Bot. Gard.* **61**:539-623.

Rosen, D. E., 1975, A Vicariance Model of Caribbean Biogeography, *Systematic Zoology* **24**:431-464.

Stoddart, D. R., 1968, Isolated Island Communities, *Science Journal,* April, pp. 32-38.

Stott, P., 1981, *Historical Plant Geography,* Allen & Unwin, London, 151p.

Tralan, H., ed., 1969, *Index Holmensis: A World Index of Plant Distribution Maps,* vol. 1, The Scientific Publishers Ltd, Stockholm, complete set 12 vol.

Wiley, E. O., 1980, Phylogenetic Systematics and Vicariance Biogeography, *Systematic Botany* **5**:194-220.

Chapter 6

Paleogeography and Biological Development

The most fundamental of all questions in paleogeography and historical biogeography deals with the past stability of continental areas and the nature of their relationships with each other.

G. G. Simpson, *Mammals and the Nature of Continents*

Present distributions of terrestrial life cannot be understood or explained without reference to the paleogeography of land masses and their former connections. Indeed, the first theories of continental drift were formulated out of observed similarities in fossil distributions over widely separated areas—distributions that could only have arisen if the land masses were at one time connected. The nature of those past connections and the geophysical mechanisms that alter them have led to differences of opinion regarding the earth's biotic history.

DRIFT AND BRIDGES

Essentially two lines of reasoning have emerged to explain observed biotic patterns. The land bridge theory, best defended by Van Steenis (1962), argues that continental locations have remained stationary, at least since Paleozoic time, but that there have been periodic land connections between them. Alternatively, it is argued that the continental plates themselves have separated and are moving in predictable directions at measurable speeds. Over thousands and millions of years, the distances achieved have led to the variety of biotic distributions observed today.

Land Bridges

In the rationale for land bridges, Van Steenis based his arguments on the idea of steady state. This argument avers that since the Mesozoic era, neither the climatic zones nor the general morphology of earth have changed dramatically. Hence, the idea of a uniformly warm planet over which a uniform biotic development took place is highly unlikely. Despite the fact that climatic belts are fixed by the axis of the earth, Van Steenis concludes that Tertiary cooling may have led to a compression of tropical belts and a widening of both temperate and polar belts, and that these fluctuations could account for some of the observed patterns of distribution. He goes on to conclude that continents may be spreading apart, but that this movement began in the Paleozoic, long before the evolution of mammals and flowering plants. Consequently, the present location of continents cannot be important as an explanation of present distributions.

Quite apart from the idea of steady state is a second argument that plants have not changed their thermoecological potential since their origin. In other words, tropical taxa have always been tropical and temperate taxa have always been temperate. The idea that plant life originated in a uniform coherent land mass that subsequently broke up and drifted to new climatic regions does not seem credible to Van Steenis. His rather restricted view largely denies the powerful process of adaptation to new adaptive zones.

Evidence considered to be generally in favor of the land bridge concept may be summarized as follows:

The tropical transpacific flora exhibits almost as many shared taxa as the tropical transatlantic. If North and South America have been drifting westward, one would expect to find very few tropical transpacific similarities, since these floras have presumably never been in contact. At some time in the past there must have been a transpacific land bridge.

There is living evidence of tropical land bridge relicts. For example, the plants one would expect to be present in the southwest Pacific are there and could only have been dispersed across a land connection. The island groups and guyots of the southwest Pacific are relicts of a former land bridge, and the vegetation is similarly relict. There are undoubtedly secondary invaders that have arrived along island stepping stones and as anthropochores since the bridge broke up.

The predominance of Indo-Malayan floral elements in the western Pacific suggests that this area is the relict of a once larger land mass.

Figures 6.1 a, b and c, redrawn from Van Steenis, illustrate his estimates of the location and duration of the earth's primary land bridges. From the

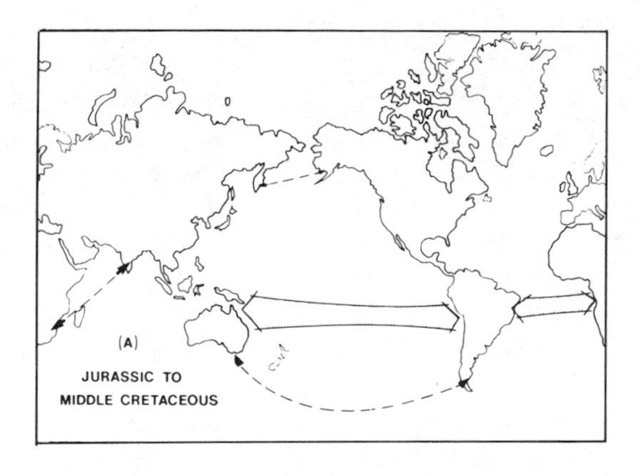

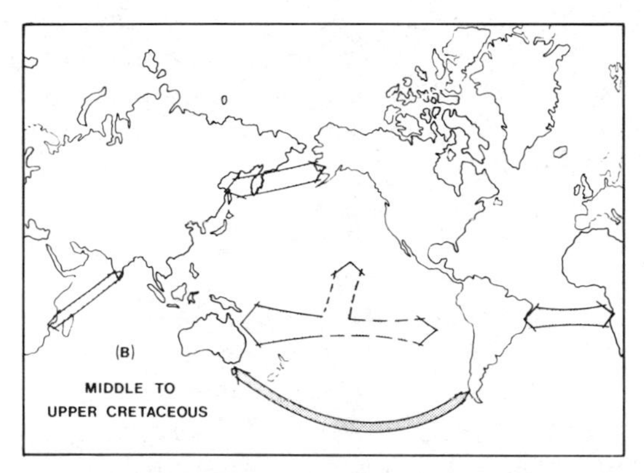

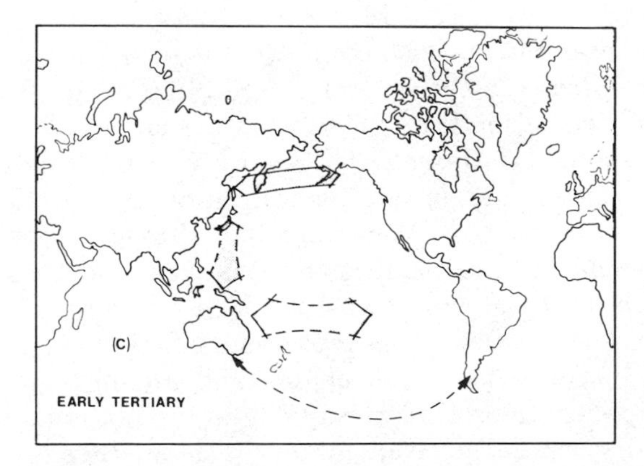

Figure 6.1. Reconstructions of presumed intercontinental land bridges according to Van Steenis. *(Redrawn from C. G. G. J. Van Steenis, 1962, The Land Bridge Theory in Botany, Blumea* **11***(2):342, 344, 345; reprinted by permission of the editor of Blumea.)*

Jurassic to the Middle Cretaceous (150 to 90 million years ago, Figure 6.1a) tropical land bridges may have connected South America with Melanesia and Africa. Narrower, or less persistent, bridges may have existed between Africa and India, between temperate South America and Australia, and between Alaska and Siberia (Beringia). The presence of these bridges would account for the transpacific and transatlantic floral similarities, as well as for some of the oldest known angiosperm distributions in Africa and India.

Middle to Upper Cretaceous land bridges (90 to 70 million years ago, Figure 6.1b) might have shown the same general pattern as those in Figure 6.1a with a few differences of degree. The tropical transpacific bridge must have begun to founder from the eastern (South American) side, along with the entire transatlantic bridge. The Afro-Indian, Beringian and transantarctic bridges, however, must have been more viable avenues of migration than was the case earlier. It would have been during this time that the southern beech *(Nothofagus)* migrated into the temperate areas of Australia, New Zealand, and South America (see also Chapter 12). The Bering land bridge is believed to have been over a thousand kilometers wide, along which there was a two-way migration. Along its northern fringes, cool temperate floral elements migrated, while to the south, warm-temperature types intermingled.

By early Tertiary times (70 to 40 million years ago, Figure 6.1c), the transatlantic bridge is shown to have disappeared along with the entire eastern half of the transpacific bridge. The Afro-Indian bridge is also gone. The transantarctic bridge is believed to have been reduced to a very narrow island archipelago. A new connection, called the Marianas bridge, is supposed to have connected Micronesia to a point just north of the Bonin Islands near Japan. Such connection is necessary to account for the major break between the holarctic vegetation of the Bonin Islands and Malaysian vegetation throughout Micronesia.

Two points should be stressed with regard to an understanding of the land bridge theory. First, the continental margins, as drawn in Figure 6.1, are not intended to imply their constancy throughout geologic time. The theory argues that their shapes have changed as a function of rising and falling sea levels and the transgressions of continental seas; but that their relative locations have not changed. Secondly, the bridges themselves are viewed as island chains and archipelagos capable of serving as stepping stones for two-way migration, but not as continuous stretches of dry land. Perceived in this fashion, these bridges most probably served as filters for selective migration of forms.

One of the most unpalatable facets of the land bridge theory, aside from the dearth of underwater remains of former islands, is that the bridges serve as a convenience mechanism to explain whatever distributions they can. Where distributions cannot be explained, science must assume it was the failure of the species to disperse.

Continental Drift (Plate Tectonics)

The general conformance of the African/South American and Australian/ Antarctic continental margins led to the speculation that they were once connected. In fact, the more one studies the continents, the more they resemble the pieces of a gigantic jigsaw puzzle. Even greater conformity can be seen if one "fits" the continental shelves at the 1,000-fathom isobath rather than at present-day coastlines. It appears from this observation that all the continents were once united and have subsequently separated. A second observation that fueled early speculations is the Glossopteris Flora. This assemblage of fossil ferns occurs in similar sediments in both South America and Africa, now separated by more than 50 degrees of longitude. More will be said about this assemblage in Chapter 10. For the present, it suggests that the two areas were once joined. Van Steenis, of course, would argue that this flora merely proves the existence of a former transatlantic bridge.

That plate tectonics has had a checkered history of acceptance stems partly from differences between the biological and geophysical records. In 1943, Simpson wrote that ". . . the drift theory offers no possible answers . . . when they [the proponents] try to apply it to Madagascar. The distribution of mammals definitely supports the hypothesis that continents were essentially stable throughout the whole time involved in mammalian history (well over 100 million years)" (p. 29). In a similar fashion, it has been difficult to explain the presence of many angiosperm families in Australia if that continent has been isolated for more than 70 million years.

Further difficulties have arisen with the theory because, until the 1960s, geophysicists could not supply a mechanism whereby continents could move. As recently as 1959, both Wilson and Kennedy, in separate articles, described the origins of continents and oceans without even mentioning continental drift. The recent shift in favor of the theory has occurred because a mechanism has been found, and because the timing of the breakup and drifting can be moved forward in geologic time to account for at least some of the major biological problems. The mechanism is a combination of *sea-floor spreading,* whereby new material is created along the trailing edge of crustal plates and older material is subducted along their zones of contact. The identification of tectonic plates, together with the articulation of a mechanism, caused a major scientific revolution that began about 1965. Since then, the theory has explained much about our earth and the foundations for present-day biological patterns. The theory, however, does not answer all questions equally well. For this reason, there may yet be a resurgence of alternative hypotheses, just as there are serious shifts between evolution and biblical creation (see Broad, 1981).

Instead of continents drifting apart on a sea of molten rock, they are viewed as being rafted or pushed apart through gradual widening of the ocean along the trailing edges of the tectonic plates. A map of these plates, simplified from Dewey (1972), is shown in Figure 6.2. The earth's surface is divided into a mosaic of these rigid plates. As they move, they either collide with each other or drift apart; thus, new material must be created and old material digested. Mountain ranges are usually formed along the lines of collision by the creation of extreme crustal pressures. Material from one plate is also subducted under the other. The so-called Pacific ring of fire, for example, refers to a volcanic zone that stretches from The Philippines through the Kuriles, Aleutians, Rockies, and the Andes. Along this zone, the North American and South American Plates are colliding with plates to the west as the sea floor spreads laterally along the Mid-Atlantic Ridge.

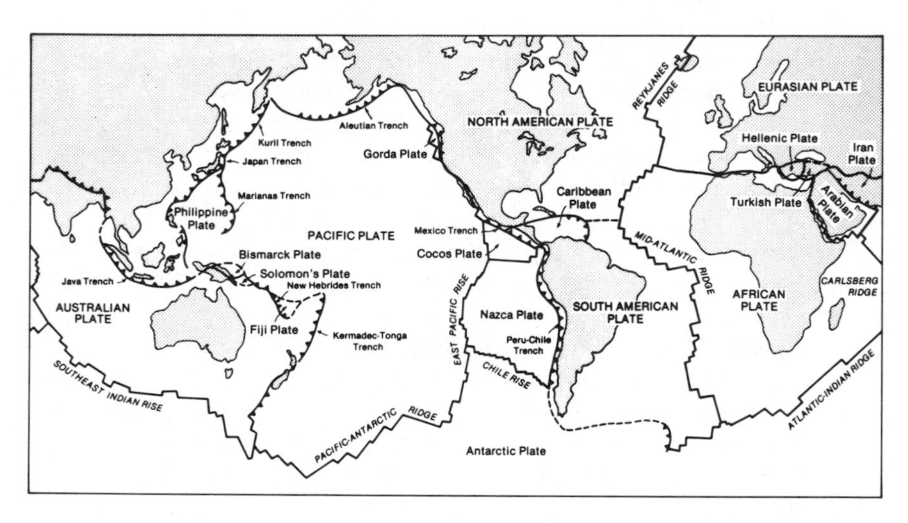

Figure 6.2. A mosaic of plates forms the earth's lithosphere. According to the theory of plate tectonics, these plates are not only rigid, but also in constant relative motion. Their boundaries are of three types: ridge axes, where plates are diverging and new oceanic floor is generated; transforms, where plates slide past each other; and subduction zones, where plates converge and one plate dives under the leading edge of its neighbor. *(From J. F. Dewey, 1972, Plate Tectonics, Scientific American,* **226***(5):56, 57; copyright © 1972 by Scientific American, Inc. All rights reserved.)*

The present location of continents has been achieved over the last 200 million years by plates moving on the order of a few centimeters per year. Pangaea is the name given to the parent Paleozoic land mass (Figure 6.3). The broad outlines are shown for North and South America, Africa, Eurasia, India, Antarctica, and Australia when they were connected. It appears likely that Pangaea drifted as a single unit during the late Paleozoic, although some scientists argue that *Gondwanaland* (the land mass comprising today's southern continents) may have drifted northward into *Laurasia* at that time.

Panthalassa was the ancestral Pacific Ocean. The Tethys Sea (the ancestral Mediterranean), formed a large bay separating Africa and Eurasia. Its last vestige is the Ganges River Valley. The relative positions of the continents, except for India, are based on best fits made by computer, using the 1,000-fathom isobath to define continental boundaries (Dietz and Holden, 1970).

Separation of Pangaea commenced in Triassic time with the creation of the Atlantic Rift and Afro-Indian Rift. These divisions gave rise to Laurasia, east Gondwana, and west Gondwana. By the end of the Jurassic Period, South America had split from west Gondwana, and by the end of the Cretaceous period, Australia had split from east Gondwana—making six basic continents in all. Madagascar and the Seychelles became distinct during the Tertiary, as did North America and Europe at Greenland. India migrated northward into Asia but continued to be a separate entity until relatively recent times.

The world, as it looks today, has been produced in the past 65 million years during the Cenozoic era. Nearly half of the ocean floor has been created in this geologically brief period. India completed its drift northward by colliding with Asia and closing the Tethys Sea. A rift separated Australia from Antarctica. The North Atlantic Rift finally entered the Arctic Ocean, completing the split between Europe and North America.

Tectonics and Biotic Development

If we assume that earth's major land masses were once in direct contact, then it is clear that the breakup of that ancient surface initiated geographic isolation on a grand scale and allowed for the subsequent diversification of life forms. Concomitantly, as the land masses became gradually more isolated by intervening climatic, oceanic, and other barriers, the stage was set for increasing endemism. In the early stages of separation, many forms were shared between adjacent areas separated by only narrow, shallow seas. On each "island," however, adaptation progressed in tune with the nature of environmental changes. The record of these changes is

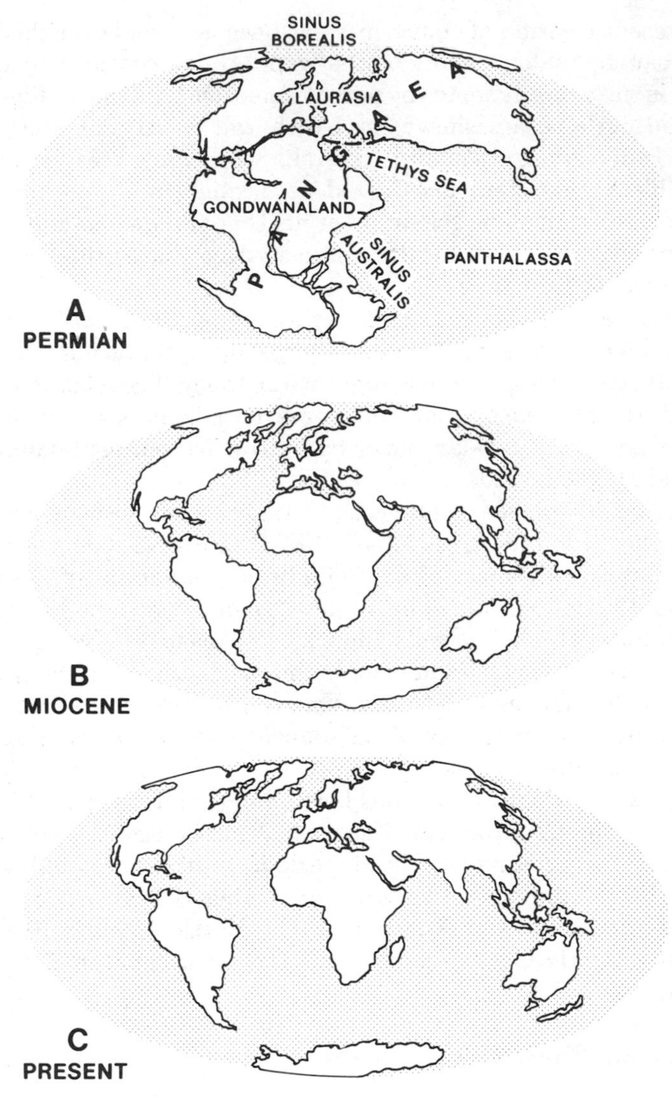

Figure 6.3. Continental positions, Permian to the present. Considering Africa as a pivotal point rotating counterclockwise, Laurasia split along the Atlantic Rift to form North America and Eurasia. Gondwanaland split along the South Atlantic Rift giving rise to west Gondwana (South America and Antarctica) and east Gondwana (Africa, India, and Australia). By Cretaceous time, India and Australia had drifted northward. India made contact with Eurasia, closing the Tethys Sea along what is now the Ganges River Valley. Australia has remained in isolation and can be regarded as the world's largest island.

contained in both the flora and fauna, but it is by no means clear and unambiguous.

Figure 6.4 shows the hypothetical reconstruction of Laurasia and Gondwanaland, as they might have existed at the end of the Jurassic period. The Tethys Sea separated these land masses and was narrower to the west, at Gibraltar, than to the east. The reptilian orders were present on both supercontinents and these included Crocodilia (crocodiles, etc.); Saurischia (bipedal predators and amphibious herbivores); Ornithischia (bipedal herbivores and armored quadrapedal herbivores); Pterosauria (flying reptiles); Chelonia (turtles); Squamata (lizards and snakes); and Choristodera (amphibious predators). Fossil evidence so far available suggests that the Crocodilia, Saurischia, and Ornithischia originated in Gondwanaland; whereas, the Chelonia, Squamata, and Choristodera

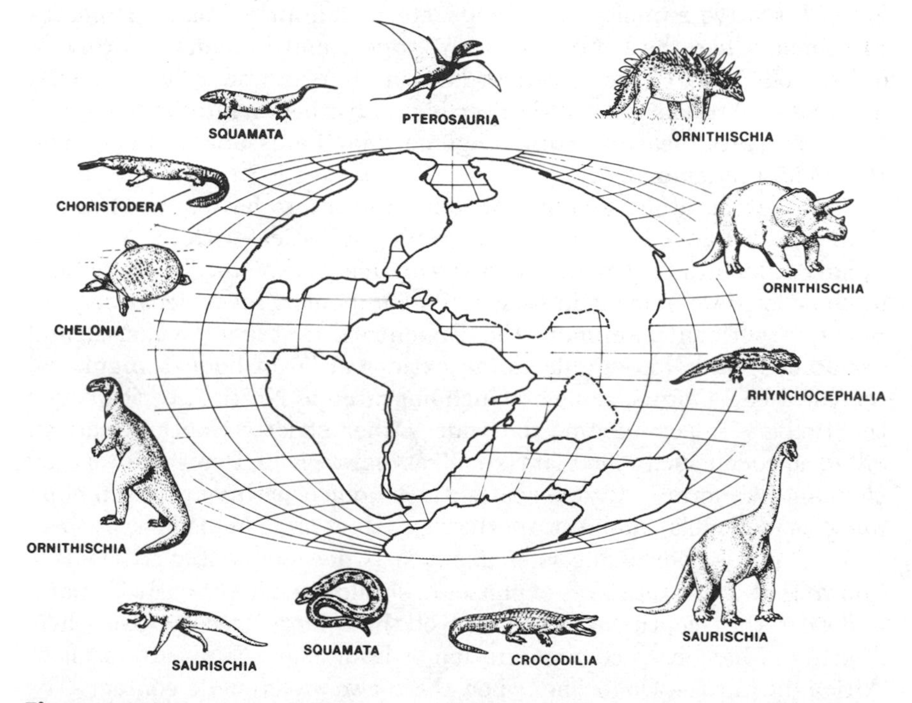

Figure 6.4. Two supercontinents of the Mesozoic era were Laurasia in the north and Gondwanaland in the south. The 12 major types of reptiles, represented by typical species, are those whose fossil remains are found in Cretaceous formations. Most of the orders inhabited both supercontinents; migrations were probably by way of a filter through Spain. *(Redrawn from B. Kurtén, 1969, Continental Drift and Evolution, Scientific American* **220***(3):55, copyright © 1969 by Scientific American, Inc. All rights reserved.)*

may have arisen in Laurasia. Pterosauria's origins are uncertain, although the oldest fossils are from present day Eurasia.

Gondwanaland was somewhat larger than Laurasia and had a more varied climate (including some glaciated parts in the far south). In contrast, Laurasia was more uniformly tropical or subtropical. Life forms probably spread from one continent to the other by way of the western Tethys Sea or along island stepping stones. All in all, at the end of the age of reptiles, forms were rather widespread over a relatively uniform environment.

Figure 6.5 illustrates major mammalian orders in Laurasia and Gondwanaland. The three present day continents that formed Laurasia are North America, Greenland, and Eurasia. These have all been in close contact as recently as the early Eocene. Within these three nuclei, a total of 16 orders of mammals is believed to have arisen, of which 9 still exist: Insectivora (moles, hedgehogs, etc.); Chiroptera (bats); Primates (prosimians like the lemurs, monkeys, apes, and humans); Carnivora (cats, dogs, bears, etc.); Perissodactyla (horses and other one-toed ungulates); Artiodactyla (cattle, deer, pigs and other even-toed ungulates); Rodentia (rats, beavers, etc.); Lagomorpha (hares and rabbits); and Pholidota (pangolins).

The fragmentation of Gondwanaland appears to have started earlier than that of Laurasia, and since the timing is critical with regard to the rapidly differentiating mammalian and angiosperm biotas, it is necessary to consider development in each of its component areas separately. In South American mammals, the Endentata (ant bears, sloths, and armadillos) and Notoungulates (an extinct order of hoofed ungulates) represent early forms, some of which migrated to North America before the Bolivar Trench became too wide. Other strictly South American orders include Paucituberculata (small rat marsupials), Pyrotheris (extinct elephant-like forms), Litopterna (extinct hoofed herbivores resembling horses and camels), and Astrapotheria (extinct large hoofed herbivores).

Very little is known for certain about early development in Africa. It is known that there were large expanses of shallow sea in the early Tertiary, and that the "continent" consisted of three large pieces. Also, there appears to have been some migration of Laurasian (Eurasian) forms to Africa during the Oligocene, when these two areas made contact. The following orders are of African origin: Proboscidea (mastodons and elephants); Hyracoidea (the canines and their extinct relatives); Embrithopoda (extinct large mammals); and Tubulidentata (the aardvarks).

Almost nothing is known of mammalian development on the remainder of Gondwanaland. In Australia, no early Tertiary mammals are known from the fossil record, though the Monotremata (platypus and spiny anteater) are among the most ancient of mammals and probably origi-

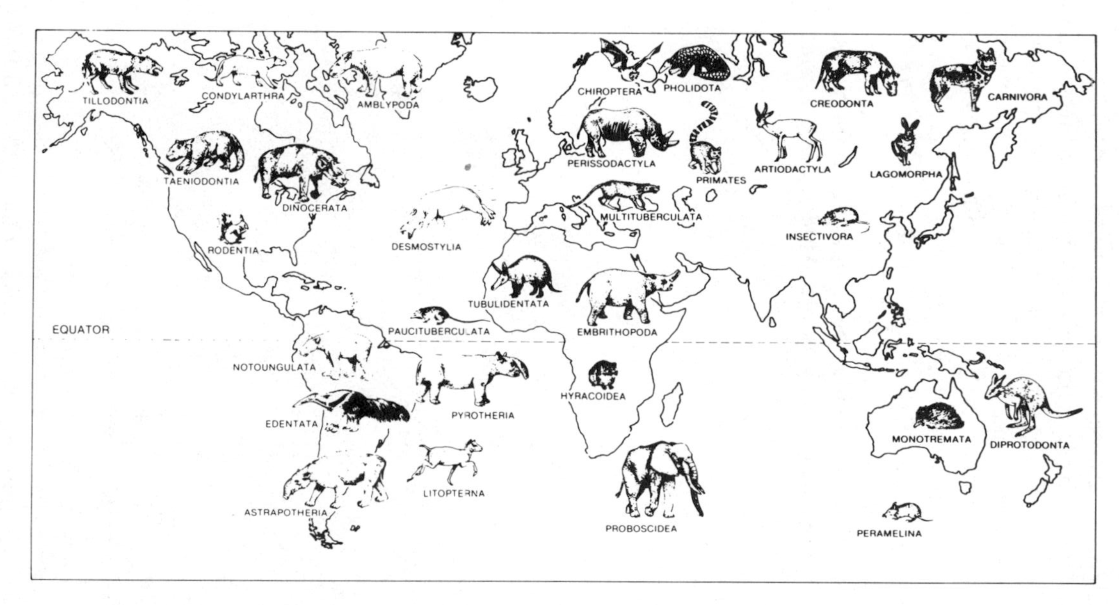

Figure 6.5. Continental drift affected the evolution of the mammals by fragmenting the two supercontinents early in the Cenozoic era. In the north, Europe and Asia remained connected with North America during part of the era. The free migration that resulted adds certainty to the origin of many orders of mammals that evolved in the north. The wider rifting of Gondwanaland allowed the evolution of unique groups of mammals in South America, Africa, and Australia. *(From B. Kurtén, 1969, Continental Drift and Evolution, Scientific American* **220***(3):62,63; copyright © 1969 by Scientific American. All rights reserved.)*

nated there. Only two other orders are believed to have originated in Australia: the Peramelina (bandicoots) and the Diprotodonta (kangaroos, wombats, etc.).

The pattern described above is concerned with the origins of mammalian orders, rather than attempting to explain the details of their distribution. Fooden (1972), among others, has begun to piece these details together. He argues that the oldest progenitor mammals (subclass Prototheria and Metatheria) arose in east Gondwanaland (Figure 6.6a) while Pangaea was still a single unit, some 200 million years ago. Prior to the end of the Triassic, 180 million years ago (Figure 6.6b), primitive marsupials in the subclass Metatheria arose in east Gondwanaland. From there they must have spread into west Gondwanaland before east and west became too separated. By the time separation between South America and Africa

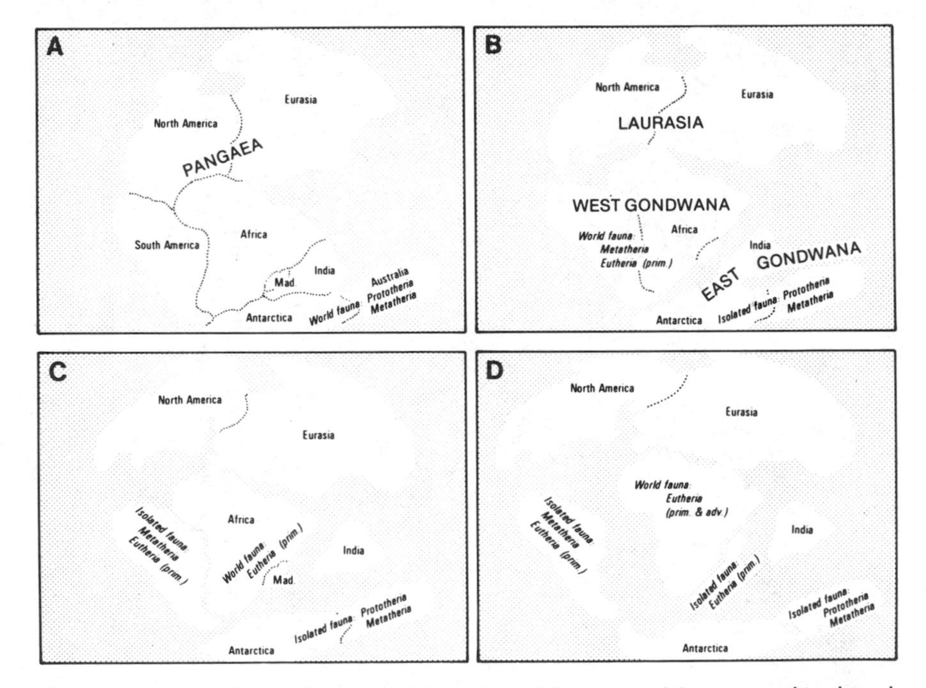

Figure 6.6. Hypothetical composition of world mammal faunas and isolated mammal faunas during successive stages of continental fragmentation: (*A*) approximately 200 million years ago; (*B*) approximately 135 million years ago; (*C*) approximately 100 million years ago; (*D*) approximately 70 million years ago. *(From J. Fooden, 1972, Breakup of Pangea and Isolation of Relict Mammals in Australia, South America and Madagascar, Science* **175:***897; copyright © 1972 by the American Association for the Advancement of Science. Reprinted with permission.)*

was initiated (Figure 6.6c), the early placental mammals in subclass Eutheria were present in both ancestral South America and Africa. The marsupials (Metatheria) were present in South America and Australia; but the monotremes (Prototheria) were present only in Australia. Thus, isolation of Prototherians and Metatherians in Australia and New Guinea may date from the Middle or Upper Cretaceous, and isolation of Eutherians in Madagascar may date from the Paleocene.

The Australian-New Guinean fauna appears to be older than that of South America which, in turn, appears to be older than that of Madagascar. In terms of the maps, it is likewise clear, and consistent with the above relative ages, that east Gondwanaland was the first to split from Pangaea, followed by South America, then Madagascar. It appears that Prototherians and Metatherians were the dominant mammal groups of Pangaea before the initial split, and that Prototherians declined in Laurasia and west Gondwanaland after that break. Eutherians were evolving and beginning to radiate at about the time South America broke away, thus giving that "island" a compliment of Metatherians and early Eutherians for evolution in isolation. Madagascar became isolated during the second stage of Eutherian evolution. The main line of Eutherian development (the dominant Afro-Asian mammalian fauna) took place on the "mainland" of Africa.

Quite apart from the reptilian and mammalian faunas, one must consider the floristic evidence in relation to the evolution of continents. The major objection to continental drift as an explanation of plant distributions has been that it occurred too early—before modern genera were in existence. More recent evidence, however, suggests that drift commenced in Permo-Triassic times and continues today. The major objection is therefore erased. In its place is the more troublesome question of how to explain the origin and rapid advance of angiosperms throughout the world. This question has stimulated the best minds, including Darwin who wrote, "I have been so astonished at the apparently sudden coming in of the higher phanerogams, that I have sometimes fancied that development might have slowly gone on for an immense period in some isolated continent or large island, perhaps near the South Pole."

Hawkes and Smith's (1965, 48) explanation is that

> The angiosperm families must have been in existence, possibly as small populations only, at least by Jurassic times, and that the group as a whole may have originated even further back in Carboniferous or Permian in the southern land mass of Gondwanaland. The apparent absence of fossil angiosperms in the Southern Hemisphere might be explained partly by assuming that the flowering plants occurred as small, rather isolated populations and that although some fossil material may exist, it has not

yet been discovered because of its relative scarcity. Other authors have discussed the possibility that the Gondwanaland flowering plants might well have exploited drier, non-fossilizing habitats (possibly montane) situated well away from the competition of the humid Glossopteris flora. . . . A possible explanation for the apparently sudden appearance of flowering plants in the Northern Hemisphere is that they evolved in Gondwanaland and were able to migrate from it only after one or more of its fragments had drifted northward and come into contact with the northern land masses of Europe, Asia and North America.

DEVELOPMENTAL TRENDS

Summary of Vertebrate Evolution

In the general scenario for the evolution of terrestrial vertebrates, several adaptive developments were essential. Vertebrate fossil remains are found in Early Silurian materials (i.e., early Paleozoic). These earliest forms are agreed to have evolved in freshwater lakes and streams, rather than the sea, at a time commensurate with the invasion of near-shore and freshwater environments by land plants. For successful upstream migration, early vertebrates had to be active swimmers to avoid being carried back to the sea. Their advance and adaptation must have been rapid because by Late Silurian and Devonian times fish were prominent dwellers of inland waters.

A new phase of development occurred with the emergence of jaws (see Figure 6.7), which permitted radiation into a new adaptive zone. By Early Devonian times, there was a variety of placoderms, Actinopterygii (ray-finned fish), Dipnoi (lungfish) and Crossopterygii (lobefin fish that were progenitors of land vertebrates). Among the latter, the coelacanths represent the sole surviving branch, and of these there is but one known surviving form (*Latimeria)* in the Indian Ocean. There is a close relationship between the Crossopterygii and the Amphibia, particularly in the Labyrinthodontia (primitive amphibia), which spanned the time range from Devonian to Triassic. One group of labyrinthodonts (the *anthracosaurs*) gave rise to the reptiles. Little is known about the history of the surviving orders of amphibians, but somewhere along the line there was a major evolutionary step from water to land.

The specific environmental condition that led to the origin of the amphibia is believed to have been seasonal drought. Numerous Permian and Triassic red beds are evidence for the prevalence of such an environment. Biological developments necessary for adaptation to drought include the origin of lungs and land limbs. Under drought conditions,

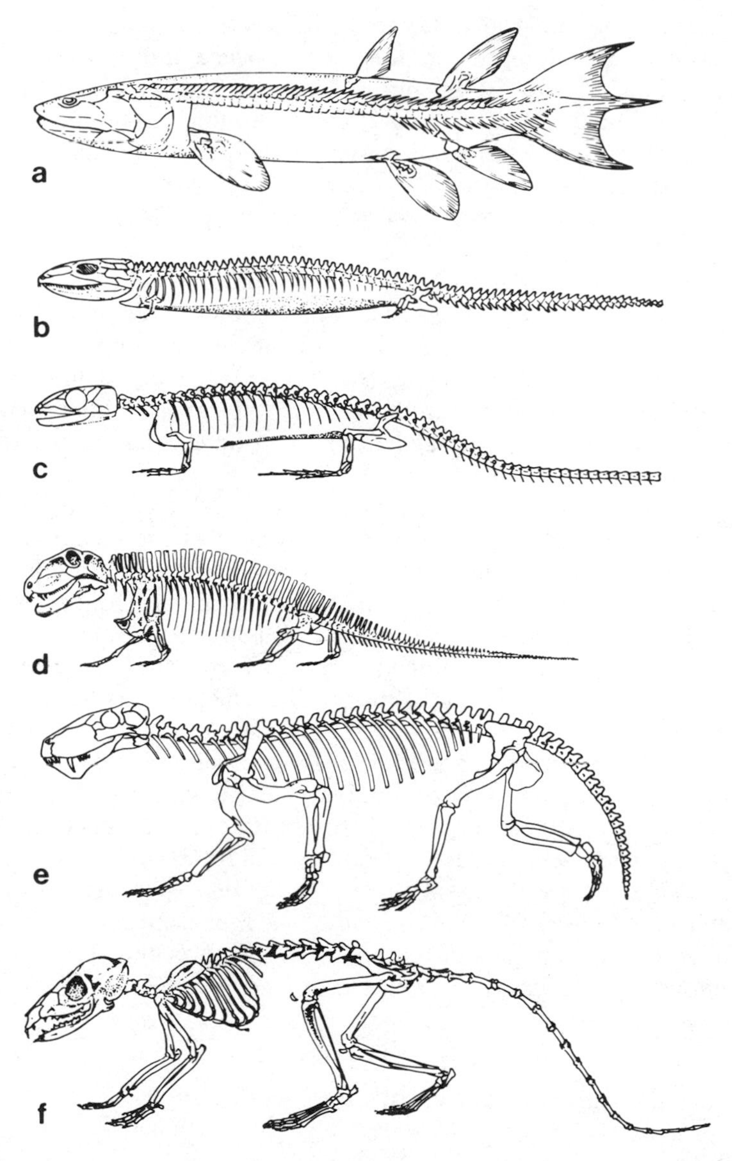

Figure 6.7. Abbreviated phylogenetic sequence from crossopterygian fish (top) to placental mammal (bottom): (*a*) early lobefin fish; (*b*) early amphibian; (*c*) primitive reptile; (*d*) early mammal-like reptile; (*e*) advanced mammal-like reptile; (*f*) modern mammal. *(From A. S. Romer, 1968, Major Steps in Vertebrate Evolution, Science* **168***(3809):1633; copyright © 1968 by the American Association for the Advancement of Science. Reprinted with permission.)*

water levels drop and the water itself becomes stagnant. It would be advantageous, therefore, for organisms to surface and breathe oxygen. Interestingly, there are five genera of true lung fish today, and they all reside in seasonally droughty areas. Limbs would be required for those instances where the drought was prolonged and the pond actually dried up. Under these circumstances, the fish would have to have some viable means of locomotion to gain access to nearby ponds; otherwise they would perish in the mud.

The distinction between reptiles and amphibians is in their mode of reproduction. The latter lay eggs in water, which later undergo metamorphosis; the former lay dry-land eggs from which a miniature replica of the adult emerges. The rise of this type of egg (the amniotic egg) provided the means for escaping the aquatic habitat, and is believed to be another adaptation to prolonged drought. Under drought conditions, most larval eggs would surely perish.

With lungs, limbs, and the amniotic egg, full terrestrial existence became possible. With the rise of insects during the Carboniferous period, a basic food supply became available as a new adaptive zone. One group of reptiles, the therapsids, is particularly interesting as the progenitor of mammals or mammal-like forms. They arose during Permian times and were unique in their narrow gait and speed. They soon gave way to the thecodonts, or bipedal "ruling" dinosaurs. In other words, mammal-like reptiles actually arose earlier than the classic dinosaurs and gave way to them only after leaving progenitors for later mammalian diversification in the Late Cretaceous. Eventually, body scales were replaced by body hair—a development that improved the ability of mammals to maintain body temperature, and one that opened up still further adaptive zones. A second development was the shift from egg laying to live birth in both the marsupials and placentals. By the end of the Mesozoic and beginning of the Cenozoic, the evolution of mammals at the class and subclass levels was complete. The Cenozoic era has been one of rapid differentiation of families and genera.

Summary of Angiosperm Evolution

Less is known about the origins of the flowering plants than is known about mammals. The process of photosynthesis originated with the earliest beginnings of life 2,000 to 3,000 million years ago. Terrestrial plants first emerged during Silurian times, according to evidence of freshwater and near-shore spores found in places like New York and Sweden (Gray, Laufeld, and Boucot, 1974), and it is known that by Carboniferous times the world was inhabited by ferns and other cryptogams. These species form the bulk of the world's coal beds. Until recently, however, there was no conclusive evidence for the pre-Cretaceous existence of angiosperms. In 1970, fossil remains of a palm-like plant from Jurassic sediments in Utah, western Colorado, France, and Greenland were reported by Tidwell and his associates. These are all Laurasian areas, and would seem to discount the prospect for a Southern Hemisphere origin for the angiosperms.

Among the many who have studied angiosperm evolution, Stebbins (1967), believes that rapid evolution may have been stimulated by environmental changes so subtle that fossil remains do not appear in the paleo record. The environment would not have caused genetic changes in species, but it would had have selective value in determining which new gene combinations survived. It is assumed that angiosperms originated in warm and humid upland tropical regions, then spread poleward during the Cretaceous (see Axelrod, 1959, 1967). The pattern of spread on a global basis is shown in Figure 6.8; a more detailed view for Australia is given in Figure 6.9. As the new genetic strains spread, they encountered drought in the form of seasonal wet and dry climates and as topographic rain-shadows. The response to drought can be observed in modern ecosystems in the latitudinal adaptation of vegetation belts. In Africa, for example, one proceeds from rain forest at the equator, through savanna and dry deciduous types, to thorn forest, and finally to tropical desert. Initial evolution probably occurred in locally dry, or xeric, sites, rather than on featureless lowlands. Evolution could have been fast because of the range of adaptation required for entering a new adaptive zone. Several known adaptations would have included reduction of leaf area, deciduousness, pubescence, extensive root systems, rapid flowering, and seed production.

Topographic variation might have stimulated evolution of adaptive characters through local isolation that created peripheral populations. Thus, deciduousness may be a favorable character for higher latitudes, but a character developed initially for drought resistance. Many northern hardwoods have evergreen relatives in lower latitudes, but their deciduous habit assists them in three ways: (1) the reduced rate of nutrient uptake during the cool season is offset by the absence of nutrient-requiring leaves; (2) the low light levels of winter are offset by the absence of photosynthesizing leaves; and (3) low moisture availability is offset by the absence of transpiring leaves, and protection from concomitant subfreezing temperatures.

SUMMARY

Superimposed upon all the mechanisms that species have evolved to ensure their active dispersal are the mechanisms for fostering passive dispersal through geographic isolation. Continental drift, shifting climatic cycles, glacial advances, and marine transgressions are a few of the better known processes by which species are relocated. The legacy of these events can be read today in the taxonomic distributions at familial and higher levels of biotic development. Thus, there is an interplay of organisms competing for resources in ecological time, while their very environments are being molded by events in geologic time. The record clearly shows that the main lines of evolutionary development in higher plants and animals were complete by Late Cretaceous time. Each of the

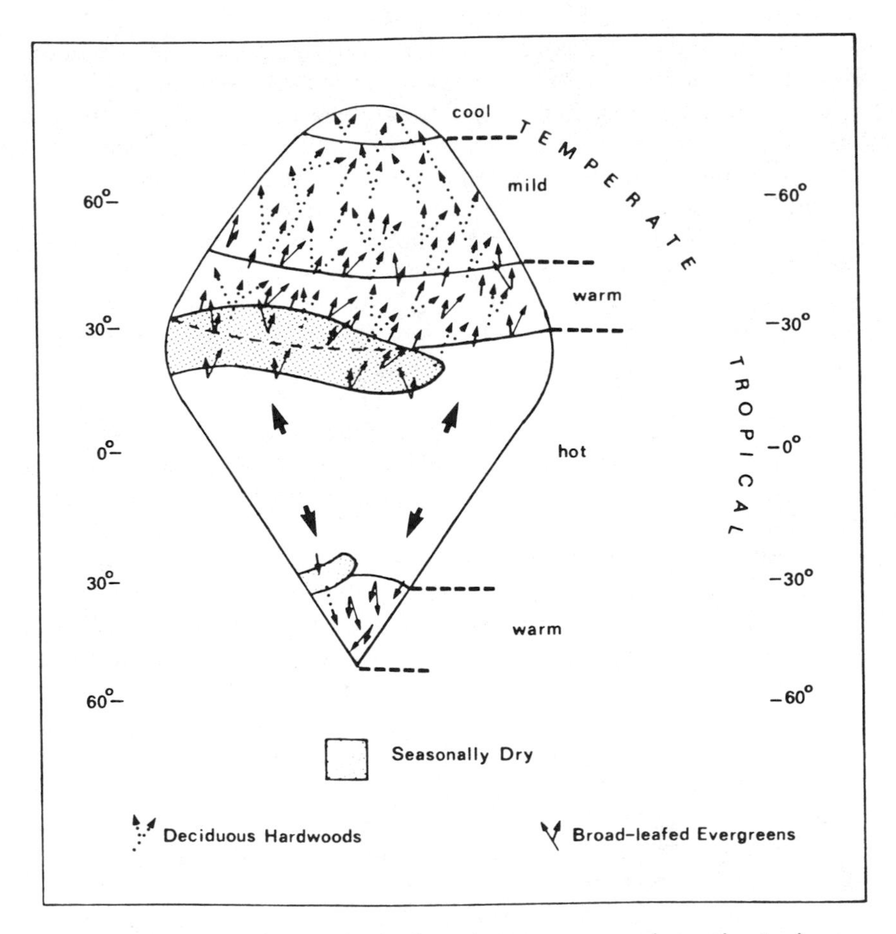

Figure 6.8. Geographic spread of angiosperms on a hypothetical supercontinent during the Late Cretaceous. *(Modified from D. I. Axelrod, 1966, Origin of Deciduous and Evergreen Habits in Temperate Forests, Evolution* **20***(1):12.)*

major continental areas of Laurasia and Gondwanaland had representatives of at least some of these forms, and these have undergone adaptive radiation at the familial, generic and specific levels during gradual separation of land areas in the last 70 million years.

In its purest form, historical biogeography is concerned with the distributional and phylogenetic histories of individual taxa. This aim is feasible on small isolated islands but becomes clouded by overwhelming numbers of species as studies of continental proportion are undertaken. It is not numbers alone, however, that fog our understanding, but the concomitant increase in habitat diversity conditioned by a complex geologic past, which almost forces us to consider broader units, like plant

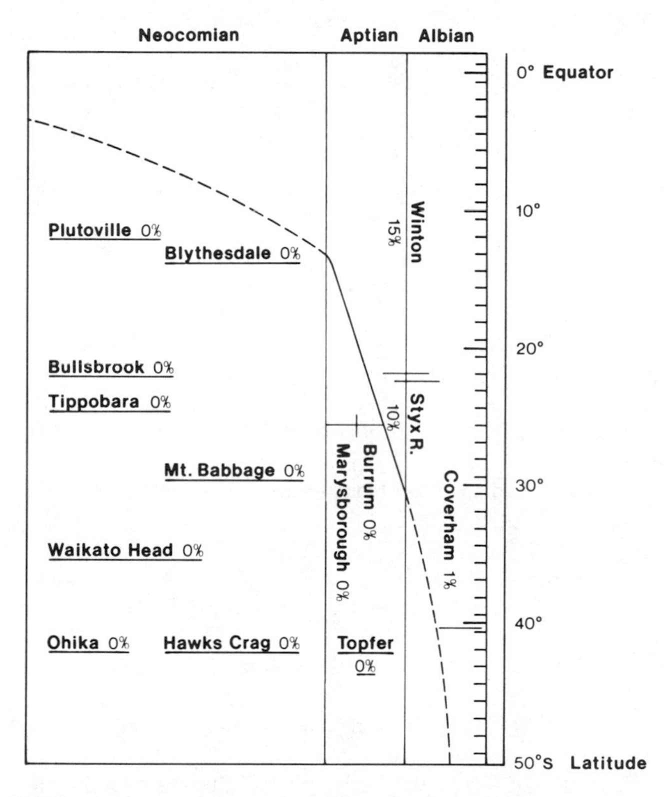

Figure 6.9. Time-space relations of Early Cretaceous floras, and the percentage of angiosperms they contain. All of these fossil floras are from Australia and New Zealand. *(Redrawn from D. I. Axelrod, 1959, Poleward Migration of Early Angiosperms, Science* **130:**203; *copyright © 1959 by the American Association for the Advancement of Science. Reprinted with permission.)*

communities, for study. The chapters that follow in Part II reflect not only the disparity of our knowledge between biotic realms, but also the need for greater generalization as geographic area increases. At some points the discussion is quite detailed as to the origins and percentages of elements comprising a more or less local biota; at other points, in favor of a more abstract discussion of mammalian and vegetational history, only passing reference is made to these concerns.

REFERENCES

Axelrod, D. I., 1952, A Theory of Angiosperm Evolution, *Evolution* **6**(1):29-60.

Axelrod, D. I., 1959, Poleward Migration of Early Angiosperm Flora, *Science* **130**:203-207.

Axelrod, D. I., 1966, Origin of Deciduous and Evergreen Habits in Temperate Forests, *Evolution* **20**(1):1-15.

Axelrod, D. I., 1967, Drought, Diastrophism and Quantum Evolution, *Evolution* **21**(2):201-209.

Broad, W. J., 1981, Creationists Limit Scope of Evolution Case, *Science* **211**:1331-1332.

Dewey, J. F., 1972, Plate Tectonics, *Scientific American* **226**(5):56-68.

Dietz, R. S., and J. C. Holden, 1970, The Break-up of Pangaea, *Scientific American* **223**(4):30-41.

Fooden, J., 1972, Breakup of Pangaea and Isolation of Relict Mammals in Australia, South America and Madagascar, *Science* **175**:894-898.

Gray, J., S. Laufeld, and A. J. Boucot, 1974, Silurian Trilete Spores and Spore Tetrads from Gotland: Their Implications for Land Plant Evolution, *Science* **185**:260-262.

Hawkes, J. G., and P. Smith, 1965, Continental Drift and the Age of Angiosperm Genera, *Nature* **207**(4992):48-50.

Kennedy, G. C., 1959, The Origin of Continents, Mountain Ranges, and Ocean Basins, *American Scientist* **47**(4):491-504.

Kurtén, B., 1969, Continental Drift and Evolution, *Scientific American* **220**(3):54-64.

Romer, A. S., 1967, Major Steps in Vertebrate Evolution, *Science* **168**(3809): 1629-1637.

Simpson, G. G., 1943, Mammals and the Nature of Continents, *Am. J. Sci.* **241**:1-31.

Stebbins, G. L., 1967, Adaptive Radiation and Trends of Evolution in Higher Plants, *Evol. Biol.* **1**:101-142.

Van Steenis, C. G. G. J., 1962, The Land Bridge Theory in Botany, *Blumea* **11**(2):235-372.

Wilson, J. T., 1959, Geophysics and Continental Growth, *American Scientist* **47**(1):1-24.

FURTHER READING

Adie, R. J., 1965, Antarctic Geology and Continental Drift, *Science Journal* Aug., pp. 65-73.

Alexander, T., 1975*a*, A Revolution Called Plate Tectonics Has Given Us a Whole New Earth, *Smithsonian,* Jan., pp. 30-40.

Alexander, T., 1975*b*, Plate Tectonics Has a Lot to Tell Us About the Present and Future Earth, *Smithsonian,* Feb., pp. 30-47.

Anonymous, 1978, Skepticism Persists as Plate Tectonic Answers Come Harder, *Science* **199**:283.

Barnett, S. A., ed., 1958, *A Century of Darwin,* Harvard University Press, Cambridge, Mass., 376p.; see especially, A. S. Romer, Darwin and the Fossil Record, pp. 130-152.

Beck, C. B., 1976, *Origin and Early Evolution of Angiosperms,* Columbia University Press, New York, 342p.

Beloussov, V. V., 1967, Against Continental Drift, *Science Journal,* Jan., pp 2-7.

Ben-Avraham, Z., 1981, The Movement of Continents, *American Scientist* **69**(3):291-299.

Bolin, B., 1977, Changes of Land Biota and Their Importance for the Carbon Cycle, *Science* **196**:613-615.

Briggs, J. C., 1966, Zoogeography and Evolution, *Evolution* **20**(3):282-289.

Camp, W. H., 1947, Distribution Patterns in Modern Plants and the Problems of Ancient Dispersals, *Ecol. Monogr.* **17**(2):161-183.

Cartmill, M., 1974, Rethinking Primate Origins, *Science* **184**:436-443.

Christian, J. J., 1970, Social Subordination, Population Density, and Mammalian Evolution, *Science* **168**:84-90.

Clemens, W. A., 1970, Mesozoic Mammalian Evolution, *Annu. Rev. Ecol. and Syst.* **1**:357-390.

Colbert, E. H., 1980, *Evolution of the Vertebrates,* 3rd ed., Wiley Professional Books, Somerset, N.J., 510p.

Cracraft, J., 1974, Continental Drift and Vertebrate Distribution, *Annu. Rev. Ecol. and Syst.* **5**:215-261.

Croizat, L., 1952, *Manual of Phytogeography,* W. Junk, The Hague, 587p.

Dobzhansky, Th., 1950, Evolution in the Tropics, *American Scientist* **38**:209-221.

Doyle, J. A., 1978, Origin of Angiosperms, *Annu. Rev. Ecol. and Syst.* **9**:497-518.

DuToit, A. L., 1944, Tertiary Mammals and Continental Drift, *Am. J. Sci.* **242**:145-163.

Elliot, D. H., 1970, Triassic Tetrapods from Antarctica: Evidence for Continental Drift, *Science* **169**:1197-1201.

Federov, An. A., 1966, The Structure of the Tropical Rainforest and Speciation in the Humid Tropics, *J. Ecol.* **54**(1):1-11.

Hallam, A., 1972, Continental Drift and the Fossil Record, *Scientific American* **227**(5):57-66.

Hartley, W., 1950, The Global Distribution of Tribes of the Gramineae in Relation to Historical and Environmental Factors, *Aust. J. Agric. Res.* **1**:355-373.

Heirtzler, J. R., 1968, Sea-Floor Spreading, *Scientific American* **219**(6):60-70.
Herbert, D. A., 1950, Present Day Distribution and the Geological Past, *Victoria Nat.* **66**:227-232.
Hurley, P. M., 1968, The Confirmation of Continental Drift, *Scientific American* **218**(4):52-64.
Keast, et al., 1972, *Evolution, Mammals and Southern Continents,* State University of New York Press, Albany, N.Y., 543p.
Kerr, R. A., 1979, How is New Ocean Crust Formed? *Science* **205**:115-118.
Kerr, R. A., 1980, The Bits and Pieces of Plate Tectonics, *Science* **207**:1059-1061.
Lillegraven, J. A., 1974, Biogeographical Consideration of the Marsupial-Placental Dichotomy, *Annu. Rev. Ecol. and Syst.* **5**:263-284.
Mayr, E., 1976, *Evolution and the Diversity of Life,* Harvard University Press, Cambridge, Mass., 721p.
McLean, D. M., 1978, Land Floras: The Major Late Phanerozoic Atmospheric Carbon Dioxide/Oxygen Control, *Science* **200**:1060-1062.
Nelson, G., and D. E. Rosen, eds., 1981, *Vicariance Biogeography,* Columbia University Press, New York, 594p.
Newell, N. D., 1967, *Revolutions in the History of Life,* United States Geological Survey, Special Publication 89, Washington, D.C., pp. 63-91.
Raven, P. H., and D. I. Axelrod, 1974, Angiosperm Biogeography and Past Continental Movements, *Ann. Missouri Bot. Gard.* **61**:539-673.
Regal, P. J., 1977, Ecology and Evolution of Flowering Plant Dominance, *Science* **196**:622-629.
Stebbins, G. L., 1956, Cytogenetics and Evolution of the Grass Family, *Am. J. Bot.* **43**:890-905.
Tidwell, W. D., S. R. Rushforth, J. L. Reveal, and H. Behumin, 1970, Palmoxylon simperi and Palmoxylon pristina: Two Pre-Cretaceous Angiosperms from Utah, *Science* **168**:835-840.
Tidwell, W. D., S. R. Rushforth, and A. D. Simper, 1970, Pre-Cretaceous Flowering Plants: Further Evidence from Utah, *Science* **170**:547-548.
Valentine, D. H., ed., 1972, *Taxonomy, Phytogeography and Evolution,* Academic Press, New York, 431p.
Windley, B. F., 1977, *The Evolving Continents,* Wiley, New York, 386p.
Woodburne, M. O., and W. J. Zinsmeister, 1982, Fossil Land Mammal from Antarctica, *Science* **218**:284-285.

PART II

Regional Biogeography

Chapter 7

The Nearctic Realm

The Nearctic Realm includes all of North America and Greenland, extending along the Sierra Madre of Mexico and Central America to about Tegucigalpa in Honduras. All of the United States, excluding the Everglades in extreme southern Florida, and Hawaii, is included in this realm. Rzedowski (1973) presents good evidence to indicate that the Mexican dry regions have closer floristic affinities to the Neotropical Realm than to the Nearctic. On this basis the Sonoran and Chihuahuan deserts of California, Arizona, New Mexico, and southwest Texas might also be excluded. Overall the latitudinal range extends from about 15°N to 85°N, with a longitudinal extent from Greenland at about 20°W to the western terminus of the Aleutian Islands at 170°E. In all, there are more than 24,531,000 square kilometers of land over which species can evolve, migrate, concentrate, and diverge. Superimposed on this expanse is an almost complete spectrum of life zones and habitats, which throughout history have acted as stimuli for biological progression.

For the mammals and angiosperms, Pleistocene glaciation and Holocene aridity are the two most important environmental influences on the developmental histories of the realm. In fact, from the Upper Great Lakes and Northern Rockies northward throughout Canada and Alaska, very little is known about biotic development, because these areas are only now recovering from the last glacial transgression. As far as plant life is concerned, the Canadian sector in particular is too "complicated and confused, or in the case of the Arctic, too inadequately investigated, to permit any clear generalizations" (Ritchie, 1965, 68). Current estimates of Holocene biotic recovery over much of the Nearctic are derived from palynology and studies of paleoecology, whose interpretations are, in many cases, controversial or inconclusive.

The history of glacial activity in North America began about 1.5 million years ago with the Nebraskan Ice Age. The times and durations of subsequent events are shown in Figure 7.1. Lower temperatures and drier conditions have led to at least four ice transgressions, each separated by an interglacial stage during which plant and animal life migrated out of refugia behind the receding ice. During the peak of the most recent full glacial advance 18,000 years ago, equatorial ocean temperatures were as much as 6°C cooler, and terrestrial temperatures south of the ice sheets were as much as 15°C cooler (Hammond, 1976). At this time continental Wisconsin ice spread as far south as central Ohio and Long Island in the eastern half of North America, while in the Pacific Northwest, alpine glaciers coalesced to cover most of Washington (Stuiver, Heusser, and Yang, 1978). Figure 7.2 presents a general view of this ice sheet and the major ice-free refugia.

The net effect of these advancing and retreating ice sheets was to obliterate existing plant and animal communities and force the more adaptive and competitive forms into protected areas. For several degrees of latitude in the van of the ice front, cooler and drier climatic conditions stimulated the migration of life forms toward the equator. The more or less long history of uniform tropical and semitropical conditions that resulted in the development of a lush and widespread Tertiary flora thus came to a rather abrupt end at the beginning of Pleistocene time. Existing biotic patterns observed in the Nearctic are almost entirely molded by events that have unfolded in the last 70th of angiosperm and mammalian history. As one moves southward into the Neotropical Realm, a complex pattern of plant and animal communities is encountered that has been less affected by severe climatic changes but which, nevertheless, shows inherited characteristics from younger migrations superimposed on Tertiary assemblages (see Graham, 1973).

NEARCTIC VEGETATION

Although the shape and topography of North America have changed considerably in the last 70 million years, the climates and vegetational patterns have been fairly stable through all but the last few million years. Relatively mild, moist climates dominated the realm and the early Tertiary flora was rich in *Magnolia, Sequoia, Rhododendron, Taxodium, Sequoiadendron,* and related genera. Today, many of these forms are restricted to the Pacific Northwest or portions of the southeastern United States. Their vicarious distributions are living testimony

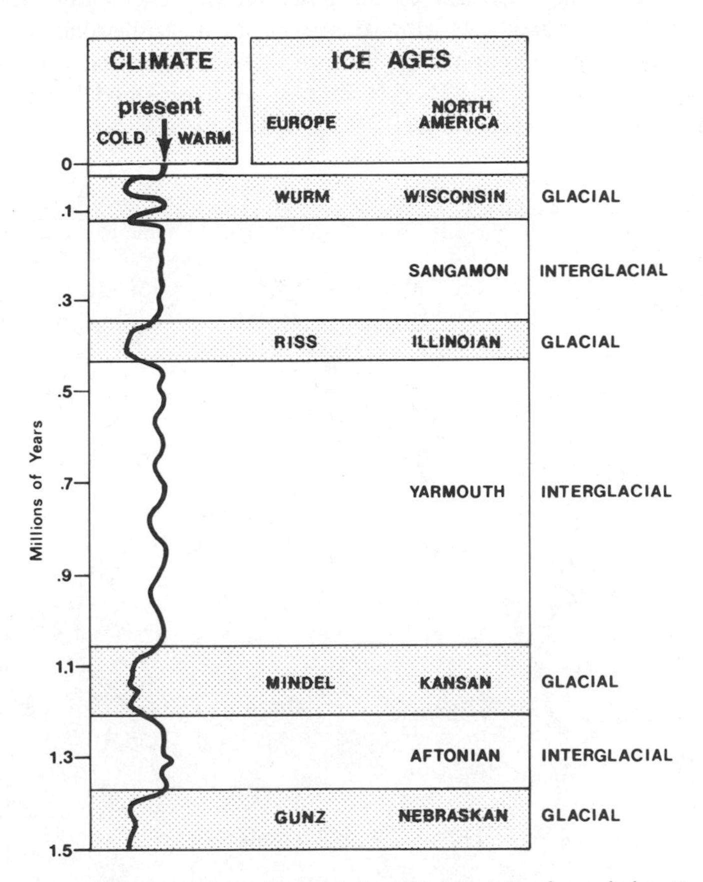

Figure 7.1. Pleistocene glacial and interglacial episodes of the Northern Hemisphere. Glacial transgressions have been relatively short by comparison to the somewhat warmer interglacial episodes. The oscillations of these warmer and cooler cycles are mirrored by geographic displacements of plant and animal life across most of the Nearctic and Palearctic Realms. As one progresses into higher latitudes, the length of recovery time since the last ice age decreases. *(From G. de Q. Robin, 1966, Origin of the Ice Ages, Science Journal Reprint no. 66/166, p. 5; copyright © 1966 by Science Journal. Reprinted with permission of the publisher.)*

to a once more uniform and equable environment, and their fossil remains attest to their former arctic extent (Figure 7.3).

By Miocene time several events were under way that stimulated a long period of floristic and mammalian evolution and adaptation. The Rocky Mountains began uplifting in Late Cretaceous time and by the mid-Tertiary were high enough to influence Pacific Maritime and Polar Continental air masses. During this same time, the broad expanse of Mesozoic seas that occupied the interior of North America receded and, because of the continentality of climate, were replaced by featureless savanna plains that experienced periodic droughts.

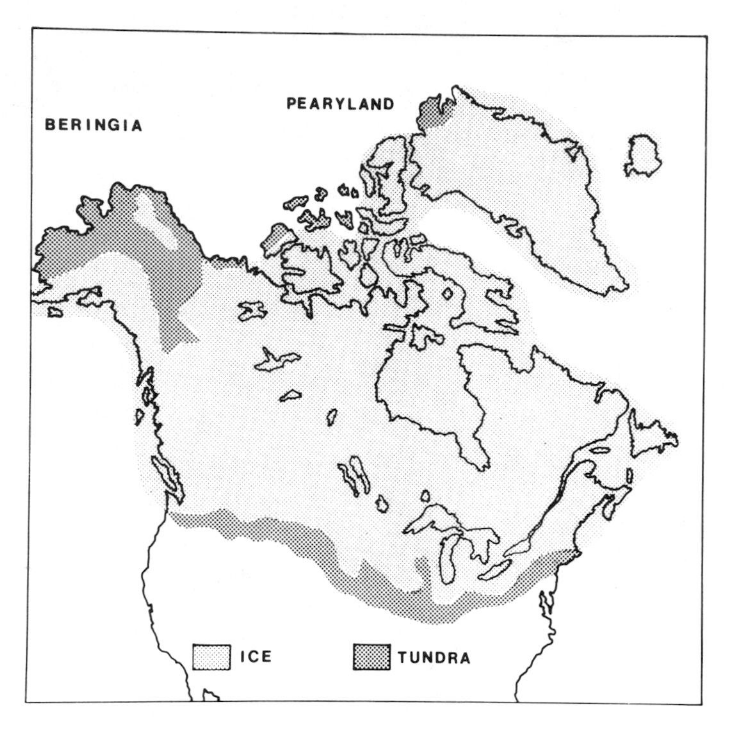

Figure 7.2. Extent of Wisconsin ice about 18,000 years before present. Important ice-free areas were located in Beringia (portions of Alaska and Siberia, which at that time were connected because of lowered sea level), Pearyland, in northern Greenland, and along a narrow zone from the Mackenzie Delta to the Elizabeth Islands. A more or less continuous zone of tundra on the southern margin of the ice is hypothesized. *(From A. H. MacPherson, 1965, The Origin and Diversity of Mammals of the Canadian Arctic Tundra, Systematic Zoology* **14***(3):155; copyright © 1965 by Systematic Zoology. Reprinted with permission of the publisher.)*

Pliocene vegetation in North America can be summarized as consisting of a western and eastern forest, both of which extended quite far to the north, and a central region of open tree savanna grading eastward from the Rocky Mountain front into the more continuous eastern forest. This eastern forest was probably characterized by tropical and subtropical genera that extended well into Canada, but many of the species were probably differentiating from their southern relatives along the lines of deciduousness and physiologic drought. By the end of the Pliocene epoch, two million years ago, many of the generic adaptations to altitude and latitude had probably been initiated in the preglacial flora.

The stimulus to further sifting and differentiation at the species level came as a result of major climatic shifts and isolations caused by the southward flow of ice. The complex history of subsequent glacial and interglacial episodes, major longitudinal mountain barriers (along the Pacific, Rocky Mountain, and Appalachian systems), Holocene aridity in the American Southwest, and the sudden massive impact of *Homo sapiens* all led to the present segregation of biotic elements. The result, however, is one of rather gradual ecotones from the Rocky Mountain front eastward, and more abrupt transitions westward. To the east, glacial and interglacial episodes forced a north-south migration of plant

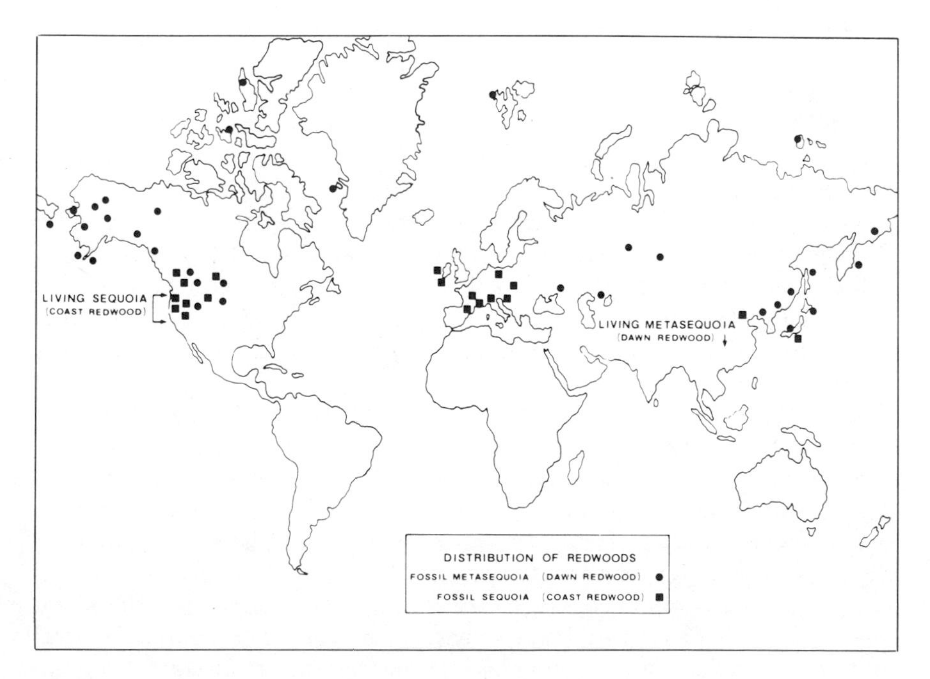

Figure 7.3. Distribution of fossil redwood (*Sequoia* and *Metasequoia*).

and animal life. Refugia were often rather broad in extent and more or less latitudinally arranged. Geographic distributions were compressed during the glacials and expanded during the interglacials. Between the Rockies and the Sierra Nevada/Cascade Ranges, however, quite the opposite process emerged. Here, the glacial, or pluvial, episodes permitted the downslope expansion and convergence of vegetational zones, while the warmer and drier interglacials forced biota into isolated mountain refugia. One can speculate, therefore, that the periods of greatest plant differentiation occurred during interglacial episodes in the west, and during glacial episodes in the east.

The Eastern Forests

Too little remains of the pristine eastern forest to construct a good picture of its development. Indian and European settlement have so altered the forest's pattern that it is almost impossible to reconstruct its history. Pollen analyses suggest that subarctic vegetation dominated the upper midwest and eastern United States after the retreat of Wisconsin ice. The absence of large numbers of trees, as judged by the quantity of pollen, further suggests that an arctic tree-line environment existed there as recently as 12,000 years ago. About 11,500 years B.P. the climate apparently warmed, and forest vegetation dominated by spruce (*Picea*) and other boreal species began to develop. By 8,000 B.P. temperate deciduous trees including maple (*Acer*), birch (*Betula*), oak (*Quercus*), beech (*Fagus*), and hickory (*Carya*) were dominant. At this stage there appears to be good floristic correspondence between the United States and northern Europe, suggesting that the two regions recovered along similar lines after glacial retreat. The increase in conifers in New England during the past few thousand years is now attributed to reforestation after clearing by people, and is not the result of climatic change. Concomitantly, the popular view that the current abundance of oak, maple, and hickory owes its origin to historical human activity appears to be wrong.

Two features of the eastern forest are noteworthy for their implications on biogeographic development. The "driftless area" of southwest Wisconsin, adjacent Minnesota, Iowa, and Illinois was never glaciated but was bounded on the north, east, and west by ice. It has long been considered a refugium for plants during glacial periods and a center for dispersal during ice retreat. If so, it could be argued that these refugial species survived glacial episodes by adapting to boreal conditions and were better prepared to advance outward as the ice retreated. Several endemic plants occur there, suggesting their periodic isolation during the Pleistocene, but the genetics of these endemics are too poorly understood to allow more than speculation about their age and place of origin. Their original home could still have been the tropics, and their presence in the Great Lakes region might be explained in terms of an early wave of migration northward during preglacial or interglacial times.

A second point of interest is the "prairie peninsula"—a wedge of grassland penetrating eastward into Minnesota and Iowa. Its origin and persistence have been hotly debated, and it has been variously explained as a naturally occurring, climatically controlled area, on the one hand; and as fire induced on the other. Gleason (1922) proposed two major episodes of prairie expansion since the retreat of Wisconsin ice. The first of these episodes was stimulated by a warm, dry trend that forced the retreat of forests to the east. The second episode possibly resulted from prairie fires set by Indians. We can observe the same trends today in the Prairie Province, as Küchler (1972) has demonstrated. Although the setting of fires by humans may explain why the grassland has persisted, it fails to explain why grasslands do not extend farther into the east. Certainly the setting of fires was not confined solely to the Great Plains, and recent evidence by Bormann and Likens (1979) indicates that fire has been a natural and regular visitor to the eastern forest ecosystem.

As a general proposition, development of the eastern forest has been compressed into a short period of time since the retreat of the last glacier. Many believe that recovery is not complete and that the complex pattern merely reflects different aspects of the recovery, highly modified by the impact of humans. As the ice retreated, environments became available for recolonization, which meant that successional patterns, competition, tolerance levels, and isolating mechanisms all came into play. The situation appears to have been perfect for adaptation and speciation, especially in areas of rugged terrain like the Appalachian and Ozark regions, where local habitat parameters might restrict the flow of species or their genetic material.

The modern arboreal flora of Tennessee may well reflect the intricacies of habitats recovering not only from a major glacial advance to within a few hundred miles, but also from the effects of human disturbance. In all, 184 arborescent species are recorded for the state, 156 of which are native and 28 of which are introduced (Tennessee Department of Conservation, 1964). Table 7.1 lists some of the major genera and shows their percentage contribution to the overall tree flora; Figure 7.4 shows the complex manner in which some of these species are segregated into local communities according to slope and aspect.

While glaciers spread over the north, pine and spruce forests dominated the southeastern United States as far south as northern Florida and Texas (Dillon, 1956). By late Wisconsin time (15,000-10,000 B.P.) the Chesapeake Bay area, as far south as Georgia, was rich in spruce, pine, fir

Table 7.1. Important Genera in the Arborescent Fraction of the Tennessee Flora

Common name	*Latin*	*No. of species*	*% of total trees*
Birch	*Betula*	3	2
Elm	*Ulmus*	6	4
Gum	*Nyssa*	3	2
Hickory	*Carya*	8	5
Magnolia	*Magnolia*	4	3
Maple	*Acer*	8	5
Oak	*Quercus*	21	13
Pine	*Pinus*	6	4
Plum	*Prunus*	6	4

(*Abies*) and birch. The dominance then began shifting to pine, birch, and alder (*Alnus*); and, by the close of the glacial episode farther north, the southern forest was dominated by oak, hickory, and other species. In postglacial times (10,000 B.P.-present) there has been a gradual transformation in the Carolinas to oak, hickory, and maple, with cypress (*Cupressus*) and gum (*Nyssa*) dominating the swampier regions. In Arkansas and eastern Oklahoma, oak and hickory have dominated the pattern since preglacial times and some believe that this was the original home of the eastern oaks.

The situation farther south in Florida and along the Gulf Coast as far as eastern Texas is not at all well known. The prime question is whether Pleistocene influences ". . . extended sufficiently westward to allow temperate elements of the eastern deciduous forest to transgress the present desert barrier in south Texas and northern Mexico" (Graham, 1973, 305). Some compression of vegetation types probably occurred as the glaciers moved toward the south, but any such compression would have become less and less pronounced as the tropics were approached. The vegetation of southern Florida is truly tropical, including tropical species of pine, so it is unlikely that any major shifts of vegetation have taken place there in several million years.

The Central Mountains and Plains

The present-day North American prairies are unique in their geographic position and history. During most of the Cretaceous, the region was submerged in a shallow inland sea that served as an effective barrier for biotic interchange between east and west. As the sea gradually receded, the land became occupied by a characteristic Tertiary flora mixed with increasing numbers of graminae extending as far as 60° to 65° north

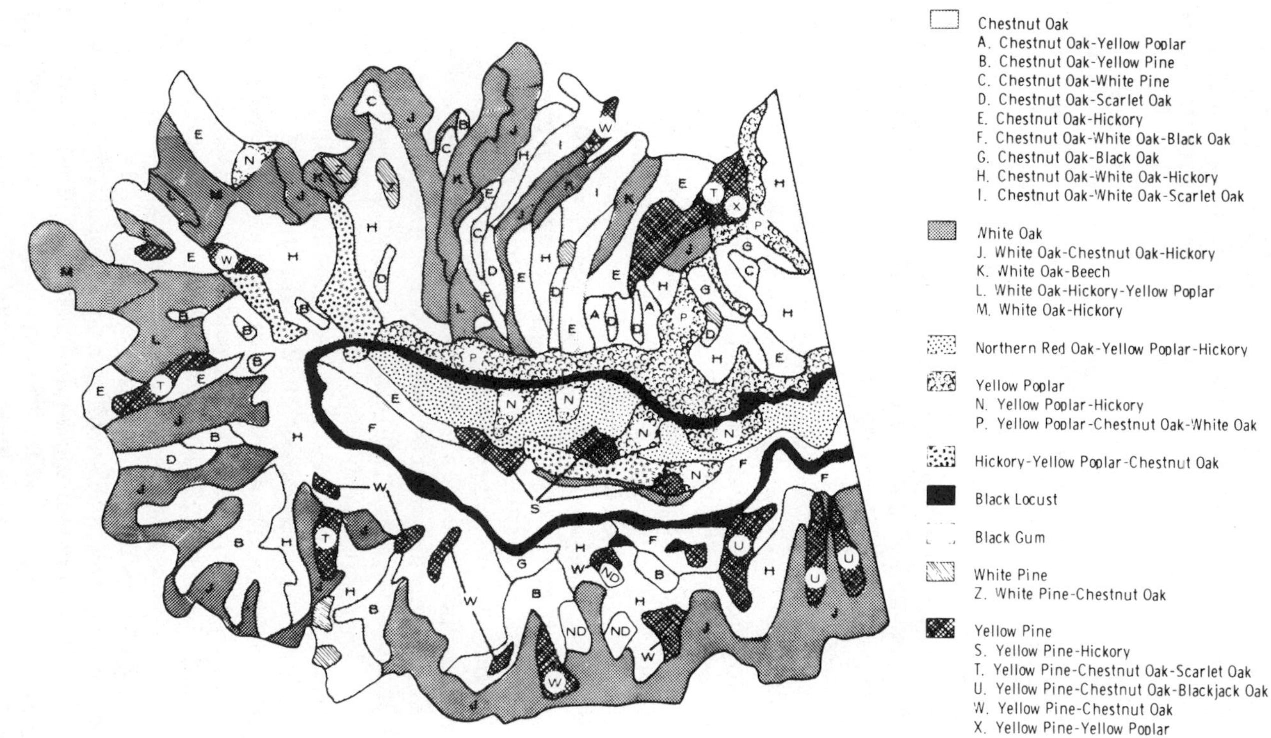
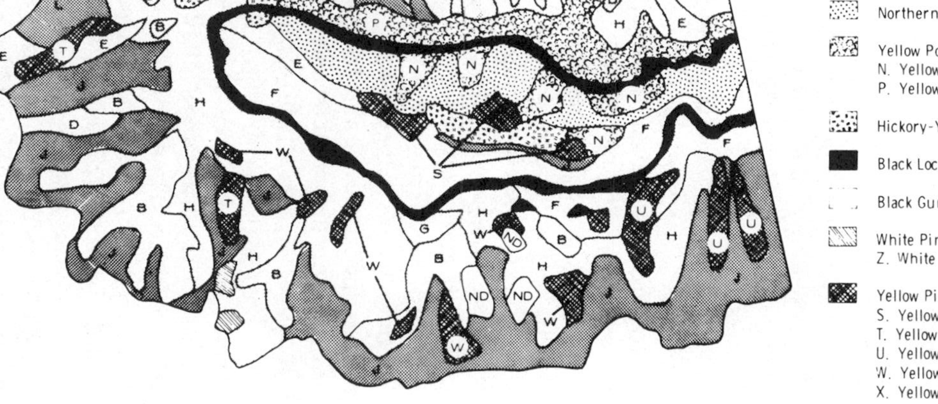

Figure 7.4. An example of vegetational sorting as a function of slope, aspect, and topographic position.

latitude. By late Pliocene or early Pleistocene time the great interior lakes had disappeared and an equable climate extended across most of North America. The longevity and continuity of the Tertiary flora was consequently relatively short in the interior prior to the first refrigeration of the Nebraskan Ice Age.

The gradual onset of Pleistocene glacial events forced the northern life zones southward, compressing the tropical Tertiary flora in their van. Though the ice ages were relatively short by comparison to the interglacials, the warming periods never again permitted a tropical flora to return to latitudes much above 45°N. Some boreal species persisted in the higher elevations of what is now the Grasslands Province, and the temperate flora occupied much of what had formerly been tropical. The Kansan and Illinoian glacial episodes forced successive migrations deep into the province, and on each occasion fewer and fewer of the arborescent species returned.

The origin of prairies is believed to derive from floristic enrichment of tundra regions (see Carpenter, 1940; Thompson, 1939). With each succeeding ice advance, tundra and boreal species transgressed over the heart of America. The development of subhumid climates in the interior of the continent allowed species that were adapted to high evapotranspiration and physiologic drought to extend rather far southward by comparison to their expansion in the more or less narrow tundra zone of New England. Broad-scale leveling of terrain, combined with the interior climate, thus promoted large expanses of tundra, and later of grassland.

The postglacial advance of forests on the eastern margin of the prairies was, and is, confined to river valleys and sloping terrain. The more xeric uplands have remained treeless and today are known ecologically as balds. In rolling or highly dissected terrain like that of eastern Kansas, the balds are very small and the forest nearly continuous, but as one progresses westward, trees disappear rapidly and the open prairie emerges. The persistence of these prairies in Recent time may stem from rapidly moving range fires, but their origin cannot be attributed to fire alone.

The Rockies have formed a major north-south highway for vegetational elements moving north from the Andes and south from the Asiatic arctic. The river systems also act as corridors, while the divides between them serve as effective east-west barriers. Currently, the mountain system is effectively isolated from similar alpine areas by steppes and lowlands. In early Tertiary times, as the Rockies continued to be uplifted, subtropical and tropical species comprised the flora of Colorado. Species included sequoia, willow, elm, and basswood. The environment became warmer and drier by late Oligocene times and many of the broadleaf forms died out. By Mio-Pliocene times only a few relict areas apparently remained.

This trend toward eliminating the broadleaf forms continued in response to drying.

The Southwestern Deserts

Figures 7.5 and 7.6 show the full-glacial and modern vegetation of the American Southwest. It is clear by comparing the figures that spruce-fir forest used to be more widespread and that the yellow-pine forest of 18,000 years ago has yielded slowly to grassland. The Sonoran Desert has expanded northward to replace the retreating sagebrush-chaparral communities that occupy today's Great Basin Desert. Similarly, species of the Chihuahuan Desert have replaced the piñon-juniper type in southern New Mexico and Texas. The explanation for all of these changes is based on glacial episodes in the north. During glacial advances, suitable climatic regimes apparently prevailed that would permit the expansion of types both downslope and southward. As the glaciers retreated, conditions became drier, and desert species, radiating out of the Mexican dry regions, had their opportunity to reinvade northward. The majority of these invaders have Neotropical or tropical affinities compared to those in the more mesic higher elevations, which seem to be descendant from preadaptive forms of northern origin.

Rzedowski (1973) has provided interesting summary data on the floristic similarities between the major dry regions of the southwestern United States and those of Mexico. At the generic level it appears that the Sonoran Desert, which includes Baja California, differs somewhat from the Chihuahuan, San Luis Potosi, and Hidalgan Deserts (Figure 7.6 and Table 7.2). Generally, fewer than 50 percent of the genera are common to both the Sonoran and any of the other deserts. Of the 12 most frequent and characteristic plants of the Sonoran and Chihuahuan Deserts, for example, only three are common to both (*Larrea, Fouquieria,* and *Prosopis*). In contrast, similarity seems higher between the other three deserts; generally, in excess of two-thirds of the genera are shared. This pattern suggests that a north-south avenue for migrating genera was slightly more important than an east-west movement.

Another interesting feature of the southwest deserts is the relationship between the vegetation and the Pleistocene mammal assemblage. It is believed that many plant species developed resinous and unpalatable attributes in response to heavy grazing by native horses, camels, elephants, antelope, and bison. An equilibrium is believed to have existed between forage and browsing populations, such that the sudden extinction of these animals after almost a thousand years of glacial recovery cannot easily be explained. Oddly, these animals disappeared within 2,000 years

Table 7.2. Floristic Similarity of Four Mexican Deserts

	Sonora (272 genera)	*Chihuahua* (258 genera)	*San Luis Potosi* (177 genera)	*Hidalgo* (189 genera)
Sonora	—	51.9%	45.4%	41.0%
Chihuahua		—	71.8%	67.8%
San Luis Potosi			—	87.0%
Hidalgo				—

Source: J. Rzedowski, 1973, "Geographical Relationships of the Flora of Mexican Dry Regions," in *Vegetation and Vegetational History of Northern Latin America,* A. Graham, ed., Elsevier, Amsterdam, p.65.

Note: The values represent percentages of genera shared between indicated deserts.

of human arrival from the north, leading several authors to suggest that it was *Homo sapiens* who triggered the so-called Pleistocene megafaunal extinction in North America.

The Californian Province

California is perhaps the most unique floristic region in the Nearctic Realm. By various calculations, 30 to 40 percent of the almost 5,000 native species are endemic. The origin of these natives, and their geographic locations, have intrigued biogeographers and botanists for over 200 years.

The endemics, by and large, can be subdivided into those that are relict (*paleoendemics*) and those that are more recent (*neoendemics*). These two types are not confined to California, but this area is favorable for their study because of the close juxtaposition of environments. The backdrop for the floristic uniqueness of California lies in its wide climatic and topographic diversity. This diversity includes hot deserts in the southeast that have been shown to contain species ranging from Arizona, Texas, and Mexico; and the maritime forests of the northwest coast, containing species that range throughout Alaska into boreal Asia, and parts of most of the remaining life zones of the world, except the humid tropics. These are primarily climatic belts, within which the rough, broken topography adds still more complexity to the pattern of habitats.

The accurate delineation of the patchwork of climatic types and topographic diversity makes biogeographic mapping of the area almost impossible. The map shown as Figure 7.7 is only one approach to the problem. Endemism as a whole is highest along the central coast, in Southern California, and in the Great Basin; but it is also high in the Sierra, and proportionately high in the Central Valley. In general, high

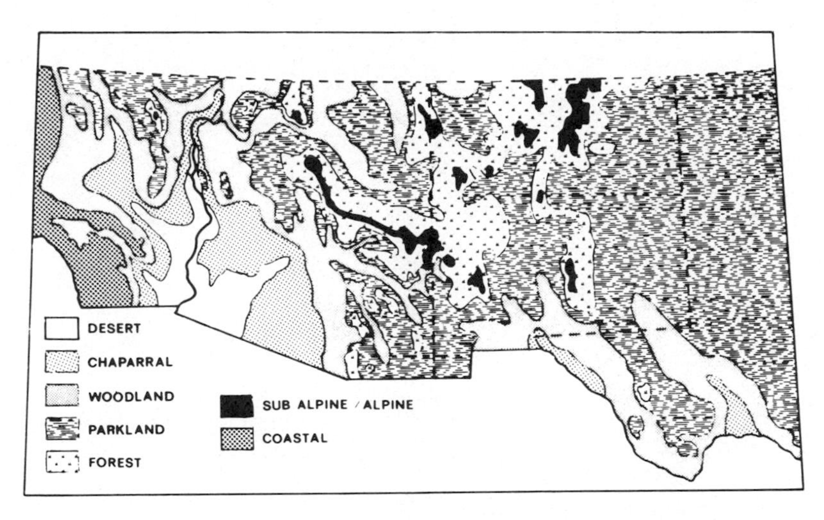

Figure 7.5. Full glacial vegetation of the Southwest 17,000–23,000 years ago, based mainly on carbon-dated fossil-pollen spectra. In the modern vegetation, the sagebrush-chaparral type has been replaced by grassland.

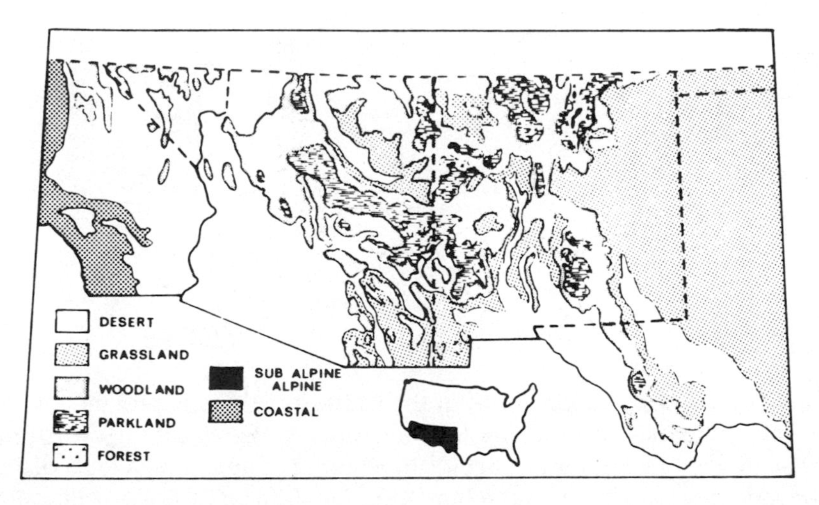

Figure 7.6. Modern vegetation of the Southwest. *(Based on H. L. Shantz and R. Zon, 1924, Atlas of American Natural Vegetation, U. S. Department of Agriculture, Washington, D. C., Fig. 2.)*

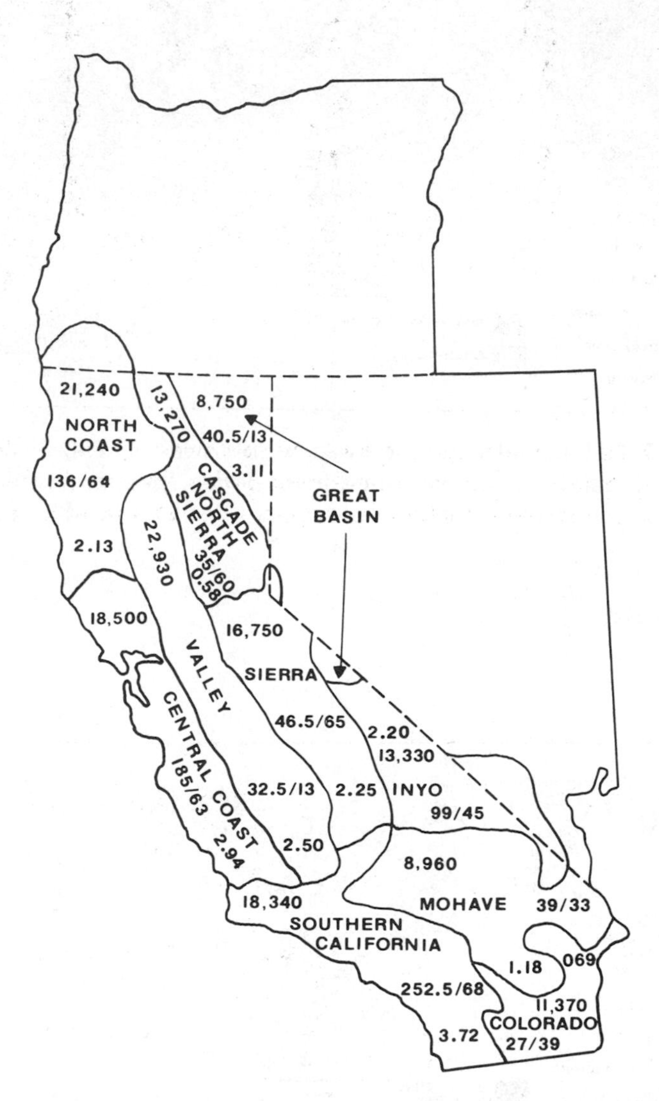

Figure 7.7. Map of subdivisions of the California floristic province. Using the North Coast as an example, the numbers represent 21,240 square miles in size, 136 endemic species in 64 relict genera, and a ratio of 2.13 between these endemic and relict categories. *(From G. L. Stebbins and J. Major, 1965, Endemism and Speciation in the California Flora, Ecological Monograph* **35***(1):7; copyright © 1965, the Ecological Society of America. Reprinted with permission of the publisher.)*

endemism is found in areas of greatest habitat diversity and in areas not subjected to Pleistocene glaciation.

There are several recognized types of endemics. The paleoendemic relicts were left by extinction of their close relatives. The neoendemics are newly evolved taxa usually restricted to a small area. In practice, this simple dichotomy is not entirely satisfactory because there are no means for distinguishing between the two types. A more detailed classification, shown below, has been proposed by Stebbins and Major (1965), based on cytological characters. By adopting it, one can compare the distribution of the older relict species with the more youthful element.

Paleoendemics are ancient isolated taxa with little genetic variability. They are ecologically specialized and perhaps becoming extinct. They do not necessarily occupy the same area where they arose and they may be polyploids. Examples include coast redwood (*Sequoia*), giant redwood (*Sequoiadendron*), and incense cedar (*Libocedrus*).

Schizoendemics are the result of gradual speciation or more or less simultaneous divergence from a parent. They usually have the same number of chromosomes.

Patroendemics are diploid endemics that have given rise to more widespread polyploids. Most of them seem to be gramineous or herbaceous. An example might be wild buckwheat (*Eriogogonum fasciculatum* var. *typicum*) located near Santa Barbara. It has a diploid chromosome number of 40 but has given rise to *E. fasciculatum foliolosum,* which is widespread over Southern California and has a chromosome count of 80.

Apoendemics are polyploid endemics that have arisen from widespread diploid parents. Great Basin sagebrush (*Artemisia tridentata,* spp. *parishii*), for example, has 36 chromosomes and is confined to Antelope Valley near Los Angeles. It has derived from *A. tridentata typica,* which has 18 chromosomes and is widespread in the Basin and Range Province.

The geography of paleo- and schizoendemics shows that they once had a more widespread distribution. Figure 7.8 shows their frequency of occurrence in the present flora. The two areas of concentration are in the northwest coastal zone and the southeast desert zone. Most of the genera in the north have their nearest relatives in the southeastern United States or in temperate Eurasia. The most important category is represented by forest species with relatives in Eurasia. This group includes the redwood, giant redwood, and incense cedar already mentioned. These, among others, form part of the so-called *Arcto-Tertiary flora,* a mesic coniferous and deciduous forest that was widespread in the Northern Hemisphere during Mio-Pliocene times.

In the Southern California desert, the majority of relicts have their closest relatives in other parts of southwestern North America, including Mexico. This group is referred to as part of the *Madro-Tertiary Flora.* The great similarity of North American species to those of South America, and in some cases Africa, suggests that they have evolved very little since these vicarious elements became separated. They exist today in essentially the same form as when the Madro-Tertiary Flora developed in the middle Tertiary. Common examples include the Washingtonia palm (*Washingtonia filifera*), smoke bush (*Daleya tesota*), and creosote bush (*Larrea divaricata*).

Patro- and apoendemics represent the youthful element in the California flora and display a rather different geography. Regions containing large numbers of endemic polyploids with nearby related diploids can be regarded as areas of active speciation. In contrast, regions with a

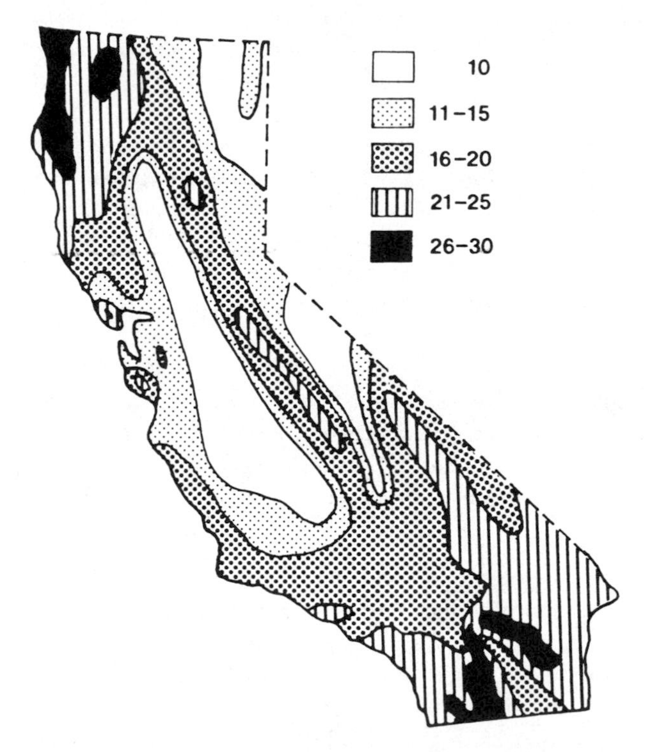

Figure 7.8. Map of the frequency of paleo- and schizoendemic species in California. *(From G. L. Stebbins and J. Major, 1965, Endemism and Speciation in the California Flora, Ecological Monograph* **35***(1):10; copyright © 1965, the Ecological Society of America. Reprinted with permission of the publisher.)*

concentration of diploid forms that have more widespread polyploid relatives are recognized as recessional areas. The maps in Figures 7.9 and 7.10 show the frequency distributions of patroendemics and apoendemics. The former are concentrated in the central coast region, primarily in the area high in summer fog. A second concentration is found in the middle elevations of the Sierra Nevada between 1,500 and 3,000 meters. Apoendemics are less frequent in the coastal region and more common in the interior coast ranges and higher elevations of the Sierra Nevada, particularly on the leeward side.

PLEISTOCENE MAMMALIAN HISTORY

The Canadian Arctic section of the Nearctic Realm is comparatively poor in mammalian species diversity. Including *Homo sapiens,*

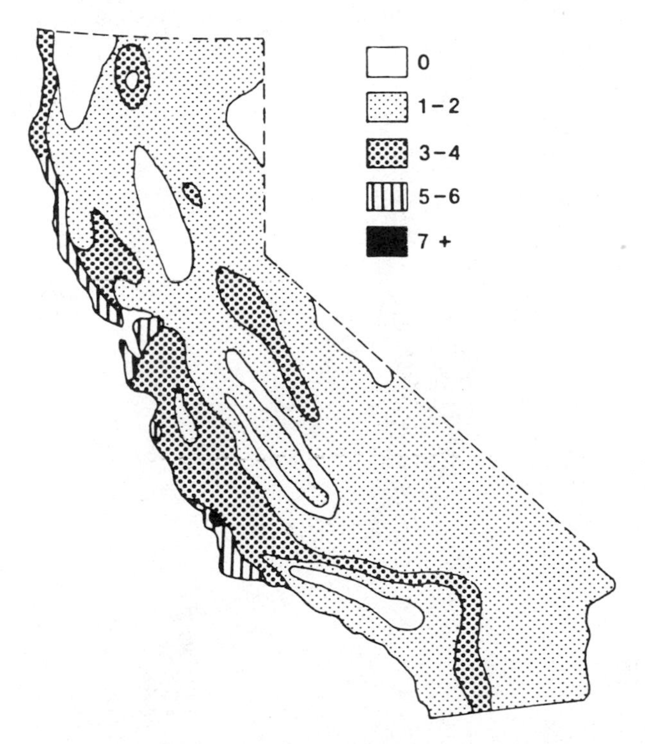

Figure 7.9. Map of the distribution frequency of patroendemics in California. *(From G. L. Stebbins and J. Major, 1965, Endemism and Speciation in the California Flora, Ecological Monograph* **35***(1):19; copyright © 1965, the Ecological Society of America. Reprinted with permission of the publisher.)*

only 32 species are reported by MacPherson (1965), but these 32 have apparently given rise to an additional 40 subspecies owing their origins to refugia within the Wisconsin ice sheet. The arctic hare (*Lepus arcticus*), along with the lemmings (*Dicrostonyx torquatus* and *Lemmus sibiricus*) have been among the more genetically and geographically variable species. Other forms, like the polar bear (*Ursus maritimus*) and tundra vole (*Microtus oeconomus*), display more ubiquitous characters across the realm.

Range distribution maps of all of the species and subspecies suggest that pre-Wisconsin populations became gradually isolated from each other as ice and snow fields grew in size and began to coalesce into a continental sheet. At its height 18,000 years B.P., Wisconsin ice covered all of the Canadian Arctic except for Beringia, Pearyland, and a narrow

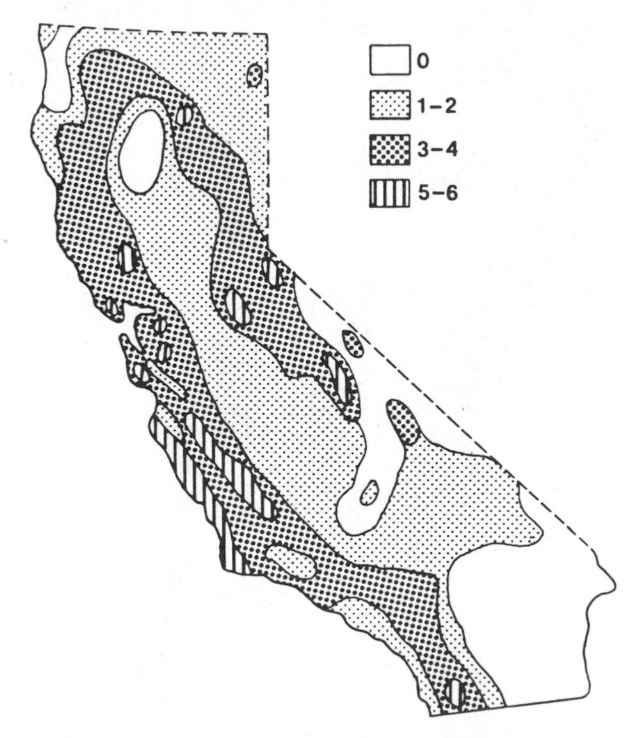

Figure 7.10. Map of the distribution frequency of apoendemics in California. *(From G. L. Stebbins and J. Major, 1965, Endemism and Speciation in the California Flora, Ecological Monograph* **35***(1):19; copyright © 1965, the Ecological Society of America. Reprinted with permission of the publisher.)*

zone from the Mackenzie Delta to the Elizabeth Islands (see Figure 7.2). These exceptions served as havens for mammalian and tundra species during the advance of ice and as source regions for recolonization during ice retreat. Among these species, during their confinement, genetic drift probably took place that led to the variety of forms observed today and listed in Table 7.3.

The refugium located on the southern margin of the ice sheet apparently fostered fewer subspecies than either Beringia or Pearyland. Only 10 subspecies are recorded here as compared to 19 for Beringia and 11 for Pearyland. Conditions were less isolated in the south because of the connections with more southerly life zones and because of the relative ease of east-west movement across more than 50° of longitude (see Dillon, 1956). Zones of overlap and population interspersion could well have hindered the origin of new genetic stock. South of the southern refugium, in what is now the United States, the Wisconsin glacial advance has been viewed as having stimulated little the origin of new species. The effect of the advancing ice was one of displacement, not isolation of forms.

Figure 7.11 shows the locations of major faunal fossil sites in the conterminous United States. The basic data for reconstructing the history of Quaternary mammals in the Nearctic Realm consists of fossil remains for some 170 genera (Hibbard et al., 1965). This fauna represents the best known Quaternary group of animals on the continent. Their chronology can be related to glacial-interglacial climatic changes in much the same manner as was described for the vegetation. This relationship is best documented in the southern Great Plains, particularly in southwest Kansas.

Although there is no general agreement on the location of the Plio-Pleistocene stratigraphic boundary in North America, it is often placed at 2 to 3 million years B.P. The upper boundary of the Pleistocene is equally ambiguous but can be placed at somewhere between 4,000 and 10,000 years B.P. In the earlier part of this time frame, speciation of mammals appears to have been fairly rapid. It culminated 18,000 years B.P. in the extinction (or retreat) of several of the more important forms, such as elephants, mastodons, camels, giant bison, giant beaver, ground sloths, and horses.

The tabulation provided by Hibbard et al., (1965) has been reformatted in Table 7.4 in the form of a matrix showing the times of first and last records for each of the 170 genera. Two observations can be drawn from the matrix. First, it is evident that about half (52 percent) of all the genera that have roamed North America since the Pliocene are still present. Of those that have disappeared, almost half again (46 percent) terminated their stay during Wisconsin time, perhaps after the arrival of humans. Secondly, it is interesting to note that periods of glacial advance

Table 7.3. Species and Subspecies of Canadian Arctic Mammals Listed by Refugia of Their Suspected Origin (see also Figure 7.2)

Common Name	*Beringia*	*Pearyland*	*South*
Arctic hare	*Lepus othus* *L. arcticus* banksicola andersoni north	*L. arcticus* *porsildi* *groenlandicus* *monstrabilis* *hubbardi*	*L. articus* *labradorius*
Varying lemming	*Dicrostonyx torquatus* *kilangmiutak* *lentus*	*D. torquatus* *groenlandicus* *clarus*	*D. torquatus* *richardsoni* *D. hudsonius*
Muskox	— —	*Ovibos moschatus* *wardi*	*O. moschatus* *moschatus*
Caribou	*Rangifer tarandus* *groenlandicus*	*R. tarandus?* *pearyi*	*R. tarandus?* *caribou*
Gray wolf	— —	*Canis lupus* *orion* *arctos*	*C. lupus* *labradorius* *manningi*
Ermine	*Mustela erminea* *arctica* *semplei*	*M. erminea* *polaris*	*M. erminea* *richardsoni*
Masked shrew	*Sorex cinereus* *ugyunak*	—	*S. cinereus* *cinereus*
Red-backed mouse	*Clethrionomys rutilus* *dawsoni* *washburni*—north *platycephalus*—east	—	*C. gapperi*
Ground squirrel	*Spermophilus undulatus* *kennicotti* *plesius* *parryi*	—	*S. columbianus*
Wolverine	—	—	*Gulo gulo* *luscus*
Red fox	—	—	*Vulpes vulpes*
Brown lemming	*Lemmus sibiricus* *alascensis* *subarcticus* *yukonensis* *phaiocephalus*—north *trimucronatus*—east	—	*L. sibiricus* *helvolus* —
Tundra vole	*Microtus oeconomus*	—	—
Brown bear	*Ursus arctos*	—	—
Man	*Homo sapiens*	—	—

Source: A. H. MacPherson, 1965, "The Origin and Diversity in Mammals of the Arctic Tundra," *Syst. Zool.* **14**(3):170.

show greater numbers of genera making their first appearance: Nebraskan, 6 genera; Kansan, 35 genera; Illinoian, 14 genera; and, Wisconsin, 31 genera. This breakdown represents about 87 percent of all the genera making their first appearance in the Quaternary. Only 13 percent made their first appearance during the interglacial episodes. The view expressed by Dillon (1956) that the Wisconsin episode did not substantially influence speciation must therefore be modified and placed into perspective with other Quaternary events. In contrast, the interglacials, including the present, seem to exhibit somewhat higher incidence of last records. Although the number of Recent extinctions is not indicated in the table, it is known from other evidence (Chap. 4, *Man the Extinctor*) that wholesale extermination is now taking place.

Other considerations that should accompany the interpretation of Table 7.4 include faunal shifts, paleogeography, and evolution. All influence the definition of first and last appearances, and can significantly alter our understanding of Pleistocene patterns. Faunal shifts represent changes in distribution that are associated with ecological changes but display little taxonomic differentiation. This phenomenon was discussed earlier in the context of plant migrations. The first major southward faunal shift came in late Kansan time and was repeated to a greater or lesser extent in Illinoian and Wisconsin time. The stratigraphic sequence in southwest Kansas, northwest Oklahoma, and Texas clearly shows this fluctuation of environments and the dominant mammalian forms present

Figure 7.11. Most of the well-known sites for Pleistocene mammalian assemblages in North America.

Table 7.4. Numbers and Percentages of North American Mammalian Genera Making Their First and Last Appearances in the Quaternary Fossil Record

Pleistocene Episodes	*Total number—first appearance*	*Numbers (%) of last appearance*								
		Plio	*Neb.*	*Aft.*	*Kan.*	*Yar.*	*Ill.*	*Sang.*	*Wis.*	*Living*
Pliocene	71	12(17)	1(1)	13(18)	5(7)	2 (3)			5 (7)	33(46)
Nebraskan	6			2(33)					2(33)	2(33)
Aftonian	5								2(40)	3(60)
Kansan	35					4(11)	1 (3)	1(3)	16(46)	13(37)
Yarmouthian	4					2(50)			1(25)	1(25)
Illinoian	14						2(14)		2(14)	10(72)
Sangamonian	4								1(25)	3(75)
Wisconsin	31								8(26)	23(74)
Totals	170	12(7)	1(.5)	15 (9)	5(3)	8 (5)	3 (2)	1(.5)	37(22)	88(52)

at each major interval. This sequence of events can be traced through complex patterns of present day distributions, as well as through fossil occurrences that extend far beyond the range of living populations of the same species.

Paleogeographic changes differ from faunal shifts in that resulting distributions are accompanied by taxonomic differentiation. The Nearctic Realm is inhabited today by 111 living genera of placental mammals (Hibbard, et al., 1965). Most endemism is at the family level, suggesting rather recent evolution. Nevertheless, at least 21 species are common to both North America and Eurasia, indicating that some of the forms evolved fairly early and have been successful in extending their range. Throughout late Pliocene and Pleistocene time, the spread of mammals from the Old World to the New could only have been via the Bering Strait. Among the best known families to make the crossing were the elephants and the bovines. Recent evidence from the Seward Peninsula in Alaska shows that in late Pleistocene time, tundra environments were favorable for the passage of mammoth, bison, and horse.

Besides the Old and New World connections there are many species common to both the Nearctic and Neotropical Realms, and one can assume that these regions had still another pattern of development. Some South American forms of rodents and tapirs extended northward into Central America and Mexico sometime during Plio-Pleistocene time. In some cases, glacial episodes may have had little effect on these forms, aside from stimulating changes in their ranges. In contrast, there are cases like the peccaries that spread northward out of the southern Nearctic region and became extinct in their home range. This pattern gives the false impression that they are Neotropical in origin.

Finally, in considering the question of evolution, evidence seems to point to a modernization of the Quaternary Nearctic mammalian fauna through two processes: new genera and species evolved from families already present in the region; and Neotropical and Palearctic stock immigrated. "Modernization" apparently has progressed also by absolute extinction and by retreat, through the influence of early *Homo,* to refugia outside the Nearctic.

The generalization that mammalian *species* have arisen during the Quaternary is regarded as essentially valid. However, the fossil record is so scanty in key areas like Alaska that one cannot yet tell just how much of the mammals' speciation actually took place in the Nearctic Realm and how much predated their arrival on the continent.

SUMMARY

The broad vegetational patterns of the Nearctic contain vestiges of a once widespread Tertiary flora that extended across the continent into boreal latitudes. By the close of the Pliocene epoch this flora had been disrupted by changing climatic regimes into distinctive eastern and western forests that were separated by tree savannas in the Central Plains. The Rocky Mountains and west coast ranges probably continued to serve as north-south migration routes linking the eastern Palearctic with the Western Hemisphere. Glacial episodes commencing in the Nebraskan cold period of the Pleistocene began a series of compressions southward in the eastern United States and into isolated refugia in what is now the Canadian Arctic. West of the Rocky Mountains, ice was confined primarily to mountain environments, and this pattern resulted in the downslope movement and spread of biota.

Each of the ice transgressions and their related interglacial periods heightened the complexity of biotic relationships. In addition to faunal and floral shifts in distribution, there were rearrangements of migration routes, leading to local and regional extinctions. Modernization of the present day biota has therefore been a progressive phenomenon and should not be summarized as being merely a result of the most recent glacial advance. Humankind has been the prime cause of continued environmental disruption since the retreat of Wisconsin ice, and most of that has occurred in the last 100 years.

REFERENCES

Bormann, F. H., and G. E. Likens, 1979, Catastrophic Disturbances and the Steady State in Northern Hardwood Forests, *American Scientist* **67**(6):660-669.

Carpenter, J. R., 1940, The Grassland Biome, *Ecol. Monogr.* **10**(4):618-684.

Dillon, L. S., 1956, Wisconsin Climate and Life Zones in North America, *Science* **123**(3188):167-176.

Gleason, H., 1922, The Vegetational History of the Middle West, *Assoc. Am. Geog. Ann.* **12**:39-85.

Graham, A., 1973, History of the Arborescent Temperate Element in the Northern Latin American Biota, in *Vegetation and Vegetational History of Northern Latin America,* A. Graham, ed., Elsevier, New York, pp. 301-314.

Hammond, A. L., 1976, Paleoclimate: Ice Age Earth Was Cool and Dry, *Science* **191**:445.

Hibbard, C. W., D. E. Ray, D. E. Savage, D. W. Taylor, and J. E. Guilday, 1965, Quaternary Mammals of North America, in *The Quaternary of the United States,* H. E. Wright, Jr., and D. G. Frey, eds., Princeton University Press, Princeton, N.J., pp. 509-525.

Küchler, A. W., 1972, The Oscillations of the Mixed Prairie in Kansas, *Erdkunda* **26**:120-129.

MacPherson, A. H., 1965, The Origin of Diversity in Mammals of the Canadian Arctic Tundra, *Syst. Zool.* **14**(3):153-173.

Ritchie, J. C., 1965, Aspects of Late Pleistocene History of the Canadian Flora, in *The Evolution of Canada's Flora,* R. L. Taylor and R. A. Ludwig, eds., University of Toronto Press, Toronto, pp. 66-80.

Robin, G. de Q., 1966, Origin of the Ice Ages, *Science Journal* Reprint no. 66/116, 8p.

Rzedowski, J., 1973, Geographical Relationships of the Flora of Mexican Dry Regions, in *Vegetation and Vegetational History of Northern Latin America,* A. Graham, ed., Elsevier, New York, pp. 61-72.

Shantz, H. L. and R. Zon, 1924, *Atlas of American Agriculture Natural Vegetation,* U.S. Department of Agriculture, Washington, D.C., 28p.

Stebbins, G. L., and J. Major, 1965, Endemism and Speciation in the California Flora, *Ecol. Monogr.* **35**(1):1-35.

Stuiver, M., C. J. Heusser, and I. C. Yang, 1978, North American Glacial History Extended to 75,000 Years Ago, *Science* **200**:16-22.

Tennessee Department of Conservation, 1964, *Common Forest Trees of Tennessee,* Division of Forestry, Nashville, Tenn., 72p.

Thompson, I., 1939, Geographical Affinities of the Flora of Ohio, *Am. Midland Naturalist* **21**(3):730-751.

FURTHER READING

Alford, J., 1971, A Geographic Appraisal of Pleistocene Overkill in North America, *Proc. Assoc. Am. Geogr.* **3**:10-14.

Axelrod, D. I., 1952, A Theory of Angiosperm Evolution, *Evolution* **6**(1):29-60.

Axelrod, D. I., 1966, Origin of Deciduous and Evergreen Habits in Temperate Forests, *Evolution* **20**(1):1-15.

Axelrod, D. I., 1967, Drought, Diastrophism and Quantum Evolution, *Evolution* **21**(2):201-209.

Bada, J. L., R. A. Schroeder, and G. F. Carter, 1974, New Evidence for the Antiquity of Man in North America Deduced from Aspartic Acid Racemization, *Science* **184**:791-793.

Bare, J. E., and R. L. McGregor, 1970, An Introduction to the Phytogeography of Kansas, *Univ. Kansas Sci. Bull.* **48**(26):869-949.

Briggs, J. C., 1966, Zoogeography and Evolution, *Evolution* **20**(3):282-289.

Bryson, R. A., 1974, A Perspective on Climatic Change, *Science* **184**:753-760.

Cain, S., 1944, *Foundations of Plant Geography,* Harper and Row, New York, 556p.

Cole, K., 1982, Late Quaternary Zonation of Vegetation in the Eastern Grand Canyon, *Science* **217**:1142-1145.

Dawson, M. R., R. M. West, W. Langston, Jr., and J. H. Hutchison, 1976, Paleogene Terrestrial Vertebrates: Northernmost Occurrence, Ellesmere Island, Canada, *Science* **192**:781-782.

Denton, G. H., and T. J. Hughes, eds., 1981, *The Last Great Ice Sheets,* Wiley-Interscience, New York, 484p.

Echlin, P., 1966, Origins of Photosynthesis, *Science Journal,* April, pp. 2-7.

England, J., and R. S. Bradley, 1978, Past Glacial Activity in the Canadian High Arctic, *Science* **200**:265-270.
Hall, E. R., 1980, *The Mammals of North America,* 2nd ed., Wiley Professional Books, Somerset, N.J., 624p.
Ives, J. D., 1974, Biological Refugia and the Nunatak Hypothesis, in *Arctic and Alpine Environments,* J. D. Ives and R. G. Barry, eds., Methuen, London, pp. 605-636.
Judd, W. W., and J. M. Speirs, 1972, *A Naturalist's Guide to Ontario,* University of Toronto Press, Toronto, 210p.
Küchler, A. W., 1971, A Biogeographical Boundary: The Tatschl Line, *Kansas Acad. Sci. Trans.* **73**(3):298-301.
Löve, A., and D. Löve, 1974, Origin and Evolution of the Arctic and Alpine Floras, in *Arctic and Alpine Environments,* J. D. Ives and R. G. Barry, eds., Methuen, London, pp. 571-604.
Marshall, L. G., R. F. Butler, R. E. Drake, G. H. Curtis, and R. H. Tedford, 1979, Calibration of the Great American Interchange, *Science* **204**:272-279.
Packer, J. G., 1969, Polyploidy in the Canadian Arctic Archipelago, *Arct. and Alp. Res.* **1**(1):15-38.
Pound, R., and F. E. Clements, 1900, *The Phytogeography of Nebraska,* University of Nebraska Press, Lincoln, Nebr., 2nd ed., 442p.
Radforth, N. W., and W. H. Camp, 1947, Origin and Development of Natural Floristic Areas with Special Reference to North America, *Ecol. Monogr.* **17**(2):125-234.
Romer, A. S., 1967, Major Steps in Vertebrate Evolution, *Science* **168**:1629-1637.
Van Devender, T. R., 1977, Holocene Woodlands in the Southwestern Deserts, *Science* **198**:189-192.
Van Devender, T. R., and W. G. Spaulding, 1979, Development of Vegetation and Climate in the Southwestern United States, *Science* **204**:701-710.
Waring, R. H., and J. F. Franklin, 1979, Evergreen Coniferous Forests of the Pacific Northwest, *Science* **204**:1380-1386.
Watts, W. A., and M. Stuiver, 1980, Late Wisconsin Climate of Northern Florida and the Origin of Species-Rich Deciduous Forest, *Science* **210**:325-327.
Webb, S. D., 1969, Extinction-Origination Equilibrium in Late Cenozoic Land Mammals of North America, *Evolution* **23**:688-702.
Webb, S. D., 1977, A History of Savanna Vertebrates in the New World. Part I: North America, *Annu. Rev. Ecol. Land Syst* **8**:355-380.
Webb, T., 1981, The Past 11,000 Years of Vegetational Change in Eastern North America, *BioScience* **31**(7):501-506.
Weber, W. A., 1965, Plant Geography in the Southern Rocky Mountains, in *The Quaternary of the United States,* H. E. Wright, Jr., and D. G. Frey, eds., Princeton University Press, Princeton, N. J., pp. 453-468.
Wells, P. V., 1970, Postglacial Vegetational History of the Great Plains, *Science* **167**:1574-1582.
Woodard, J., 1924, Origin of Prairies in Illinois, *Bot. Gaz.* **77**(3): 241-261.
Wright, H. E., Jr., 1981, Vegetation East of the Rocky Mountains 18,000 Years Ago, *Quaternary Res.* **15**(2):113-125.

Chapter 8

The Neotropical Realm

The Neotropical Realm is synonymous with the term *Latin America*. Biogeographically, it is an immensely interesting region because, alone among the major continents of ancient Pangaea, it shows the effects of long-term isolation followed by recent and direct contact with a former sister. Neither Australia, which has yet to make direct contact, nor the Indian subcontinent, which has made contact but does not compare in size, reveals the same history as South America. Superimposed upon the biotic alterations stimulated by its reunion with the Nearctic Realm are all of the local and subregional changes induced by Pleistocene climatic shifts. Even the barest outline of Neotropical history must therefore take note of both the early and more recent assemblages.

Physically, the Neotropical Realm extends from the coastal lowlands of Mexico, south of the Tropic of Cancer and below about 1,500 meters elevation, through Central America and the Caribbean, and on to the southern tip of Tierra del Fuego, at 55° south latitude. East-west, the realm stretches from about 34° to 110° west longitude, but the area north of the Gulf of Panama protrudes sharply westward and narrows rapidly. The overwhelming bulk of the ancient land mass currently lies between 40°W and 80°W and has its widest extent between 0° and 10° south latitude. The total area south of the Panama Canal is 17,800,000 square kilometers, or about two-thirds the size of the Nearctic Realm.

Along the western margin of the continent the Andean Cordillera rises to peaks in excess of 6,100 meters. Although these mountains become less elevated in the far south and to the north in Central America, they are everywhere a barrier to east-west movement and an avenue for north-south migration. East of the cordillera are the vast lowlands carved out by the Orinoco, Amazon, and Paraná Rivers and their associated highlands in Venezuela, the Guianas, and Brazil.

NEOTROPICAL VEGETATION

The view expressed by Darlington (1965) that South American vegetation has remained relatively stable since the Pleistocene epoch is being challenged by the view that significant changes have been stimulated by repeated climatic fluctuations (see B. B. Simpson, 1974; G. G. Simpson, 1980; Johnson and Raven, 1973). Figure 8.1 shows the generalized configuration of ice and glacial lakes during the most recent glacial advance. In the Neotropical Realm there were no continental glaciers as such, but alpine ice coalesced latitudinally and moved downslope compressing the vegetational zones in its van. Expansion of lakes, particularly in Amazonia, also compressed vegetation, but migrational areas were provided by the consequent lowering of sea level. In a very general sense, one can recognize five regions that serve to describe these vegetational shifts: the *Andean*, the *Patagonian*, the *Eastern Highlands*, the *Eastern Montane Slopes*, and the *Tropical Lowlands*.

Andean Region

The Andes can be subdivided into three components: southern, central, and northern. In the southern part, comprised of Argentina and southern Chile, geologists and biogeographers long ago saw a resemblance to the history of glacial episodes in Europe. It is reasonable to wonder whether the respective biotas have also been altered along similar lines. Studies suggest, for example, that the southern beech (*Nothofagus*) forests retreated toward the equator during the Pleistocene in a manner similar to the shift toward the equator of European forests. Today there are relict, disjunct patches of such forest as far north as 30°S.

South of 44-45°, the southern Andes were covered by ice during the Pleistocene glacial episodes. However, in the forested elevations there are several disjunct areas that have closely related plant taxa. These areas suggest that there were "islands" in the ice that acted as refugia. Of course, the presence of ice in the higher elevations forced the various

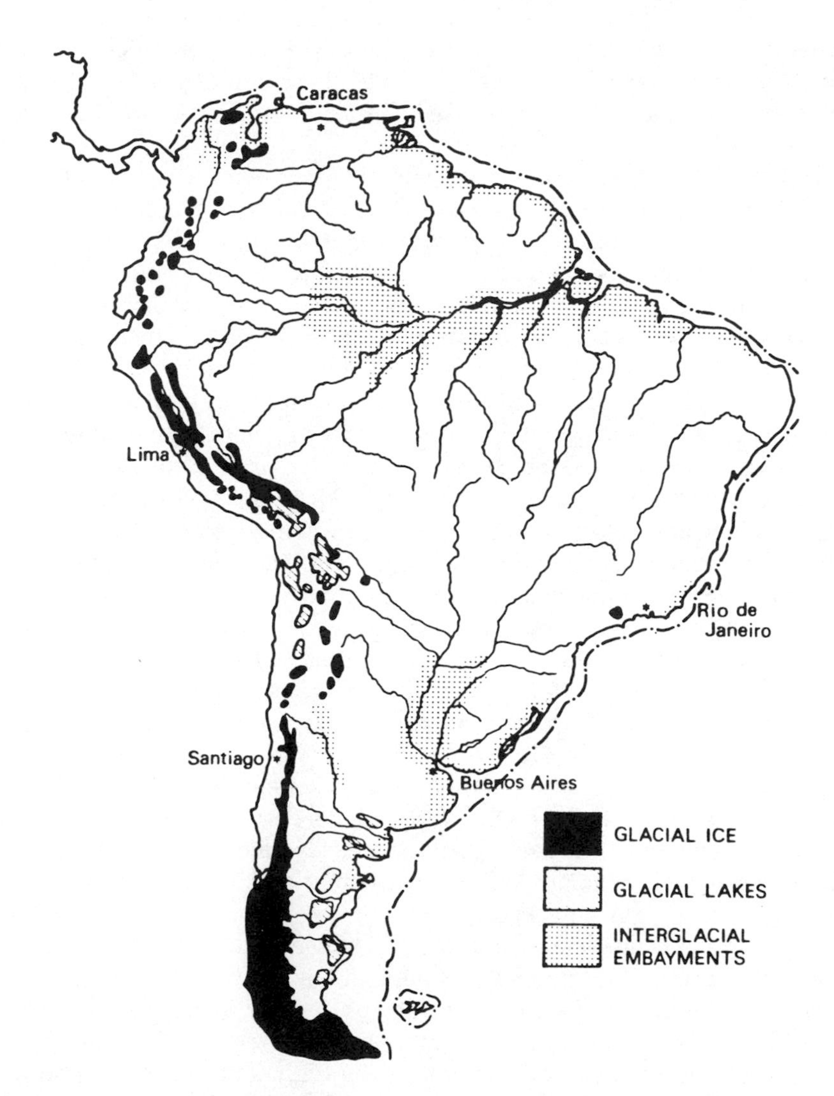

Figure 8.1. Features of the Würm glaciation together with an interglacial episode. Black areas represent glacial ice, dashed areas indicate glacial lakes, and the broken line that follows the 100-meter isobath illustrates the probable glacial coastline. Dotted areas show the locations of interglacial sea transgressions and the large, inland, freshwater lake in western Argentina. *(Redrawn from B. S. Vuilleumier, 1971, Pleistocene Changes in the Fauna and Flora of South America, Science* **173:***773; copyright © 1971 by the American Association for the Advancement of Science. Reprinted with permission.)*

vegetational zones to move downslope, and this shift provided a means for the alpine floral element to spread. The dominant elements in today's forests of southern Chile are *Nothofagus dombeyi, N. obliqua,* and *Araucaria araucana.*

In the central Andes, the vegetation of the Andean Altiplano of Peru, Bolivia, and Chile is called *puna* grassland. It consists primarily of fescue (*Festuca*) and couch grass (*Calamagrostis*). During glacial surges, these species spread downslope and outward on both the east and west flanks of the mountains in a fashion similar to that observed farther south. When one examines this general trend in detail, however, it becomes clear that glacial tongues and other irregularities represented local, physical, and ecological barriers to gene flow. One such area is located in the Andes east of Lake Titicaca where a lobe of ice apparently split the puna vegetation into two isolated parts. A few of the bird and plant species in each part display complex patterns of hybridization, suggesting that the ice lobe retreated before speciation was complete. In the present picture, puna grasslands are restricted to higher elevations and are effectively separated from one another by rain-shadow desert valleys.

Until very recently, glacial activity was not believed to have been important in the northern Andes from Venezuela to Ecuador. In this area, the vegetational zone above timberline is referred to as *paramo* grassland. In addition to the typical puna grass genera, one also finds species of *Hypericum, Geranium, Potentilla, Lupinus, Spirea,* and *Espeletia.* All of the biotic patterns in this zone have developed during the Quaternary, because the mountains have only been uplifted to their present height since the early Pleistocene. The map in Figure 8.2 shows the locations of present paramo grasslands and their probable lower elevation during glacial surges.

Vuilleumier, in a revealing study (1971), reasoned that in their present state, the grasslands represent biological islands and should therefore reflect general trends of immigration and extinction, which were discussed with thc models of island biogeography in Chapter 3. Since most of the bird species are derived from stock in the high central Andes, and not from the forested regions of Central America, any departures from predicted and observed numbers of breeding species for the 15 paramo islands should suggest the original route of migration. The data are shown in the table accompanying Figure 8.2. Obviously, there is considerable agreement between theory and observation in the numbers of species present. Departures from a perfect fit are attributed to the fact that immigration and extinction rates were probably not constant because climatic changes altered the size, shape, and arrangement of the islands through time. In general terms, the attenuation of species diversity from

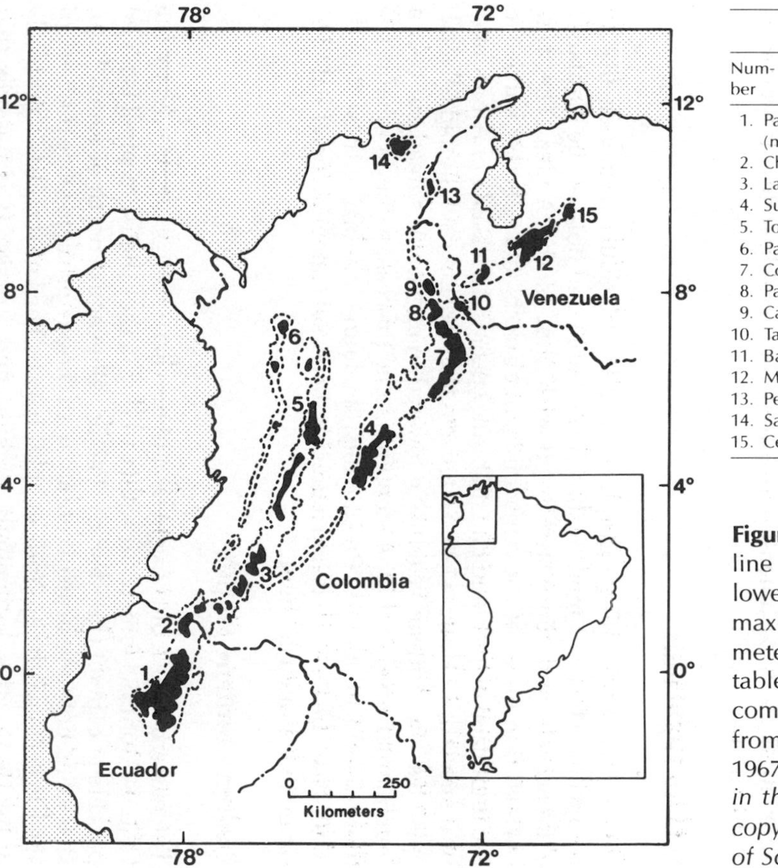

Paramo island		Species observed (no.)	Species predicted (no.)
Number	Name		
1.	Paramo 1 (northern Ecuador)	65	66
2.	Chiles	36	32
3.	Las Papas-Coconuco	30	30
4.	Sumpapaz	37	38
5.	Tolima-Quindia	35	32
6.	Paramillo	11	14
7.	Cocuy	21	25
8.	Pamplona	11	12
9.	Cachira	13	15
10.	Tama	17	20
11.	Batallon	13	13
12.	Mérida	29	23
13.	Perija	4	3
14.	Santa Marta	18	20
15.	Cendé	15	12

Figure 8.2. The present distribution (black) of land above the tree-line in northern South America. The dashed line indicates the lowest elevation reached by paramo grassland in times of glacial maxima, assumed to have been at an elevation of 2,000 to 5,000 meters. Arabic numerals indicate paramo "islands" as listed in the table above. The diversity of endemic birds (species observed) is compared to predicted diversity (species predicted) using concepts from the *Theory of Island Biogeography* (MacArthur and Wilson, 1967). *(Redrawn from B. S. Vuilleumier, 1971, Pleistocene Changes in the Fauna and Flora of South America, Science* **173:***776,778; copyright © 1971 by the American Association for the Advancement of Science. Reprinted with permission.)*

south to north suggests that the bird fauna of the northern Andes is derived from stock that migrated northward from the central Andes.

Patagonia

During the lowering of sea level accompanying glacial episodes, a huge area of the Argentine continental shelf was exposed and became available for colonization (see Figure 8.1). It was probably over this land bridge that the Falkland Islands became occupied during the Pleistocene. Indeed, the presence of partially differentiated bird and plant populations suggests that the Falkland Islands acted as a refuge when ice transgressed the Fuego-Patagonia area.

Despite the lowering of vegetational zones in the Andes and the spread of *Nothofagus* forest, this type of forest apparently never spread eastward across Patagonia. Instead, it was confined mostly to the eastern mountain front, advancing and retreating only slightly from it. Most of Patagonia appears to have been a boggy environment dissected by glacial outwash streams and lakes. It is believed that this environment allowed for the differentiation of peripheral populations observable today in numerous plant species.

Eastern Highlands

One of the great biogeographical riddles of South America lies in reconciling the close relationship of the fauna and flora of the highlands of Venezuela and southeast Brazil with that of the Andes. The wide lowland rain forest of the Amazon is an obvious and effective barrier to the migration of highland and montane species. In a study of Amazonian bird speciation, Haffer (1969) suggests that during cool, humid (glacial) episodes, the Andean vegetation lowered in elevation, spread out over a much larger area, and by a stepping-stone process, came to inhabit the more distant highlands. Presumably, the fauna inhabiting these forests, or at least the birds that had a greater dispersal capacity, increased their range in a similar fashion.

Eastern Montane Slopes

The eastern slopes of the central and northern Andes are covered by the Ceja de Montaña forest. An oddity of this is that there are small isolated pockets of ceja on the western (Pacific) slopes just north of Lima, Peru, that could not have been in contact with the eastern ceja since the close of the Pliocene, when the Andean uplift began. The forest consists primarily of heath, myrtles, bromeliads, orchids, and mosses. It is possible that these forms have migrated from the north by species crossing the Andes in stepping-stone fashion. If so, however, one would expect each of the patches to be slightly different in species composition. Furthermore, the species themselves should display considerable variation from patch to patch because of their mutual isolation and because each population should have been undergoing local adaptation and selection. In truth, the patches are so similar that the stepping-stone idea seems invalid. The next best hypothesis is that the patches are recent relicts of a once more continuous western ceja forest. This explanation, of course, does not account satisfactorily for their similarity with eastern ceja forests. Perhaps they truly are relicts of the Pliocene and once had a much more continuous distribution.

Tropical Lowlands

The tropical lowlands of Amazonia were long regarded as biogeographically stable. The difficulty is that one cannot easily explain how such great diversity, disjunction, and hybridization came to exist in a static environment. In concept, the situation is not conducive to geographic speciation. Haffer's study on bird speciation (1969) devastates the idea of stability by showing that Pleistocene climatic cycles in the region stimulated corresponding expansions and contractions of the rain forest and its resident fauna. During dry phases, only small isolated patches of rain forest could exist and these patches acted as refugia in which speciation could take place. In humid phases, refugia expanded to create overlapping ranges, which in turn stimulated hybridization. Figure 8.3 presents Haffer's view of the location of these refugia.

In summary, Pleistocene and Recent events have played a major role in molding present floristic distributions in South America. Glacial episodes in the Andean region resulted in rising and falling vegetational belts. When lowered, the puna and paramo areas expanded, later to retreat to high mountain tops during the interglacials. In the lowlands, a series of humid-arid cycles, without much change in temperature, molded the rain forest and savanna patterns and stimulated their differentiation. It is suggested that part of the great species diversity found in South America is due to two historical factors: (1) the lack of wholesale extinction; and (2) the repeated isolation of refugia, which provided ample opportunity for speciation.

NEOTROPICAL FAUNA

In the larger region of Latin America, three interrelated units can be recognized. Central America, the first unit, provided an avenue, or corridor, along which animals moved southward, and a sweepstakes route along which a few moved northward during the Plio-Pleistocene. This unit remains by and large unstudied biogeographically, yet it is fundamental to an understanding of events connecting the history of the Nearctic and Neotropical realms. The West Indies (unit 2) are regarded by Simpson (1980) as essentially a series of evolutionary traps. The third

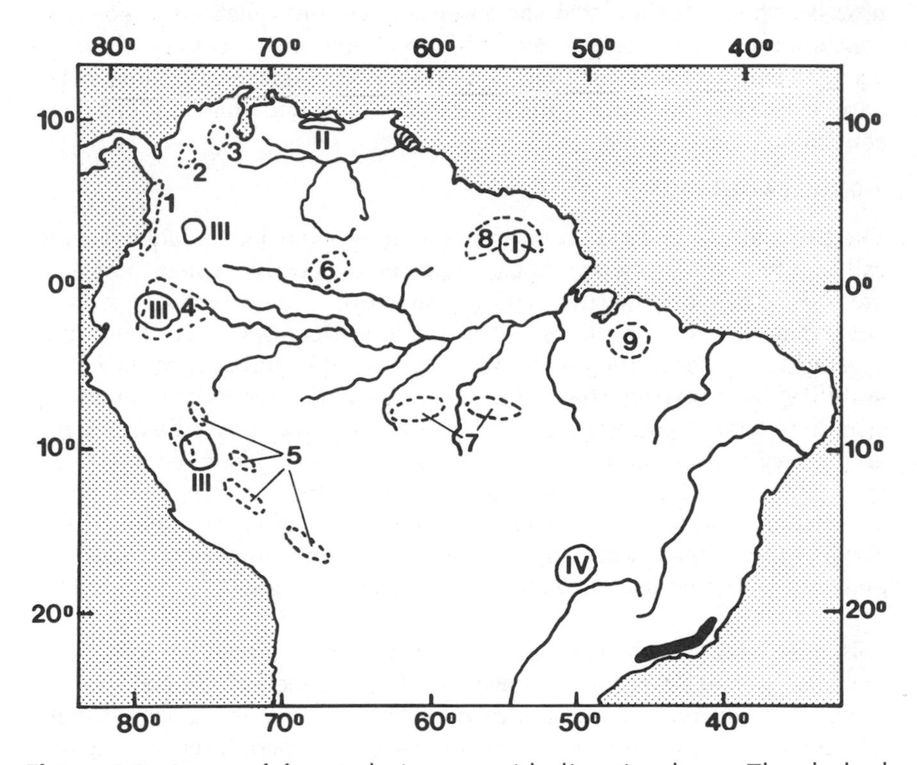

Figure 8.3. Areas of forest during an arid climatic phase. The dashed lines and arabic numerals refer to Haffer's refugia: *(1)* Chocó; *(2)* Nechí; *(3)* Catatbumbo; *(4)* Napo; *(5)* eastern Peru; *(6)* Imeri; *(7)* Madeira-Tapajó; *(8)* Guiana; and *(9)* Belém. The solid lines and Roman numerals indicate the refugia derived from the core areas recognized by Vanzolini and Williams (1970): *(I)* Serra do Tumuc Humac; *(II)* Cordillera de la Costa, Venezuela; *(III)* lower eastern slopes of the Andes; and *(IV)* southern escarpments of the central planalto, especially the Mato Grosso de Goias. *(Redrawn from B. S. Vuilleumier, 1971, Pleistocene Changes in the Fauna and Flora of South America, Science* **173:***177; copyright © 1971 by the American Association for the Advancement of Science. Reprinted with permission.)*

unit, South America, is regarded as having been an island continent through the early part of the Tertiary, after its separation from Gondwanaland and prior to its reunion with the Nearctic Realm via the Isthmus of Panama.

Central America

Ryan (1963) has delineated ten mammalian provinces in Central America based on the similarity of lists of species encountered along a transect stretching from the southern Mexican border to Colombia (Table 8.1). Provinces were delineated by separating areas of low similarity which, in effect, are ecotones between mammalian communities. The ten provinces are shown in Figure 8.4.

The validity of these provinces is suggested by the rates of endemism found in each. Table 8.2 summarizes the percentages of endemic rodents, carnivores, bats, and monkeys in each province. From these data, and

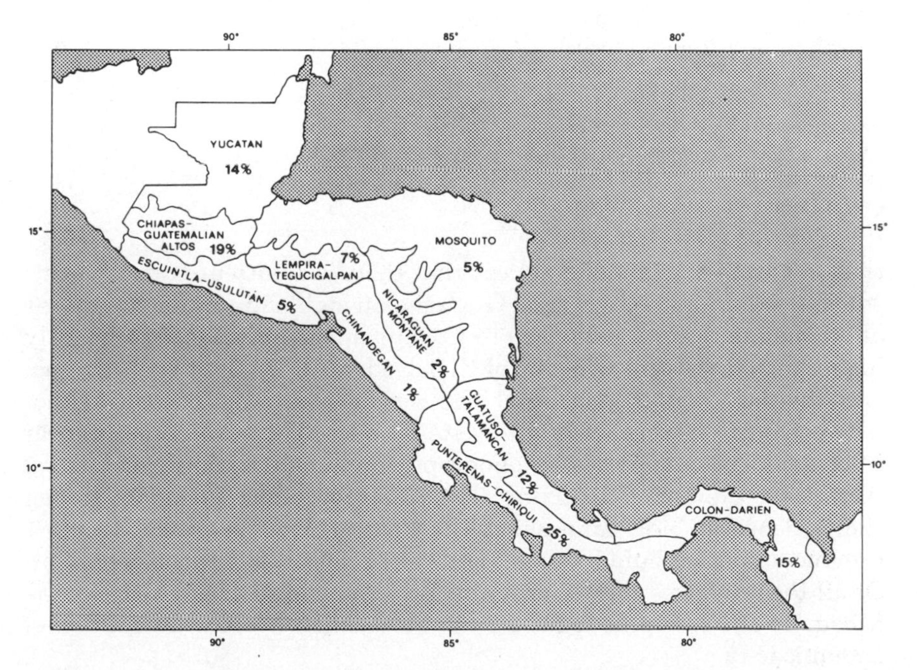

Figure 8.4. Ten mammalian provinces of Central America. Numbers indicate percent of endemism. *(From R. M. Ryan, 1963, The Biotic Provinces of Central America, Acta Zoologica Mexicana* **6***(2–3):33; reprinted with permission of Acta Zoologica Mexicana.)*

Table 8.1. Mammalian Endemism in Central America

	No. of endemics	Total no.	Percent
Yucatan	21	148	14
Chiapas-Guatemalan	18	95	19
Escuintla-Usulután	5	111	5
Lempira-Tegucigalpan	8	106	7
Chinandegan	1	109	1
Nicaraguan Montane	3	124	2
Mosquito	7	130	5
Puntarenas-Chiriquí	44	179	25
Guatuso-Talamancan	17	136	12
Colón-Darién	23	150	15

Table 8.2. Major Components (%) of Mammalian Endemism in Central America

	Total %	Rodentia	Carnivora	Chiroptera	Primate
Yucatan	92	67	10	5	10
Chiapas-Guatemalan	77	77	0	0	0
Escuintla-Usulután	100	80	20	0	0
Lempira-Tegucigalpan	100	75	13	12	0
Mosquito	90	43	0	33	14
Nicaraguan Montane	67	67	0	0	0
Chinandegan	100	100	0	0	0
Guatuso-Talamancan	92	80	12	0	0
Puntarenas-Chiriquí	93	79	2	0	12
Colón-Darién	90	80	5	5	0

that in Figure 8.4, it would appear that there is an attenuation of forms southeastward from the central Guatemalan highlands to the mountains of Nicaragua. From Costa Rica to Colombia, endemism is significantly higher than in the central region of El Salvador, Honduras, and Nicaragua, and this fact may suggest that the central area is a zone of contact between northwestern and southeastern nodes. There would not appear to be sufficient differences in topography or climate throughout Central America to prevent such mixing. Table 8.2 shows that endemism among the Rodentia is high throughout the Central American area. In most provinces, they account for more than two-thirds of the endemics present. Of all the provinces shown, only the Yucatan, and to a lesser extent the Mosquito and Lempira-Tegucigalpan areas, show a wide distribution of endemic taxa.

One must remember that throughout most of the Tertiary period, Central America existed as an island archipelago just as the Caribbean region does today. The rise and fall of islands in a tropical environment provided adequate opportunity for species to evolve, but their radiation was highly dependent on their competitive abilities as island hoppers. Figure 8.5 gives a general impression of the nature of this archipelago at the opening of Tertiary time. The Panama Bridge apparently came into existence during the late Pliocene, because it was in full operation as a migration route 2 to 3 million years ago. Prior to its linkage as a complete bridge, relatively few mammalian forms migrated from north to south as island hoppers, but even fewer migrated from south to north.

The primary interchange of Nearctic and Neotropical mammals occurred only about a million years ago, and the main elements of the pattern are shown in Figure 8.6. The overall effect on South America was devastating because the native ungulates were replaced by migrants like the llama (*Lama*, spp.) and deer, or eliminated by placental carnivores. A few forms like the armadillo (*Dasypus novemcinctus*), opossum (*Didelphis virginiana*), and porcupine (*Erethizon dorsatum*) migrated northward, but it is clear that the Panama Bridge was a sweepstakes route for northward movement.

According to Woodring (1966) the emergence of the land bridge interrupted the east-west contact of marine forms in the Tertiary Caribbean Province. Miocene North American land mammals have been found in

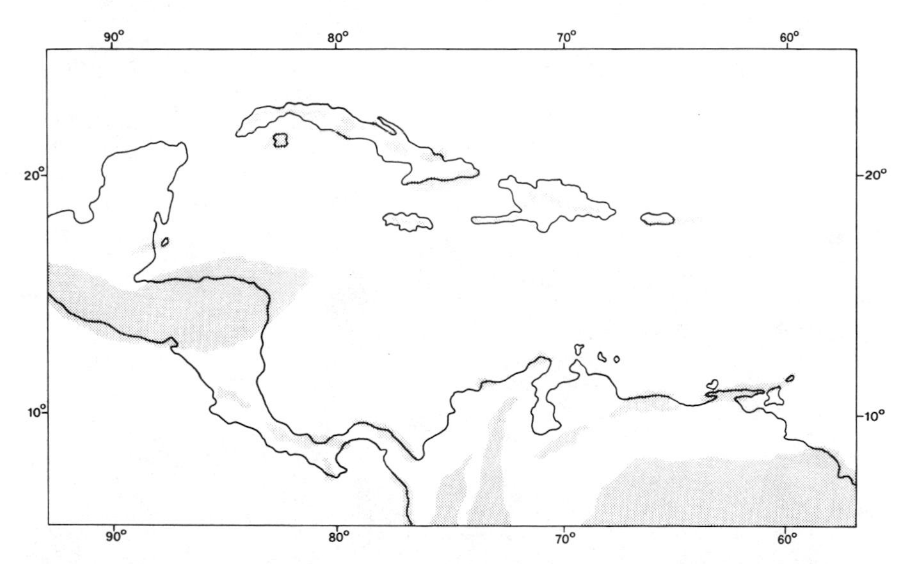

Figure 8.5. The insular nature of Central America and northern South America (shaded) some 60 to 70 million years ago allowed a unique fauna to evolve. The current coastline emerged about a million years ago. *(Modified from A. S. Romer, 1959, Darwin and the Fossil Record, Natural History* **68***(8):467; copyright © 1959 by the American Museum of Natural History. Reproduced with permission from Natural History.)*

the Canal Zone. They, or their ancestors, migrated from northern Central America, but were stopped by the still submerged Atrato Trough southeast of the Canal Zone. The presence of this trough so far south may account in part for the relative imbalance of southern forms moving northward compared to northern forms moving south. In a sense, the northern forms were poised for rapid invasion into South America once the trough emerged, whereas the southern forms were contained in their southern ranges until relatively recently.

The extent to which mammalian forms existed in southern Central America can be realized by reviewing the list of Costa Rican mammals given by Goodwin (1946). This list includes some 37 families, 121 genera, and 262 species. In all of Central America, Hall and Kelson (1959) indicate 508 species and subspecies. Table 8.3, derived from Goodwin,

Figure 8.6. Interchange of mammals across the Panama Bridge: (above) representative forms separated by the Atrato Trough; (below) representative forms after migration and adjustment. Many South American marsupials were eliminated. *(From A. S. Romer, 1959, Darwin and the Fossil Record, Natural History* **68***(8):469; copyright © 1959 by the American Museum of Natural History. Reproduced with permission from Natural History.)*

Table 8.3. Composition of Costa Rican Mammals

Taxon	*Common name*	*Families*	*Genera*	*Species*
Marsupialia	Marsupials	1	6	9
Insectivora	Insect eaters	1	1	4
Chiroptera	Bats	8	59	86
Primates	Monkeys	3	4	8
Edentata	Armadillos, sloths	4	7	8
Lagomorpha	Hares and rabbits	1	1	3
Rodentia	Rodents	9	23	105
Carnivora	Raccoons, otters, weasels, cats, dogs, etc.	5	13	30
Pinnipedia	Hair Seals	1	1	1
Sirenia	Sea cows	1	1	1
Artiodactyla	Deer, etc.	2	3	5
Perissodactyla	Tapirs, etc.	1	2	2

Source: G. G. Goodwin, 1946, Mammals of Costa Rica *Am. Mus. Nat. Hist. Bull.* **87**(5).

shows the relative importance of mammalian orders in Costa Rica. The most important constituents include the bats, rodents, and carnivores with the next most numerous being the opossums, monkeys, and sloths. Forty percent of the mammalian fauna consists of the rodents (squirrels, pocket mice, pocket gophers, rats, porcupines, agoutis, spring rats, and pacas). If the bats are added to this list, almost 75 percent of the Costa Rican mammalian fauna is accounted for. It is evident that even today, there are few megafaunal forms located in southern Central America.

Galapagos Islands

The Galapagos Islands represent the only area in South America not directly affected by Pleistocene vegetational migrations. Their fauna can be divided into two elements: (1) that having an alliance with Central America and the Antilles and (2) a less differentiated group with affinities to the Andean Pacific area. Although some authors have argued that the Galapagos have been connected in the past to the continent, the degree of differentiation among the two faunal elements, together with the known geological history of the islands, seem to obviate this. It seems more probable to Müller (1973) that the Galapagos fauna arrived by sea. During the early Tertiary, sea currents connected the Caribbean to the Pacific prior to the Panama Bridge. This fact may explain why the older fauna has affinities to Central America. With the completion of the Panama Bridge in Pliocene time the Peru current became dominant—hence the influx of Andean Pacific species. One can see from a map of world ocean currents that the Galapagos are bathed today by currents moving southwestward out of the Gulf of Panama and the southern coast of Central America.

South America

Biogeographically, South America is fascinating because its flora is similar to that of the Old World tropics, while its fauna is decidedly North American and its geology reminiscent of Africa. The general relationship between South America and the rest of the "World Continent" is given in Chapter 6. In broadest outline, it appears that South America separated from the rest of Gondwanaland sometime prior to the Tertiary, was isolated as a very large island, and then reconnected with North America during the Pleistocene. This sequence of events can be more or less traced in the faunal record, or as Simpson (1965, 1980) puts it, the *faunal stratification.* The latest series of adjustments to changing environments is represented by the rapid speciation in the youngest element of this stratification.

Three distinct strata of the South American fauna can be recognized: the so-called ancient immigrants; old island hoppers; and late immigrants. *Ancient immigrants* is a term used to designate forms that were in South America prior to its isolation. They would be Paleocene or older in age. A curious feature of this element is that it consisted of a strange assemblage of placental herbivores and marsupial carnivores. Since marsupials were apparently never an important element in either North America or Africa, their presence in South America is difficult to explain. Simpson (1965), in fact, discussed three hypotheses to account for their occurrence.

One of his ideas suggests that marsupials came from Australia, where even today they are an important element of the fauna. It seems odd, however, that these carnivores would have migrated without their ecological associates—their marsupial herbivorous prey. A second idea is that South America began its mammalian history with both elements (placental and marsupial carnivores) from the World Continent, but that the placental carnivores became immediately extinct. Unfortunately, there is little current evidence that marsupial carnivores were ever important on the World Continent; nor is there evidence for the presence of placental carnivores from the early Tertiary of South America. The last hypothesis is that the progenitors of both the marsupial and placental carnivores inhabited South America, Australia, and the remainder of the World Continent, and became differentiated after separation and isolation. The marsupials diverged and became important in South America and Australia, while the placentals dominated the North American and Old World regions. This is essentially the pattern diagrammed by Fooden (1972) that was presented in Chapter 6.

The important constituents of the early Tertiary ancient immigrant fauna of South America included (at the order level) Marsupialia, of which the opossum is the only surviving member; Edentata, like the armadillo and ground sloth; Condylarthra, which were the progenitors of the present hooved animals and were extinct by the end of the Pliocene; and Notoungulata, which were also hooved herbivores. Each of these orders has affinities with North America. However, none is known from Africa, and only the marsupials are known from Australia. This pattern suggests that North America is the source of the South American fauna. It is equally possible, however, that South America is the seat of origin for a subsequent fauna to invade the World Continent.

The term *old island hoppers* refers to a rather small group of animals consisting mainly of rodents and New World monkeys that arrived in South America during the Oligocene epoch. It is believed that these groups migrated over a sweepstakes route from North America. The fact that no New World monkeys have yet been discovered in the fossil record of North America is explained by their supposed evolution in Central America.

The ancient immigrants and first wave of island hoppers provided the stock for an evolutionary explosion to fill niches. The marsupial carnivores were the main predators, but their species diversity and, hence, their ability to hold prey in check, was low. By the end of the Pliocene, and perhaps under the impetus of Nearctic glacial events, a southward migration of forms from North America began. By this time, Central America may have provided a continuous land bridge for this movement. Although the bulk of the movement was from north to south, a few of the more adaptive forms, like the opossum and armadillo, managed to break out of South America and extend their ranges northward. Among the *late immigrants* moving south were horses, mastodons, camels, deer, cats, bears, rodents, weasels, and pigs. This exchange brought about an initial enrichment of the faunal diversity, which led eventually to competition for niches, selection of the more adaptive forms, and elimination of the less competitive. Among the most dramatic of the turnovers was the elimination of the marsupial carnivores.

The importance of Quaternary climatic fluctuations has already been stressed. In the Amazon region, dry phases resulted in the breakup of the rain forest into smaller isolated patches, followed by their reunion during humid phases. Present faunal diversity is largely a result of differentiation during these phases. In reconstructing Quaternary speciation it is necessary to locate the probable forest refuges. Haffer (1969) has approached the problem by reviewing the present day pattern of rainfall and matching the areas of maximum precipitation with distribution patterns of Amazonian birds.

Figures 8.7 and 8.8 show the distributions of modern rain forest and annual rainfall. The most important feature of the two is the location of rainfall greater than 2,500 millimeters. Haffer assumes that during the

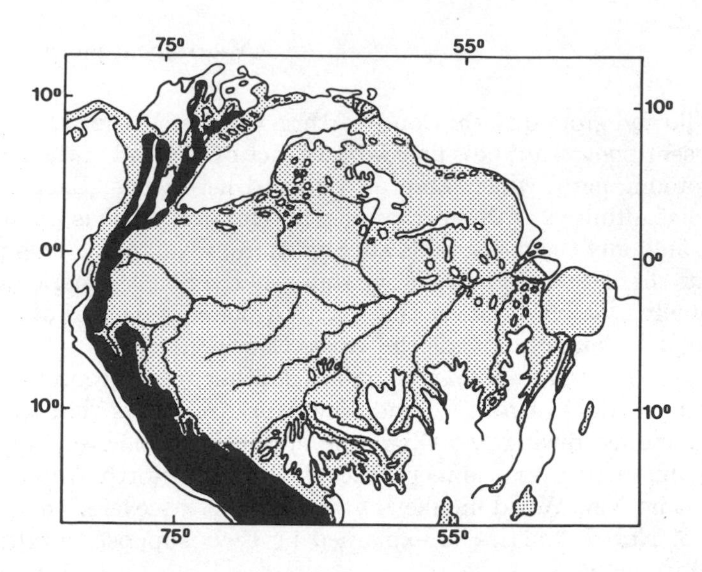

Figure 8.7. Distribution of humid tropical lowland forests (stippled) and savanna regions (white) in central and northern South America. Forests surrounding savanna regions are mostly semideciduous. The area shown in black represents the Andes Mountains above 1,000 meters. *(Redrawn from J. Haffer, 1969, Speciation in Amazonian Forest Birds, Science* **165:***132; copyright © 1969 by the American Association for the Advancement of Science. Reprinted with permission.)*

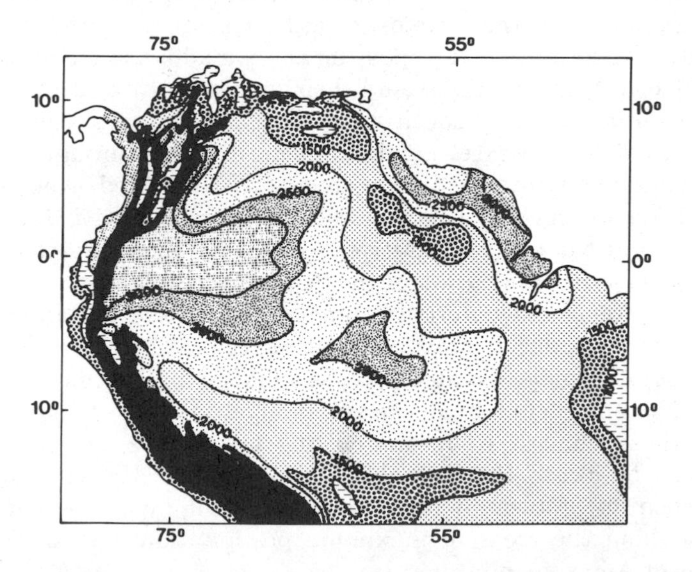

Figure 8.8. Total annual rainfall (in millimeters) in northern and central South America. *(Redrawn from J. Haffer, 1969, Speciation in Amazonian Forest Birds, Science* **165:***132; copyright © 1969 by the American Association for the Advancement of Science. Reprinted with permission.)*

dry phases, these wet spots would most likely have been refuges for the rain forest. This and other information has led to the identification of nine refuges, as illustrated in Figure 8.9. Of those shown, the Napo refuge in the upper Amazon may have been the refuge of a great many Amazonian animals, including many birds.

During the dry phases, species from southeastern Brazil are believed to have migrated northward. Some moved along the eastern Andean front and many crossed the Andes in the vicinity of Cajamarca in northern Peru. Major migration barriers at these times would have been high mountains, wide rivers and inundated areas such as the Amazonian embayment. During humid phases, an opposite pattern of adjustment and speciation would have existed, namely, savanna islands in a sea of rain forest. Thus, one can visualize a continuous process of differentiation

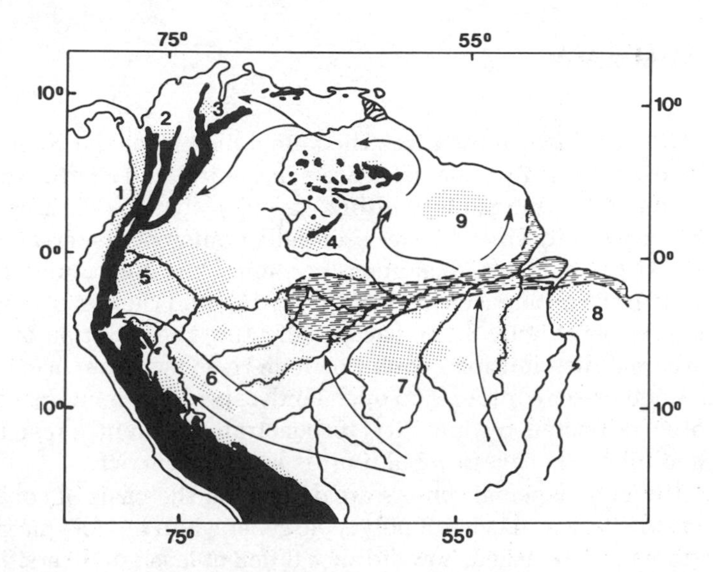

Figure 8.9. Presumed forest refuges in central and northern South America during the warm dry climatic period of the Pleistocene. The arrows indicate northward-advancing nonforest faunas of central Brazil. *(1)* Chocó refuge; *(2)* Nechí refuge; *(3)* Catatumbo refuge; *(4)* Imerí refuge; *(5)* Napo refuge; *(6)* east Peruvian refuges; *(7)* Madeira-Tapajós refuge; *(8)* Belém refuge; *(9)* Guiana refuge. The hatched area shows the interglacial Amazonian embayment (sea level raised by about 50 meters); the black areas indicate elevations above 1,000 meters. *(Redrawn from J. Haffer, 1969, Speciation in Amazonian Forest Birds, Science* **165:***134; copyright © 1969 by the American Association for the Advancement of Science. Reprinted with permission.)*

of species throughout the period of climatic shift. One can also visualize that during the expansion phases of both the rain forest and the savanna, there was the possibility for both geographic overlap and hybridization.

Haffer suggests that the major rivers of Amazonia, while not a causal factor in bird speciation, may have molded the distribution of birds by limiting species dispersal. Perhaps other animals were also limited in this way, but birds show the effect more readily because tropical species generally have very narrow niches and small day-to-day ranges. It is partly because of these attributes that Haffer believes tropical bird speciation may take only 20,000 to 30,000 years to complete. If this estimate is even approximately accurate, it is clear that most of the Amazonian bird fauna (if not the bulk of all the fauna) has differentiated in the past one to two million years.

SUMMARY

Müller (1973) summarizes the faunal history of the Neotropical Realm by describing 40 dispersal centers in Central and South America. These centers are arranged into affinity groups with the striking result that three basic units emerge: the essentially nonforest dispersal centers lying below 1,500 meters—campos, cerrados, savannas, and deserts (Figure 8.10); the arboreal centers (Figure 8.11); and the areas above tree line in the Andes (Figure 8.12). Factors affecting the isolation of faunas in the arboreal distribution centers include river systems, arms of the sea, the 1,500-meter contour, and open landscapes. The nonforest faunas seem to be restricted by inability to penetrate the rain forest biome, rivers, and inlets. The elevational limit is less of a barrier.

Since Müller's dispersal centers are defined on the basis of congruent ranges of subspecies, species of polytypic genera, and monotypic genera, the question must be asked, how did such different levels of diversification get to occupy the same area? The answer in part may be that these areas acted, from time to time, as refugia during glacial surges and periods of higher sea level, and as dispersal centers during postglacial stages. During the regressive stages, isolation of forms resulted in the emergence of subspecies.

One must assume that the regressive phases of rain forest are correlated with expansive phases of nonforest, so that one gets the mood of a pulsating, dynamic process of life-form differentiation. The overall situation can be contrasted with eastern North America, where glacial episodes literally drove all species southward into refugia. In Latin

America, glacial episodes moved species to lower elevations out of the Andes but left a vast low-lying territory over most of the continental mass. This territory was segregated into many refugia on the basis of moisture availability and soil type. In this respect the recent biotic history of South America is like that of Africa and Australia.

Figure 8.10. Nonforest dispersal centers as defined by the congruent ranges of subspecies, species of polytypic genera, and monotypic genera. *(From P. Müller, 1973, The Dispersal Centers of Terrestrial Vertebrates in the Neotropical Realm, W. Junk, The Hague, p. 175; copyright © 1973 by W. Junk. Reprinted with permission of the publisher.)*

REFERENCES

Darlington, P. J., 1965, *Biogeography of the Southern End of the World*, McGraw-Hill, New York, 236p.

Fooden, J., 1972, Breakup of Pangaea and Isolation of Relict Mammals in Australia, South America and Madagascar, *Science* **175**:894-898.

Goodwin, G. G., 1946, Mammals of Costa Rica, *Am. Mus. Nat. Hist. Bull.* **87**(5):275-473.

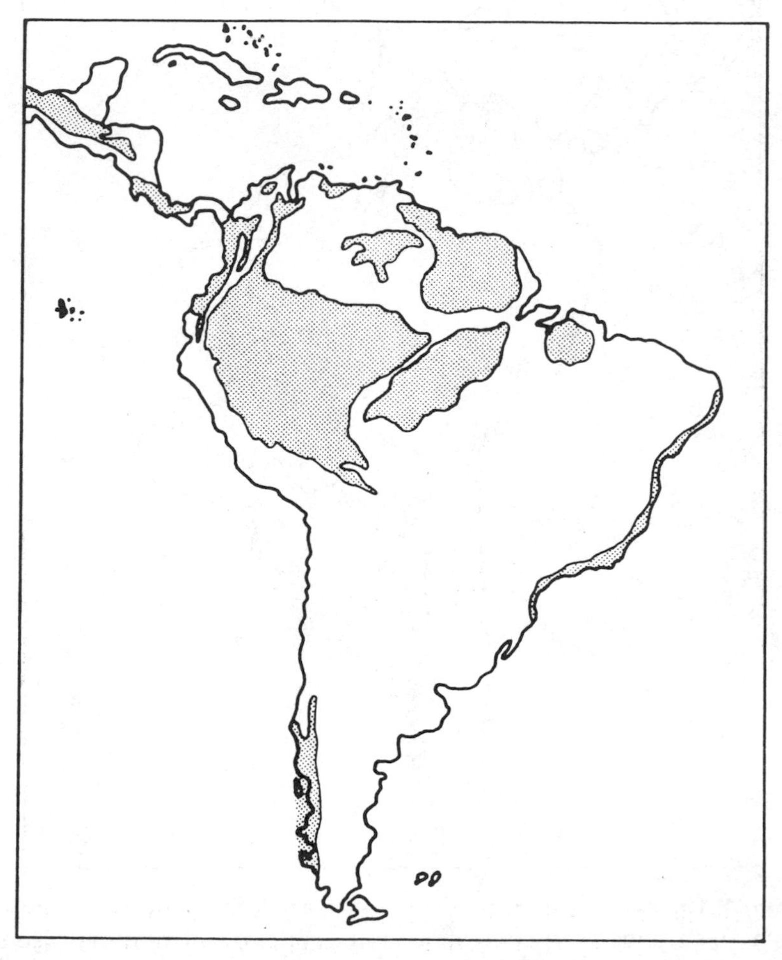

Figure 8.11. Arboreal dispersal centers. *(From P. Müller, 1973, The Dispersal Centers of Terrestrial Vertebrates in the Neotropical Realm, W. Junk, The Hague, p. 174; copyright © 1973 by W. Junk. Reprinted with permission of the publisher.)*

Haffer, J., 1969, Speciation in Amazonian Forest Birds, *Science* **165**:131-137.
Hall, R. E., and K. R. Kelson, 1959, *The Mammals of North America,* Ronald Press, New York, 2 vols.
Johnson, M. P., and P. H. Raven, 1973, Species Number and Endemism: The Galapagos Archipelago Revisited, *Science* **179**:893-895.
MacArthur, R. H., and E. O. Wilson, 1967, *The Theory of Island Biogeography,* Princeton University Press, Princeton, N. J., 203p.

Figure 8.12. Dispersal centers above tree line. *(From P. Müller, 1973, The Dispersal Centers of Terrestrial Vertebrates in the Neotropical Realm, W. Junk, The Hague, p. 176; copyright © 1973 by W. Junk. Reprinted with permission of the publisher.)*

Müller, P., 1973, *The Dispersal Centers of Terrestrial Vertebrates in the Neotropical Realm,* W. Junk, The Hague, 244p.
Romer, A. S., 1959, Darwin and the Fossil Record, *Natural History* **68**(8):456-469.
Ryan, R. M., 1963, The Biotic Provinces of Central America. *Acta Zool. Mex.* **6**(2-3):1-54.
Simpson, B. B., 1974, Glacial Migrations of Plants: Island Biogeographical Evidence, *Science* **185**:698-700.
Simpson, G. G., 1965, *Geography of Evolution,* Capricorn Books, New York, 249p.
Simpson, G. G., 1980, *Splendid Isolation,* Yale University Press, New Haven, Conn. 266p.
Vanzolini, P. E., and E. E. Williams, 1970, South American Anoles: The Geographic Differentiation and Evolution of the *Anolis chrysolepis* Species Group (Sauria, Iguanidae), *Arq. Zool. São Paulo* **19**(1-2):1-298.
Vuilleumier, B. S., 1971, Pleistocene Changes in the Fauna and Flora of South America, *Science* **173**:771-780.
Woodring, W. P., 1966, The Panama Land Bridge as a Sea Barrier, *Am. Philosophical Soc. Proc.* **110**(6):425-433.

FURTHER READING

Allen, P. H., 1956, *The Rainforests of Golfo Dulce (Costa Rica),* University of Florida Press, Gainesville.
Bailey, K., 1976, Potassium-Argon Ages from the Galapagos Islands, *Science* **192**:465-467.
Brock, J. C., 1977, Man-Made Dogs, *Science* **197**:1340-1342.
Dorst, J., 1967, *South America and Central America: A Natural History,* Random House, New York, 298p.
Hershkovitz, P., 1972, The Recent Mammals of the Neotropical Region: A Zoogeographic and Ecologic Review, in *Evolution, Mammals, and Southern Continents,* A. Keast, F. C. Erk, and B. Glass, eds., State University of New York Press, Albany, N.Y., pp. 311-431.
Janzen, D. H., and P. S. Martin, 1982, Neotropical Anachronisms: The Fruits the Gomphotheres Ate, *Science* **215**:19-27.
Lerman, J. C., W. G. Mook, J. C. Vogel, and H. de Waard, 1969, Carbon-14 in Patagonian Tree Rings, *Science* **165**:1123-1125.
Marshall, L. G., R. Pascual, G. H. Curtis, and R. E. Drake, 1977, South American Geochronology: Radiometric Time Scale for Middle to Late Tertiary Mammal-Bearing Horizons in Patagonia, *Science* **195**:1325-1328.
Marshall, L.G., S. D. Webb, J. J. Sepkoski, Jr., and D. M. Raup, 1982, Mammalian Evolution and the Great American Interchange, *Science* **215**:1351-1357.
Patterson, B., and R. Pascual, 1972, The Fossil Mammal Fauna of South America, in *Evolution, Mammals, and Southern Continents,* A. Keast, F. C. Erk, and B. Glass, eds., State University of New York Press, Albany, N.Y., pp. 247-309.

Prance, G. T., ed., 1982, *Biological Diversification in the Tropics,* Columbia University Press, New York, 714p.

Pregill, G. K., and S. L. Olson, 1981, Zoogeography of West Indian Vertebrates in Relation to Pleistocene Climatic Cycles, *Annu. Rev. Ecol. and Syst.* 12:75-98.

Quintanilla, V., 1977, A Contribution to the Phytogeographical Study of Temperate Chile, in *Ecosystem Research in South America,* P. Müller, ed., W. Junk, The Hague, pp. 31-42.

Chapter 9

The Palearctic Realm

Perhaps no other region on earth has been so important to the development of human culture and civilization as the Eurasian continent—the Palearctic Realm. It is the largest of the continental land masses, and because the northern half has experienced almost complete coverage by Pleistocene ice, it is biogeographically one of the youngest. Its occupation and alteration by both oriental and occidental humans in the past 20,000 years has transformed its biota in a way exceeding that of any other realm. By virtue of these changes, reconstructing its last 2 million years of biogeographic history is often more speculative than outlining its ancient relationships.

The Palearctic Realm consists of all the peninsulas of western Europe, together with Turkey and the Near East, as far as the Indus River, and all of the territory north of the Himalayan Mountains eastward to Kamchatka and the Islands of Japan. To the south, many authors (mainly mammologists) also include the Sahara Desert and the Arabian peninsula in this realm. (The treatment here is to defer discussion of the Saharan zone to Chapter 10.)

The Palearctic Realm extends eastward from Iceland at 20° west longitude,. half a world away to Siberia at 170°W. Its southernmost latitudes range from 36°N at the Straits of Gibraltar in Europe to 24°N in the province of Kwantung, north of Hong Kong, in China. Including all of the African and Arabian parts, it is an area of some 59,525,000 square kilometers—more than the area of the Nearctic and Neotropical realms combined.

The theme of alternating glacial and interglacial episodes already described for the Western Hemisphere is repeated in Eurasia. These events are superimposed upon a complex pattern of topography that includes high mountains, desert lowlands, swamps, and well-drained uplands, not to mention the modern-day extremes of oceanic and continental climates. Inevitably, the biogeographic evidence presents a confusing picture that stymies arguments for a generalized pattern of development. Glacial lowering of sea level created the Bering Land Bridge between the Nearctic and Palearctic realms, as well as the Mediterranean Bridge between Europe and Africa. Along these avenues hundreds of plant and animal species adjusted their ranges during periods of environmental stress. At the opposite extreme, the interglacials drowned vast areas of Siberia and the northern European lowlands, cutting off access and isolating whole ecosystems. While these events unfolded, most of the mountain ranges were lifted by an average of 200 to 300 meters. It is not only the advance and retreat of glaciers and intervening warm spells, therefore, that were important in molding the present biological pattern of northern Eurasia, but also the changing patterns of coastline, drainage, and topography.

In Eurasia glacial advances and cold periods are recognized and were separated by relatively warmer and dryer periods (Table 9.1). Following a rather lengthy period of warm and moist tropical conditions in Europe, the Pliocene ended. It was a time of general drying and warming, so that the earlier forests of the region began to share their territory with extensive grasslands, savannas, and steppes. The typical Pliocene forest fauna, which consisted of many species of deer (family Cervidae), mastodonts (*Zygolophodon*), rhinoceros (*Picerorhius*), and tapirs (family Tapiridae), began to be altered by the addition of cheetahs (*Acinonyx*), hyenas (*Hyaena*), and antelopes and the bovines (both in the family Bovidae). At the beginning of the Villafranchian age 3.3 million years ago, forms unknown from the Pliocene were important in Europe. True one-toed horses of the genus *Equus* had migrated from North America, while other genera like elephants (*Archidiskodon*), mammoths (*Mammuthus*), and cattle-like bovids (*Bos* and *Leptobos*) wandered throughout the realm.

The first sign of pending glacial episodes in Europe came with these fluctuations between forest and steppe environments. During this time sustained adjustment and sifting of both the angiosperm and mammalian biotas occurred. Today, after a million years of such alteration, this biota is viewed as being in a state of recovery from the last severe glacial

Table 9.1. Comparative Chronology of Glacial and Interglacial Episodes in North America and Central Europe

Northwestern and central Europe	*North America*
Weichselian glaciation	Wisconsin glaciation
Eemian interglacial	Sangamon interglacial
Saalian glaciation	Illinoian glaciation
Holsteinian interglacial	Yarmouth interglacial
Elsterian glaciation	Kansan glaciation
Cromerian interglacial	Aftonian interglacial
Menapian cold period	Nebraskan glaciation
Waalian warm period	
Eburonian cold period	
Tiglian warm period	
Praetiglian cold period	

advance, but with the added and perhaps more devastating element of human competition.

THE VEGETATIONAL HISTORY

Temperate Eurasia

The earliest Pleistocene episode in Europe is called the *Praetiglian cold period* and occurred sometime before the Nebraskan glaciation in North America (see Table 9.1). There were three other cold periods in Europe, separated from one another by warming trends. The cold periods were followed by three full-blown glacial episodes that corresponded roughly to Kansan, Illinoian, and Wisconsin events in the United States, and that were, of course, separated by interglacial periods. Figures 9.1-9.6 present maps modified from Frenzel (1968), that show vegetational patterns as they might have been at selected times in this sequence. The first (Figure 9.1) shows conditions in the *Reuverian* at the end of Pliocene time. Conditions for the Praetiglian cold period and the two most recent glacial episodes (*Saalian* = Illinoian, and *Weichselian* = Wisconsin) are then described. The last two maps (Figures 9.5 and 9.6) show conditions leading into the most recent interglacial periods (*Holsteinian* = Yarmouth, and *Eemian* = Sangomon).

Subtropical conditions must have existed over most of Eurasia prior to the Paleocene. The Cinnamomum Flora of the London Clay, which is pre-Tertiary in age, contains early genera of the Proteaceae and Mrytaceae that apparently migrated from Southeast Asia. By middle Tertiary time this flora had been replaced by an Arcto-Tertiary type containing maple

(*Acer*), birch (*Betula*), chestnut (*Castanea*), dogwood (*Cornus*), beech (*Fagus*), ash (*Fraxinus*), walnut (*Juglans*), pine (*Pinus*), poplar (*Populus*), sycamore (*Platanus*), oak (*Quercus*), willow (*Salix*), redwood (*Sequoia*), cypress (*Cupressus*), elm (*Ulmus*), ginko (*Ginko*), and other familiar types. Even though warmer and moister than today, the Tertiary period (Paleocene to Pliocene) must have been one of general cooling.

Figure 9.1 shows that prior to the first cold period of the Pleistocene, forest vegetation extended into the very high latitudes and southward into what is today an arid region. There was no tundra vegetation, even in the high latitudes, because apparently there had never been reason for flowering plants to adapt to such cold environments. Both lines of evidence suggest conditions that were warmer and moister than today, but cooler and drier than during the Tertiary. *Warmer* here means somewhere between 5° and 10°C higher for the mean annual temperature;

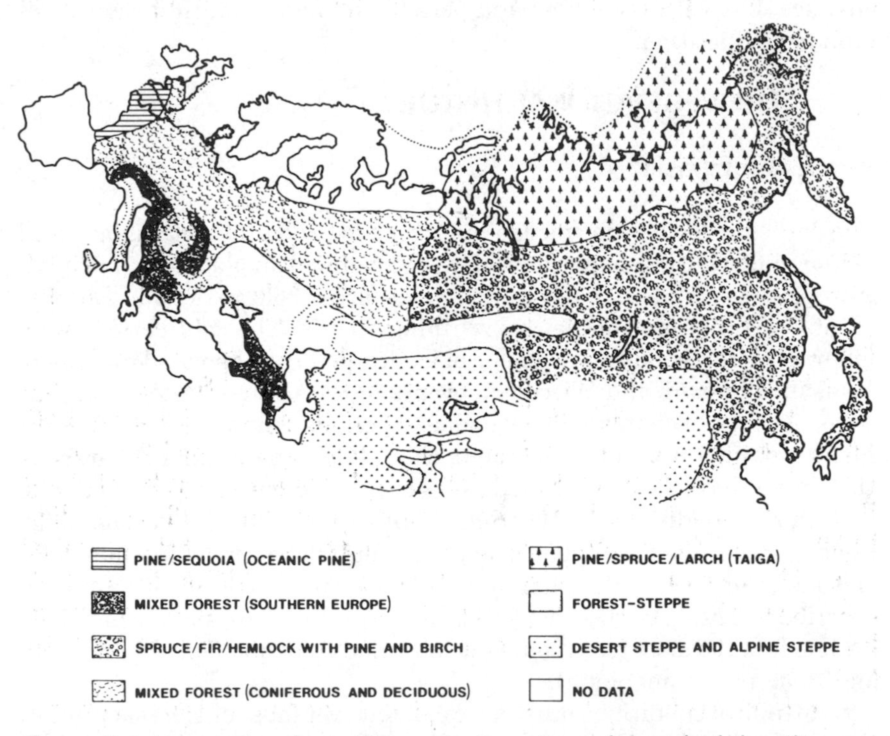

Figure 9.1. Vegetational conditions in Eurasia at the end of the Pliocene (Reuverian). *(Redrawn from B. Frenzel, 1968, The Pleistocene Vegetation of Northern Eurasia, Science* **161:**639; *copyright © 1968 by the American Association for the Advancement of Science. Reprinted with permission.)*

and *moister* means perhaps 30 centimeters more annual precipitation. This difference is roughly the same as between, say, Amarillo, Texas, and Kansas City, Missouri. While this is obviously not a vast difference in climate, the accumulated difference over several thousand years was probably enough to result in glacial and interglacial fluctuations.

It is interesting that on the map of Reuverian conditions (Figure 9.1), coniferous forests extended deep into the continent across Europe and the Balkans to the Caspian Sea. Eastern Eurasia, with the exception of China, Manchuria, Kamchatka, and Japan, was dominated by conifers. In the excepted area, one would have found a forest of redwood and cypress. The vegetation of western France was not terribly different from that of what is now Siberia.

The striking feature of the glacial maps in Figures 9.2-9.4 is the wide extent of steppe and forest/steppe vegetation. With each succeeding glacial episode, steppe conditions became more pronounced. The Praetiglian was characterized by a subarctic parkland throughout Europe and north-central Russia. Later episodes were characterized by steppe without trees or with isolated groves of extremely cold-resistant conifers

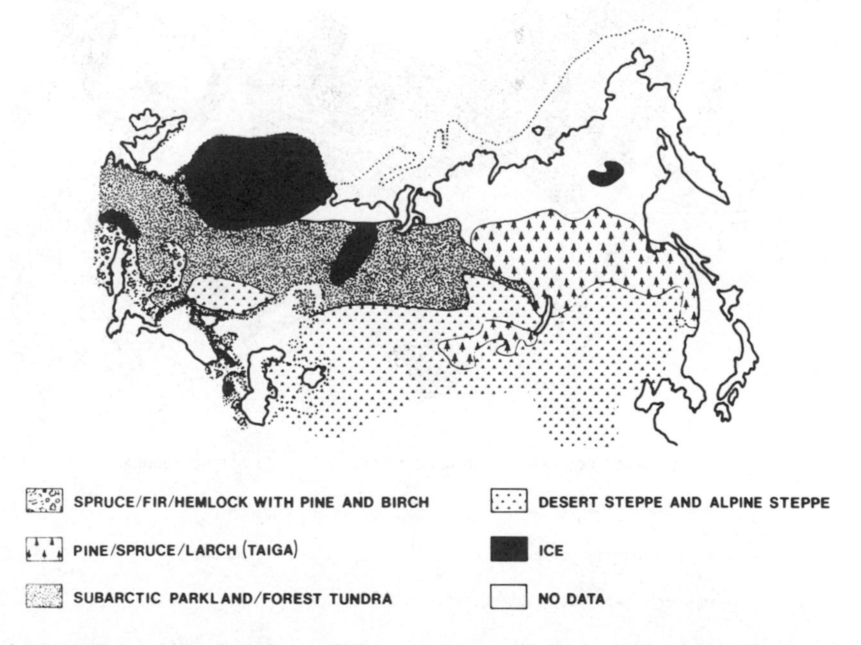

Figure 9.2. Vegetational conditions in Eurasia during the Praetiglian cold period. *(Redrawn from B. Frenzel, 1968, The Pleistocene Vegetation of Northern Eurasia, Science* **161:**640; *copyright © 1968 by the American Association for the Advancement of Science. Reprinted with permission.)*

representing populations that could adapt. These areas must have served as refugia for later vegetational advances.

Whereas the glacial episodes saw the advance of various steppe communities, the interglacial episodes were characterized by forests that spread from the Atlantic to Japan. Even as early as the first interglacial, many genera common to North America were not able to recover their former ranges in northern Eurasia. Some of these were sequoia, cypress, sweet gum, liquidambar, beech, and tulip poplar. Some, of course, have recovered more than others; but today all of them cover much less area than their ancestors did. Thus, the impoverishment of Eurasian vegetation had begun. One of the hallmarks of northern European vegetation today is its rather poor species diversity compared to previous forests. This situation has developed through limited recovery from the accumulated extinction and retreat of species. In similar fashion, the taiga of eastern Siberia has developed through the early elimination of less cold-resistant

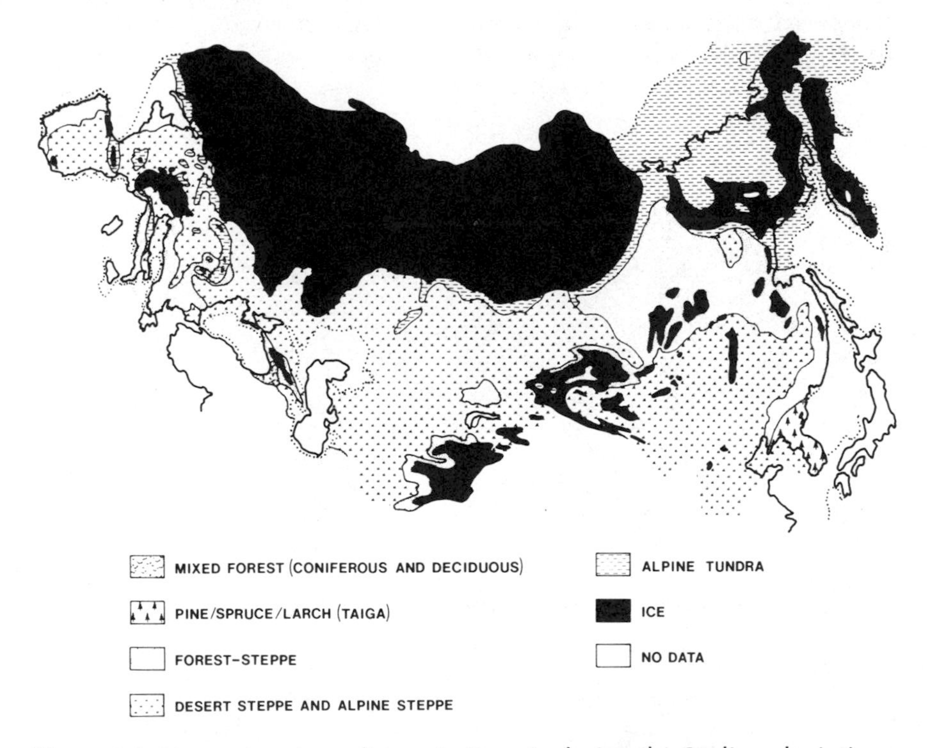

Figure 9.3. Vegetational conditions in Eurasia during the Saalian glaciation. *(Redrawn from B. Frenzel, 1968, The Pleistocene Vegetation of Northern Eurasia, Science* **161:***641; copyright © 1968 by the American Association for the Advancement of Science. Redrawn with permission.)*

species and their replacement by more cold-tolerant forms, the most diagnostic of which is the larch (*Larix*). Table 9.2 reconstructs the vegetation at several localities during these Pleistocene episodes. From this reconstruction it is possible to picture the direction and degree of changes that have taken place in the past 3 million years in northern Eurasia.

The North Polar Region

Tundra vegetation is first reported with the Saalian glacial episode in Siberia and apparently gained widespread distribution during the Weichselian. The true "home" of tundra is not known, but may have been on the slopes of mountain ranges that were "islands" in the ice. Wherever the home, it is believed that Alaska and Canada represented the retreat

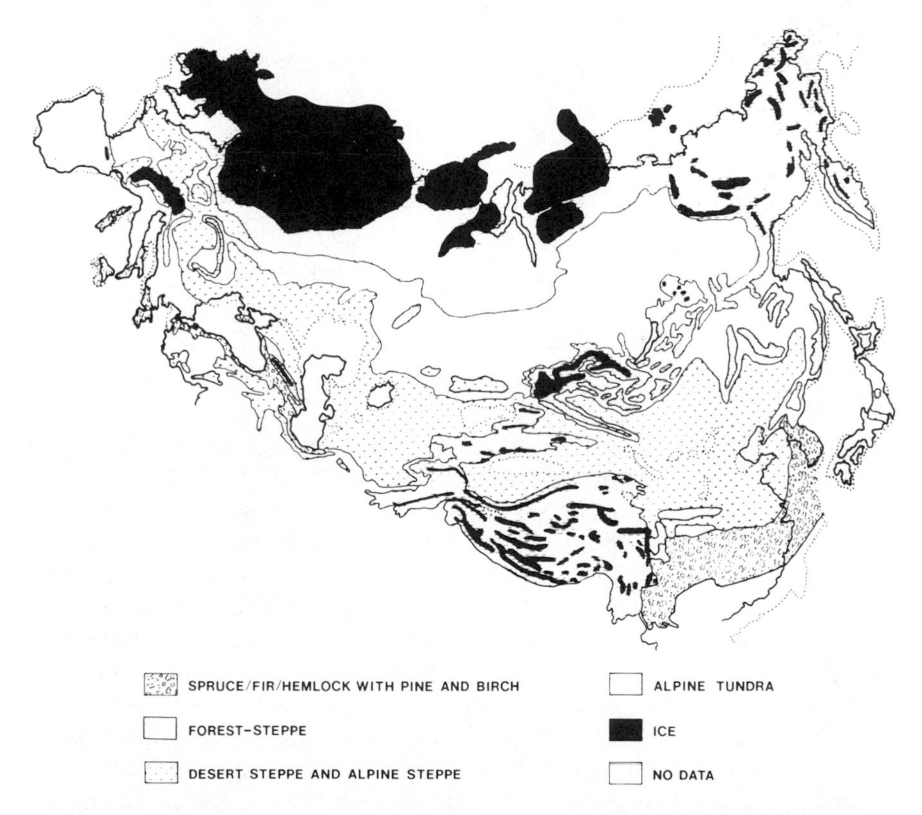

Figure 9.4. Vegetational conditions in Eurasia during Weichselian glaciation. *(Redrawn from B. Frenzel, 1968, The Pleistocene Vegetation of Northern Eurasia, Science* **161:**642; *copyright © 1968 by the American Association for the Advancement of Science. Reprinted with permission.)*

Table 9.2. Reconstruction of Vegetation at Several Localities during the Sequence of Pleistocene Episodes

Episode	*Locations*				
	France	*Denmark*	*Caspian*	*Sakahlin*	*Amur*
Weichselian	Steppe w/ tundra	ice	steppe	steppe w/ groves of hardy trees	steppe w/ groves of hardy trees
Eemian	mixed forest	oak forest	steppe	conifer in north; broad leaf decid. in south	broadleaf decid. w/ conifer
Saalian	steppe	ice	steppe	steppe w/ groves of hardy trees	steppe w/ groves of hardy trees
Holsteinian	oak/fir	pine forest	steppe	conifer in north; broadleaf decid. in south	spruce/pine w/oak & elm
Praetiglian	sub-Arctic parkland	ice	steppe	?	pine/spruce/ larch
Reuverian	pine/ sequoia	mixed forest	steppe	mixed forest as in Denmark	spruce/ hemlock

areas for tundra because the type is not a widespread constituent of the interglacial vegetation of Eurasia.

The succession of polar fossil-plant remains shows a change from gymnosperm domination in the older beds to angiosperm domination in the younger ones. Five sequential Cretaceous floras are recognized, ranging in age from Albian to Coniacian. This transition, similar to that found by Axelrod (1959) in the Southern Hemisphere, was quite rapid, suggesting that the climate on the Cretaceous coastal plain changed from warmer to cooler during a 30-million-year time span. The earliest remains are dominated by ferns and ginkgophytes (indicating warm temperatures), cycadophytes (indicating tropical to warm-temperate conditions), and conifers. Among the latter are found *Cephalotaxus, Podocarpus,* and *Juniperus,* which suggest warm-temperate to temperate conditions, and *Picea, Pinus, Sequoia, Sequoiadendron,* and *Taxodium,* which suggest temperate to cool-temperate environments.

Geologic, paleobotanical and biogeographical evidence all support the idea that tundra is the youngest of the world's major ecosystems. It probably originated as alpine tundra during the Mio-Pliocene and slowly evolved and migrated northward by late Pleistocene. The present distribution of both the circumpolar arctic tundra (see Figure 9.7) and extensive areas of alpine tundra is a product of the Pleistocene. The tundra flora probably evolved first in the highlands of central Asia and the Rockies of North America and thence spread to lower elevations and northward with continued cooling.

Arnold (1959) argues that tundra species may have evolved as early as the Permian, in concert with the earliest known glacial episodes. He envisions a mechanism almost identical to that suggested by Axelrod

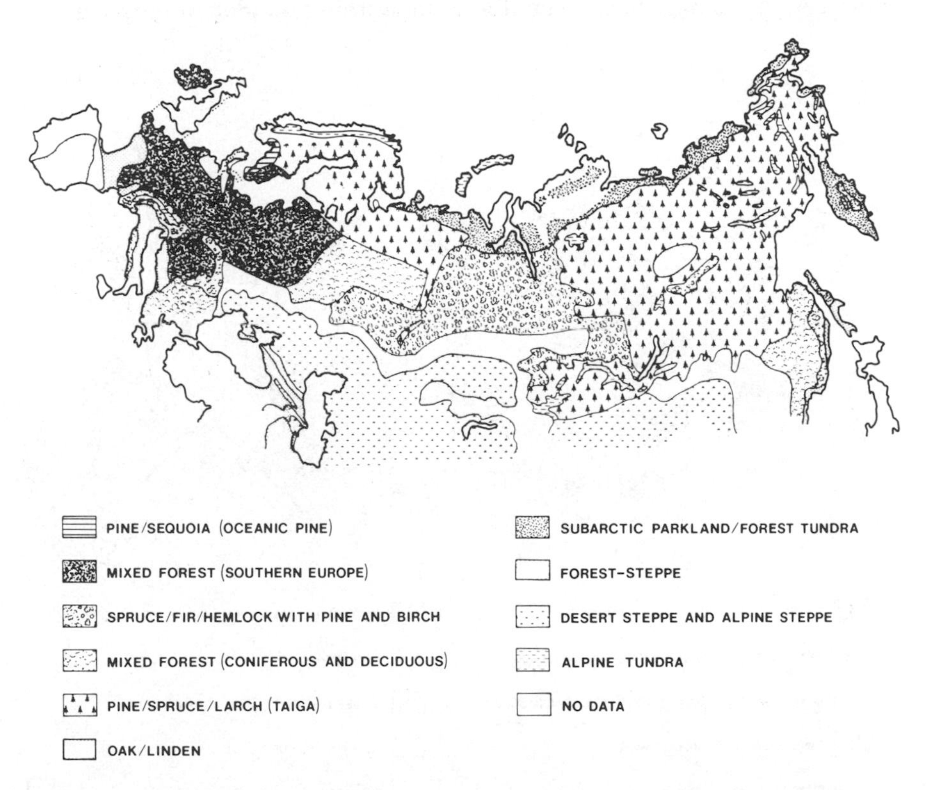

Figure 9.5. Vegetational conditions in Eurasia during the Holsteinian interglacial. *(Redrawn from B. Frenzel, 1968, The Pleistocene Vegetation of Northern Eurasia, Science* **161:***643; copyright © 1968 by the American Association for the Advancement of Science. Redrawn with permission.)*

(1966) for the development of the deciduous habit, except that cold, not drought, represented the adaptive zone. Arctic tundra species may have differentiated during the Pliocene, such that they existed in their present form by the beginning of the Pleistocene. Dwarf willows and birches may have arisen during Late Cretaceous or early Tertiary time, as did other pines and spruces. Arnold further suggests that *Metasequoia* and *Taxodium* originated in the far north during subtropical times.

Vertebrate faunas of arctic and alpine tundras share few common species, except where northern mountains extend into circumpolar tundra. Pleistocene lowland tundra faunas probably originated in central Eurasia by adaptation to cool continental environments stimulated by glacial episodes. These forms then migrated to Europe by mid-Pleistocene and to North America by the Wisconsin or post-Wisconsin. The lag time into North America may have been due to mountain and glacial barriers.

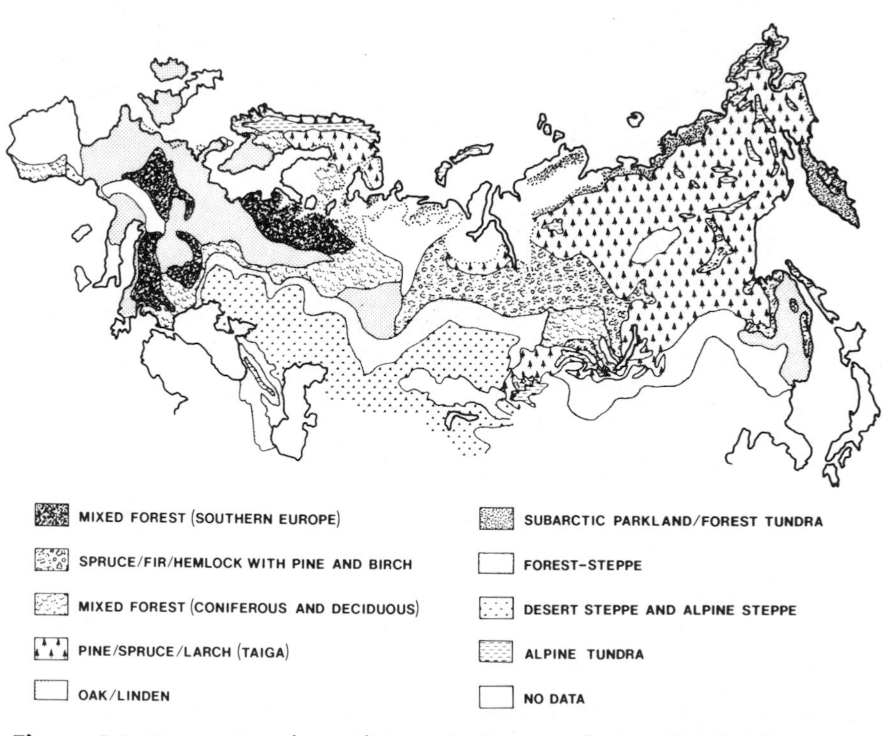

Figure 9.6. Vegetational conditions in Eurasia during the Eemian interglacial. *(Redrawn from B. Frenzel, 1968, The Pleistocene Vegetation of Northern Eurasia, Science* **161:***644; copyright © 1968 by the American Association for the Advancement of Science. Reprinted with permission.)*

The alpine vertebrate faunas, on the other hand, may have evolved earlier than the lowland tundra types and may have had as their site of origin the Mongolo-Tibetan Plateau. Most of the alpine forms apparently did not contribute to the lowland forms because they were already specialized for high altitude environments. Following their evolution, the alpine faunas are believed to have migrated to Europe and the Americas

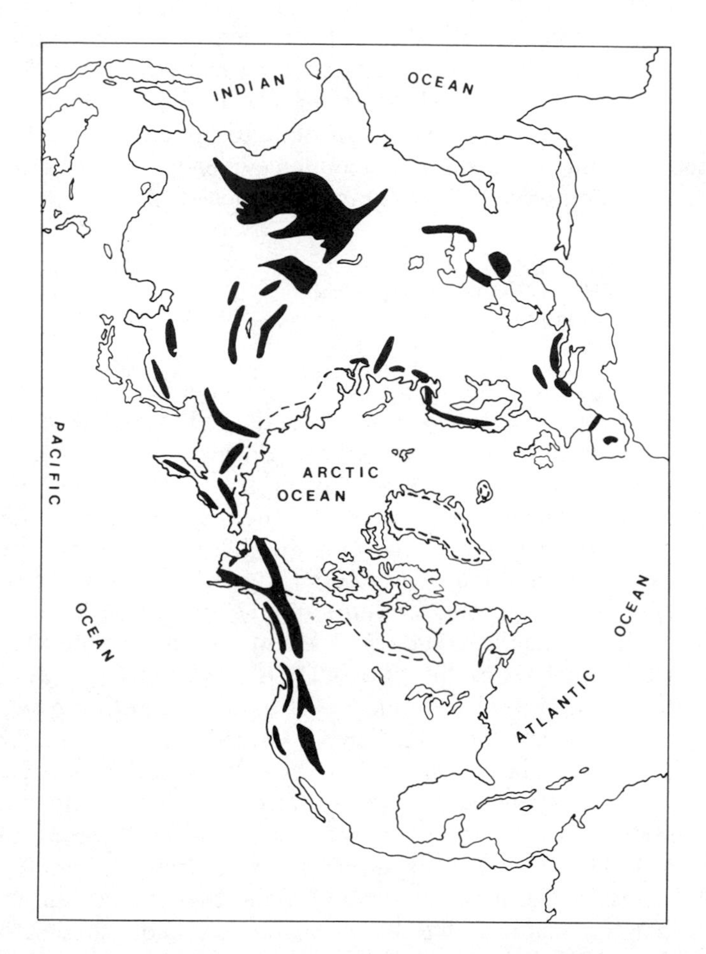

Figure 9.7. Present extent of arctic and alpine tundras in the Holarctic. *(From R. S. Hoffman and R. D. Taber, 1967, Origin and History of Holarctic Tundra Ecosystems, With Special Reference to Their Vertebrate Faunas, in Arctic and Alpine Environments, H. E. Wright, Jr., and W. H. Osburn, eds., Indiana University Press, Bloomington, Ind., p. 144; copyright © 1967 Indiana University Press. Reprinted with permission of the publisher.)*

along mountain chains, in particular the Siberian and Alaskan mountains, the Caucasus, the Rockies, and the Andes. These migration routes are believed to have acted repeatedly as refugia during interglacials.

The Bering Land Bridge could only have existed during periods of lower sea level, assuming that continental boundaries have not been changing as a result of continental drift. Glaciologists can show that during periods of intense glaciation, sea level might have dropped by several hundred feet, thereby exposing a vast region, such as that shown in Figure 9.8, between Asia and North America. Since the bridge would have existed during times of glaciation, and since it was so long and variable in its habitats, several generations of species would be required to make the crossing. These species would have to have been adapted to cold, arctic tundra climates because pollen evidence reveals that typical tundra species comprised the vegetation of those times.

PLEISTOCENE MAMMALIAN DEVELOPMENT

The most important single aspect of mammalian development in Pleistocene Europe was the rise of species adapted to cold environments. Prior to the onset of cold periods and full glacial advances there was, of course, little requirement for these adaptations, so it is remarkable that arctic biotypes occurred so suddenly in the fossil record of the Villafranchian (*Praetiglian*) and Günz stratigraphy. In the Günz, one finds the first evidence for reindeer (*Rangifer tarandus*), musk ox (*Ovibos moschatus*), several species of lemming (*Dicrostonyx*), the steppe mammoth (*Mammuthus trogontherii*) and the snow vole (*Microtus nivalis*). By Mindel times the glutton (*Gulo schlosseri*), alpine shrew (*Sorex alpinus*), and woolly rhinoceros (*Coelodonta antiquitatis*) made their first appearances, the latter being confined to Asia. Finally, by the time of the Würm glacial one encounters the polar bear (*Ursus maritimus*).

Throughout the sequence from late Pliocene to Pleistocene, a total of 278 mammalian species have been recorded in the fossil record. Most of the European sites, some 20 in number, are located in Spain, France, Italy, and southern Germany (Figure 9.9). When these species are arranged against their times of first and last appearance, a table similar to Table 7.4 can be constructed. This matrix (Table 9.3) reveals a number of interesting points. First, it is evident that the mammalian fauna of the Palearctic is young. Only 5 or 6 species extant during the Pliocene are still living. These remnants include the pigmy shrew (*Sorex minutus*), pond bat (*Myotis dasycneme*), long-winged bat (*Miniopterus schreibersi*), common field mouse (*Apodemus sylvaticus*), and steppe pika (*Mustela eversmanni*). Rates of extinction were rather extreme during the Praetiglian cold period of the Villafranchian and the Günz glacial (31 and 26 percent of the 42 Pliocene species, respectively). Moreover, of the 76 species to make their first appearance in the Villafranchian, only 4 have

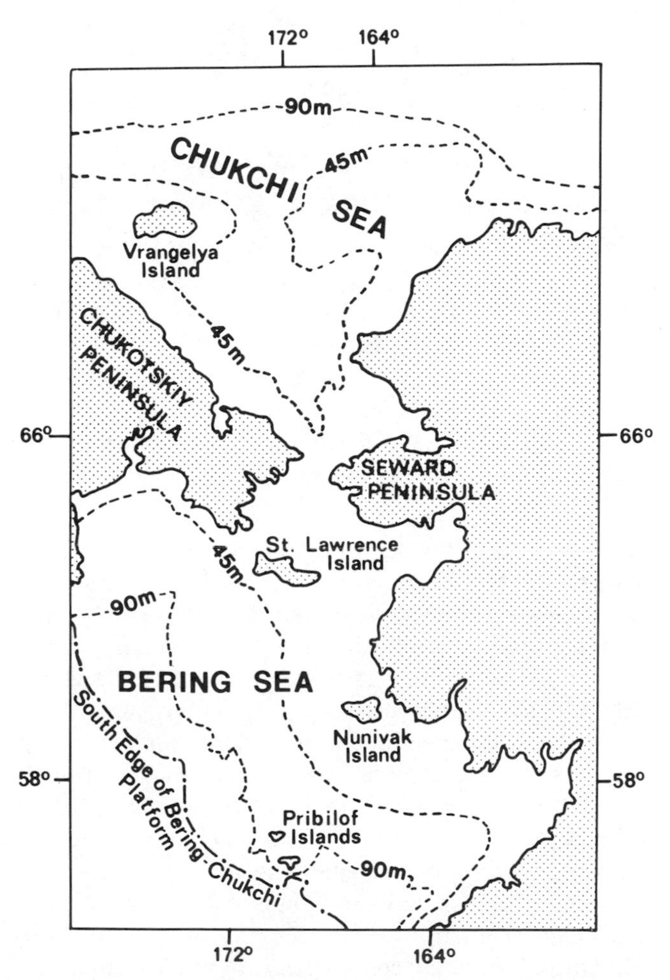

Figure 9.8. Bering and Chukchi Seas. Submarine contours in meters indicate the extent of the Bering Land Bridge when sea levels were 45 and 90 meters lower than at present. *(From D. M. Hopkins, 1959, Cenozoic History of the Bering Land Bridge, Science* **129:***1521; copyright © 1959 by the American Association for the Advancement of Science. Reprinted with permission.)*

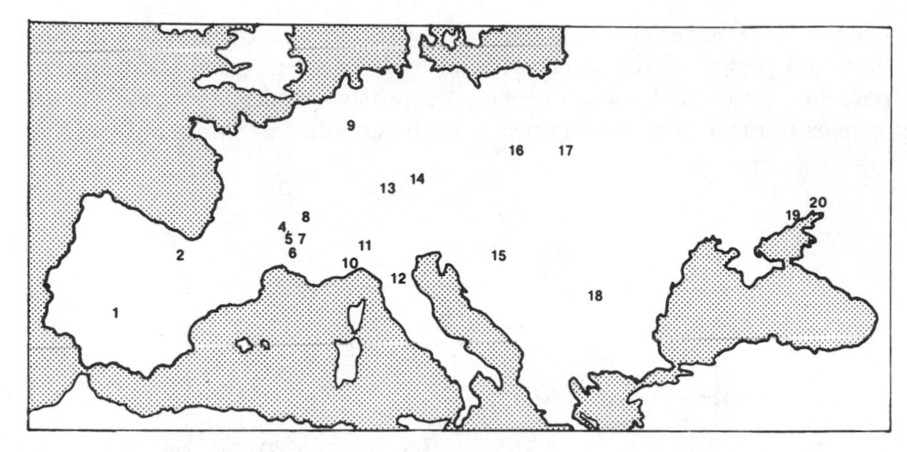

Figure 9.9. Geographic distribution of some Villafranchian mammal localities in Europe. Spain: 1, Valverde de Calatrava; 2, Villaroya. England: 3, East Anglian Crags. France: 4, Mt. Perrier; 5, Senèze; 6, Vialette; 7, Saint-Vallier; 8, Chagny. Netherlands: 9, Tegelen. Italy: 10, Villafrance d'Asti; 11, Leffe, 12, Upper Val de'Arno. Germany: 13, Erpfingen; 14, Schernfeld. Hungary: 15, Villany. Poland: 16, Rebielice; 17, Kadzielnia. Rumania: 18, Oltet. USSR: 19, Mariupol; 20, Taganrog. *(From B. Kurtén, 1968, Pleistocene Mammals of Europe, Aldine Press, Chicago, p. 11.)*

survived subsequent climatic oscillations: species of the hedgehog (*Erinaceus* spp.), the Mediterranean horseshoe bat (*Rhinolophus euryale*), one species of beaver (*Castor fiber*), and the yellow-necked field mouse (*Apodemus flavicollis*). In all, only about 17 percent of the pre-Günz fauna have survived to the present, which means that the remaining 83 percent of the 111* living species are younger than one million years old.

A second interesting aspect of Table 9.3 is revealed by comparing first appearances during glacial episodes with those of interglacial ones. Unlike the situation noted for the Nearctic, there is little difference in the numbers of new forms generated in the glacials and interglacials. (It must be remembered, however, that the data in Table 7.4 represent genera, while those in Table 9.3 are for species.) Some 83 species are first recorded in glacial sediments, and of these 83, almost 70 percent (57) are first noted in Günz strata. By comparison, 76 species are first recorded in interglacial beds, particularly those of Holsteinian and Eemian age. Remarkably, of the 159 species to have arisen in the last million years, 102 (64 percent) are still found in the Palearctic. In terms of percentage,

*Corbet (1966) and Burton (1976) list 115 and 112 species of terrestrial mammals respectively.

Table 9.3. Numbers and Percentages of European Mammalian Species Making Their First and Last Appearances in the Pleistocene Fossil Record. (Compare also Table 7.4)

	Total no. first appear-ance	Numbers (%) of last record										
		Plio.	*Villa.*	*Tigli.*	*Günz*	*Cromer.*	*Mindel*	*Holstei.*	*Riss*	*Eemian*	*Würm*	*Living*
Pliocene	42		13(31)	3(7)	11(26)	5(12)	2 (5)	2 (5)			1 (2)	5 (12)
Villafranchian	76		24(32)	7(9)	16(21)	11(14)	8(11)	1 (1)		3(4)	2 (3)	4 (5)
Tiglian	1							1(100)			—	—
Günz	57				5 (9)	4 (7)	5 (9)	6 (11)	1(2)		8(14)	28 (49)
Cromerian	13						2(15)	2 (15)			4(31)	5 (38)
Mindel	10										2(20)	8 (80)
Holsteinian	34							2 (6)	1(3)	2(6)	5(15)	24 (71)
Riss	7										3(43)	4 (57)
Eemian	22									1(5)	3(14)	18 (82)
Würm	9										1(11)	8 (89)
Living	7											7(100)
Totals	278		37(13)	10(4)	32(12)	20 (7)	17 (6)	14 (5)	2(1)	6(2)	29(10)	111 (40)

Table 9.4. Comparison of Numbers and Percentages of Species Comprising the Palearctic Mammalian Fauna

Order	Pleistocene & Recent: No. of species in record	Pleistocene & Recent: % of total Pleistocene assemblage (282)	Modern European fauna: No. of living species	Modern European fauna: % of living assemblage (113)	No. extinct since Eemian
Artiodactyla	53	19	15	13	5
Carnivora	70	25	25	22	12
Chiroptera	24	9	19	17	0
Insectivora	25	9	11	10	1
Megomorpha	9	3	5	4	0
Perissodactyla	15	5	1	1	8
Primates	6	2	2	2	1
Proboscidea	6	2	0	0	2
Rodentia	74	26	35	31	7
Totals	282	100	113	100	36

the greatest ages of extinction have been during the Cromerian interglacial and Mindel-Riss glacial episodes. It would appear from these data that *Homo sapiens* has not been as destructive an element in the Palearctic as, presumably, in the Nearctic.

Another way of assessing human impact on Europe's mammalian fauna is to compare the numbers of living species in each of the nine orders with the total number of past and present species (Table 9.4). Such a comparison gives an idea of the relative changes taking place. They can then be evaluated in terms of when they occurred. From the data in Table 9.4, it is evident that the greatest percentage decreases have occurred in the Artiodactyla and Perissodactyla. These orders are the even- and odd-toed ungulates—pigs, deer, and bovids, and horses, rhinoceros, and tapirs, respectively. Only 25 percent of these species have become extinct since Eemian time, however, so it is not likely that humanity, as a single agent, is guilty of any megafaunal extinction in Europe.

SUMMARY

The points of greatest interest in the Palearctic Realm revolve around: (1) repeated north-south migrations of forest and steppe-savanna vegetation in response to climatic oscillations; (2) the gradual impoverishment of angiosperm diversity, as southerly migrations were blocked by the Pyranees, Alps, Carpathian, and Caucasus mountain ranges during glacial surges; (3) the sudden origin of mammalian species adapted to cold environments; and (4) the relatively greater ability of modern mammalian species to withstand human competition.

Compared to the Nearctic Realm, the Palearctic is poor in both angiosperm and mammalian diversity. The reason, in part, is that unlike those of North America, European species were geographically trapped by mountain ranges and the Mediterranean Sea as they migrated southward. Only in the oriental section of the realm (which is analogous to the southeastern United States) could species find refuge in subtropical and tropical belts. Even here, the mountainous environment of southern China may have been an insurmountable barrier.

The impact of humankind throughout the realm may be hotly debated, but not ignored. Through hunting, and by indirect means, the larger game animals have been reduced in numbers, if not diversity. Offsetting this trend was the practice of domestication, a phenomenon completely lacking in the Nearctic. On the other hand, the vegetation, at least of the western European section of the realm, has been almost totally modified, whereas in the Nearctic, large areas are still described as wilderness.

REFERENCES

Arnold, C. A., 1959, Some Paleobotanical Aspects of Tundra Development, *Ecology* **40**:146-148.

Axelrod, D. I., 1959, Poleward Migration of Early Angiosperm Flora, *Science* **130**:203-207.

Axelrod, D. I., 1966, Origin of Deciduous and Evergreen Habits in Temperate Forests, *Evolution* **21**(1):1-15.

Burton, M., 1976, *Guide to the Mammals of Britain and Europe,* Elsevier-Phaidon, Oxford, 256p.

Corbet, G. B., 1966, *The Terrestrial Mammals of Western Europe,* Dufour Editions, Philadelphia, 264p.

Frenzel, B., 1968, The Pleistocene Vegetation of Northern Eurasia, *Science* **161**:637-649.

Hopkins, D. M., 1959, Cenozoic History of the Bering Land Bridge, *Science* **129**:1519-1528.

Hoffman, R. S., and R. D. Taber, 1967, Origin and History of Holarctic Tundra Ecosystems, With Special Reference to Their Vertebrate Faunas, in *Arctic and Alpine Environments,* H. E. Wright, Jr., and W. H. Osborn, eds., Indiana University Press, Bloomington, Ind., pp. 143-170.

Kurtén, B., 1968, *Pleistocene Mammals of Europe,* Aldine Press, Chicago, 317p.

FURTHER READING

Brundin, L., 1970, Antarctic Land Faunas and Their History, *Antarct. Ecol.* **1**:41-53, 54-64.

Cifelli, R. L., 1981, Patterns of Evolution Among the Artiodactyla and Perissodactyla, *Evolution* **35**(3):433-440.

Colinvaux, P. A., 1964, The Environment of the Bering Land Bridge, *Ecol. Monogr.* **34**:297-329.

Deuser, W. G., E. H. Ross, and L. S. Waterman, 1976, Glacial and Pluvial Periods: Their Relationship Revealed by Pleistocene Sediments of the Red Sea and Gulf of Aden, *Science* **191**:1168-1170.

Durham, J. W., 1963, Paleogeographic Conclusions in Light of Biological Data, in *Pacific Basin Biogeography,* J. L. Gressitt, ed., Bishop Museum Press, Honolulu, pp. 355-365.

Eisenberg, J. F., 1982, *The Mammalian Radiations: An Analysis of Trends in Evolution, Adaptation and Behavior,* University of Chicago Press, Chicago, 610p.

Giterman, R. E., and L. V. Golubeva, 1967, Vegetation of Eastern Siberia During the Anthropogene Period, in *The Bering Land Bridge,* D. M. Hopkins, ed., Stanford University Press, Menlo Park, Calif., pp. 232-244.

Hultén, E., 1937, *Outline of the History of Arctic and Boreal Biota During the Quaternary Period,* Bokförlags Abiebolaget Thule, Stockholm, 168p.

Kurtén, B., 1960, Chronology and Faunal Evolution of the Earlier European Glaciations, *Soc. Sci. Fenn.* **21**(5):1-62.

Moreau, R. E., 1955, Ecological Changes in the Palearctic Region Since the Pliocene, *Zool. Soc. London Proc.* **124**(4):253-295.

Smiley, C. J., 1966, Cretaceous Floras from Kuk River Area, Alaska: Stratigraphic and Climatic Interpretations, *Geol. Soc. Am. Bull.* **77**(1):1-14.

Stanley, S. M., 1976, Stability of Species in Geologic Time, *Science* **192**:267-269.

Chapter 10

The Africotropical Realm

CONTINENTAL AFRICA

The continent of Africa, including the Arabian Peninsula, Madagascar, and the west Indian Ocean islands, is referred to here as the Africotropical Realm. In other literature, mammologists have tended to exclude the Saharan area, or at a minimum, the Mediterranean zone north of the Atlas Mountains, because of the European flavor and depauperate aspect of its fauna. Often this somewhat reduced continental land mass is referred to as the Ethiopian Realm. As described here, and in its broader extent, the realm stretches from about 35° north latitude, to 35° south latitude, and from about 50° east longitude to better than 15° west longitude. The total area is more than 33,600,000 square kilometers. Within the Africotropical Realm's confines are two of the world's longest rivers: the Nile, running a length of 6,600 kilometers, and the Zaire, running more than 4,600 kilometers. The highest peak is Mount Kilimanjaro in Tanzania; at 5,800 meters, it is only slightly lower than Mount McKinley in Alaska, which rises 6,100 meters. The largest lake is Victoria, at about 70,000 square kilometers.

Unlike the Nearctic, Neotropical, and Palearctic Realms, the Africotropical Realm has experienced only minor alpine glacial transgression. In this respect, it is similar to the Austral and Paleotropical realms, which are discussed in later chapters. In place of ice-covered landscapes, oscillating wet and dry climates and past continental connections have promoted the change of angiosperm and mammalian distributions into their present-day form. Eight major points in the post-Cretaceous history of Africa have been stressed by Axelrod and Raven (1978).

1. During the Late Cretaceous or early Tertiary, Africa severed its ties with the southern continents and migrated some 15° north. Remnant connections through island archipelagos maintained a floristic link with other southern areas for some time.
2. Over this same time period, Africa was a low-lying continent with uniform broad expanses of tropical rain forest flanked by savanna containing small pockets of sclerophyll types.
3. Continental warping and tectonism began in the middle Tertiary. The African Rift system was initiated, and the combination of these events led to increased habitat diversity. The savannas and sclerophyll types began to spread at the expense of rain forest when seasonal rainfall became a widespread climatic regime.
4. The Tethys Sea was closed in middle Miocene time when Africa, together with the Arabian Peninsula, joined the Asian Plate at Iran. The Mediterranean Sea is a remnant of that ancient sea, as are the Tigris-Euphrates and Ganges valleys, where continental contacts were made. The rearrangement of ocean circulations at the close of Tethyan time stimulated the spread of savannas, deciduous forest, thorn forest, and sclerophyll vegetation in Africa.
5. The appearance of mountain glaciers in Antarctica provided a source of cold water, known as the Benguela Current, to bathe the southwest coast of Africa, which restricted the extent of savanna and, because of prolonged summer drought, promoted the spread of grassland, thorn forest, and semidesert.
6. During the Pleistocene glacial episodes of the Northern Hemisphere, increased aridity sponsored the development of true deserts in Africa through continued adaptations of ancient taxa already present there.
7. In modern times, a trend away from the vegetational symmetry of earlier epochs has arisen. The Sudanian savanna is different from the Zambezian, and the Sahara differs from the Namib and the Kalahari.
8. There were apparently two major periods of speciation. The first commenced in the Miocene with the rejuvenation of an isolated continent. The steppe and savanna environments promoted adaptation to open habitats and probably also stimulated coevolutionary adaptations among the mammals. A second peak in speciation may have occurred in Plio-Pleistocene time through repeated climatic shifts.

Vegetational History

Both Africa and Australia are ancient land masses in terms of their soils, relief, and geology. However, environments for fossil preservation have apparently always been very few. The tropics, in general, are poor environments for preservation because of the rapid cycling of biotic material and high oxidation rates. Part of the picture that can be pieced together, however, relates to the paleogeography of Africa, Australia, South America, India, and Antarctica. Similarities in the geology and general shape of these southern land masses indicate that, as has been discussed, they were once joined into a single continent called Gondwanaland. During the Mesozoic, until at least Early Cretaceous times, there must have been a connection between South America and Africa, as evidenced by the fossil Glossopteris Flora—a type distinguished by a genus of "tongue" ferns that occur in southeast South America, southwest Africa, and Madagascar.

The vegetational development of Africa seems to revolve around three themes: (1) a Cretaceous land mass that was connected to lands now rather distant, and from which an initial stock of forms emerged; (2) isolation following the Late Cretaceous, during which time there was extensive radiation of taxa; and (3) PostPliocene climatic and topographic changes that promoted new adaptive strategies leading to modern ecosystems (Raven and Axelrod, 1974).

As a result of all these factors, Brenan (1978) and Richards (1973), among others, describe a relative floristic poverty for the continent, particularly in the area between the two tropics. There are only something like 30,000 species in the vast 20,000,000 square kilometers between Cancer and Capricorn, compared to 40,000 for Brazil, which is less than half the size. Equally remarkable is the poverty of otherwise large and richly diversified world families. Lauraceae, for example, is a family of worldwide distribution with 32 genera and 2,000 to 2,500 species. In tropical Africa there are only 5 genera and no more than 50 species. Similarly, 17 genera and 117 species of palms occur there, but 64 genera and 837 species occur in South America, and 97 genera with 1,385 species in the Indo-Malaysian tropics.

Figure 10.1 from Axelrod and Raven (1978) summarizes the basic vegetational pattern and its changes throughout the Cenozoic. Northern Africa was covered by tropical rain forest during the early Tertiary (Figure 10.1a). At this time, the equator passed through Dakar in Senegal, as the continent drifted northward. Among the more important fossil woods and leaves are representatives of Moraceae, Proteaceae, Lauraceae, Rutaceae, Sterculiaceae, and Dipterocarpaceae. Since the highlands of East Africa did not exist at this time, it is likely that the forests ranged

Late Cretaceous-Paleocene
1. Lowland rainforest
2. Savanna-woodland
3. Montane rainforest
4. Subtropic rainforest
5. Temperate rainforest ('austral affinities')
6. Sclerophyll woodland

Oligocene-Miocene
1. Lowland rainforest
2. Savanna-woodland and thorn scrub
3. Montane rainforest
4. Subtropic laurel forest
   a. Canarian
   b. Natal
5. Temperate rainforest? ('austral affinities')
6. Sclerophyll vegetation
   a. Tethyan
   b. Cape

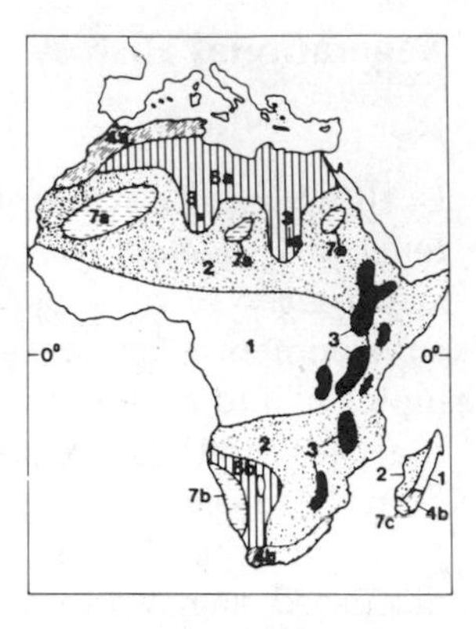

Late Miocene-Early Pliocene
1. Lowland rainforest
2. Savanna-woodland
3. Montane rainforest-afroalpine
4. Subtropic laurel forest
   a. Canarian
   b. Cape
5. (eliminated)
6. Sclerophyll vegetation
   a. Tethyan
   b. Cape
7. Thorn scrub-succulent woodland
   a. Sahelian
   b. Kalaharian
   c. Malgasan

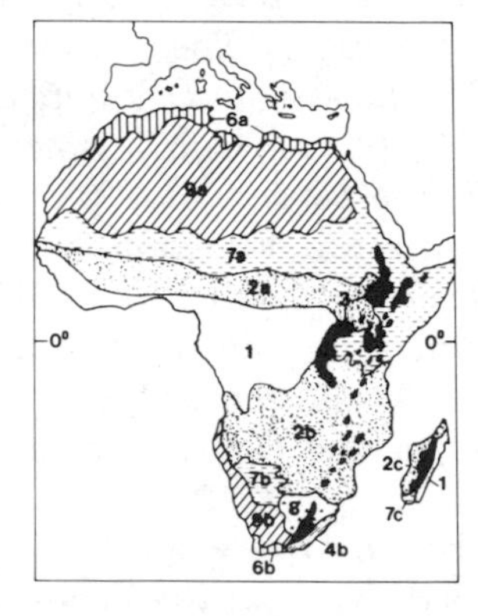

Present Vegetation
1. Lowland rainforest
2. Savanna-woodland
   a. Sudanian
   b. Zambezian
   c. Malgasan
3. Montane rainforest and afroalpine
4. Subtropic laurel forest
   a. Canarian
   b. Natal
5. (eliminated)
6. Sclerophyll vegetation
   a. Tethyan
   b. Cape
7. Thorn scrub-succulent woodland
   a. Sahelian
   b. Kalaharian
   c. Malgasan
8. Grassland
9. Desert and semi-desert
   a. Saharan-Libyan
   b. Namib-Karoo

Figure 10.1. Patterns of inferred vegetation in Africa from Late Cretaceous to present time. *(From D. I. Axelrod and P. H. Raven, 1978, Late Cretaceous and Tertiary Vegetation History of Africa, in Biogeography and Ecology of Southern Africa, M. J. A. Werger, ed., W. Junk, The Hague, p. 88; copyright © 1978 by W. Junk. Reprinted by permission of the publisher.)*

from coast to coast and extended eastward into Madagascar and India, which at that time were connected across narrow straits.

A temperate rain forest occupied the extreme southern end of Africa and had close affinities with Australia. Some of the important forms were genera allied to modern *Podocarpus, Nothofagus, Elaeocarpus,* and *Dacrydium.*

A subtropical evergreen forest extended throughout southeastern Africa during the early Tertiary and spread westward into the present Namib Desert. It effectively separated the temperate rain forest farther south from the drier savanna woodlands, which then occupied the Zaire Basin. Among the more important families represented were Anacardiaceae, Aricaceae, Burseraceae, Euphorbiaceae, Rutaceae, Sapindaceae, Tiliaceae, and Vitaceae.

Sclerophyll vegetation was highly restricted in the Paleocene and early Eocene to a few localities in the savanna. Most of the adaptations toward sclerophylly were then in their infancy and appear to have had their earliest origins in southwest Africa. For the Mediterranean climate of the Cape area, Goldblatt (1978) records some 8,550 species, 73 percent of which are endemic and probably have arisen since the Pliocene. For comparative purposes, Table 10.1 shows rates of endemism for several world locations. It is apparent from these data that only highly isolated places like Australia or insular localities like Hawaii and New Zealand match the endemism recorded for southern Africa.

By the close of Oligocene time the vegetation of Africa had acquired its near-modern aspect, but the composition and distribution of types continued to shift in response to climatic, edaphic, and topographic conditions, as the continent underwent its long period of isolation. Throughout the middle and late Tertiary, particularly the Miocene, the African continent was rejuvenated by broad regional uplifts, tectonism, and rifting. Stimulated by these processes, together with the changes they induced in soil and rainfall patterns, the vegetation reorganized itself into a more modern picture. Semideserts began to appear during the Pliocene but their full expression, along with that of true deserts, was to be a phenomenon of Pleistocene aridity.

A particularly interesting feature of Pleistocene vegetational changes is observable in Mediterranean and Saharan Africa. Quézel (1978) has made tentative analyses of the floristic composition and affinities of phytogeographic subdivisions in North Africa. From these studies he confirms an already established bioclimatic southern limit of Holarctic types and their separation from elements of strictly African origin. Figure 10.2 shows this separation together with outliers of Mediterranean species on the Ahaggar and Tibesti massifs. In general, there is an attenuation of Mediterranean species as one moves southward, and a

Table 10.1. Comparative Rates of Floral Endemism among the Genera and Species of Selected World Localities

Region	*Area km² (000)*	*Genera*	*Endemic genera (%)*	*Species*	*Endemic species (%)*
Southern Africa	2,573	1,930	560 (29.0%)	18,532	14,850 (80.3%)
Cape Floristic Region	89	957	198 (20.7)	8,550	6,252 (73.1)
Cape Peninsula	0.47	533	1 (0.2)	2,256	157 (7.0)
Australia	7,716	?	?	15,000	85%
Europe	10,000	1,340	?	10,500	3,500 (33)
California	324	795	50 (6.3)	4,452	2,125 (47.7)
British Isles	308	545	0 (0)	1,443	17 (1.2)
New Zealand	268	393	39 (9.9)	1,996	1,618 (81.1)
Hawaii	16.6	253	31 (21.3)	1,897	1,751 (92.3)
Sonoran Desert	310	746	20 (2.7)	2,441	650 (26.6)
West Tropical Africa	4,500	1,742	?	7,500	(ca. 37)

Source: P. Goldblatt, 1978, An Analysis of the Flora of Southern Africa: Its Characteristics, Relationships, and Origins, *Missouri Bot. Gard. Ann.* **65:**371.

gradual increase in Saharan and tropical African forms. Considering its total area, the Mediterranean zone is relatively rich in species (4,034) and endemics (1,038 or 26 percent) when compared to the much larger Saharan zone (1,620, 190, and 12 percent, respectively). It appears that some of the Mediterranean species arrived and others developed *in situ* during the Miocene, underwent ecological sifting during the Pliocene, and attained their present pattern from the Pleistocene onward. Contrary to the idea that the Saharan zone has served primarily as a north-south barrier, Quézel stresses that it is more appropriately regarded as a zone of transition, differentiation, and mixing of African and Mediterranean types.

Studies reported by Coe (1967) indicate the presence of alpine glaciers and lowered vegetational belts as recently as 10,000 years B.P. Temperatures are estimated to have been some 3 to 4 degrees cooler. Under such conditions, a more or less continuous montane forest might well have connected the now isolated mountain peaks of East Africa. That they have not had recent interconnection, however, is suggested by their various, but high (81 percent), rates of endemism (Killick, 1978). Mount Elgon, for example, has 23 endemic species, compared to Mounts Kenya and Kilimanjaro with 13 each, Aberdares with 7, and Meru with 3.

As is often the case, the nonendemic elements are of equal, if not greater, interest. These species show which areas seem to have the greatest influence on developmental history. In the case of the afroalpine type, the most important source areas for plant immigration have been from the southern Cape and Southern Hemisphere temperate floras. These account for about 70 percent of the nonendemic forms. The

European sector of the Palearctic is the next most important with a contribution of 15 percent, followed by Mediterranean (7 percent), Himalayan (2 percent), and Cosmopolitan (3 percent) taxa. Among the more important taxa present are *Thalictrum, Swertia, Lobelia, Helichrysum, Myrica,* and *Hagenia* from sources within Africa; and *Lithospermum, Juncus, Vaccinium, Ranunculus, Clematis,* and *Dianthus* from Europe.

Mammalian History

The African and Oriental regions are the two main faunal areas of the Old World tropics. Neither has barriers of climate or topography to match those of, say, Oceania or the Neotropics. Some authors, for example, Darlington (1957), believe that mammalian radiation may have come out of these areas. Although the present day fauna of Africa is distinctive enough to warrant its separation into a unique realm, it is clear that past

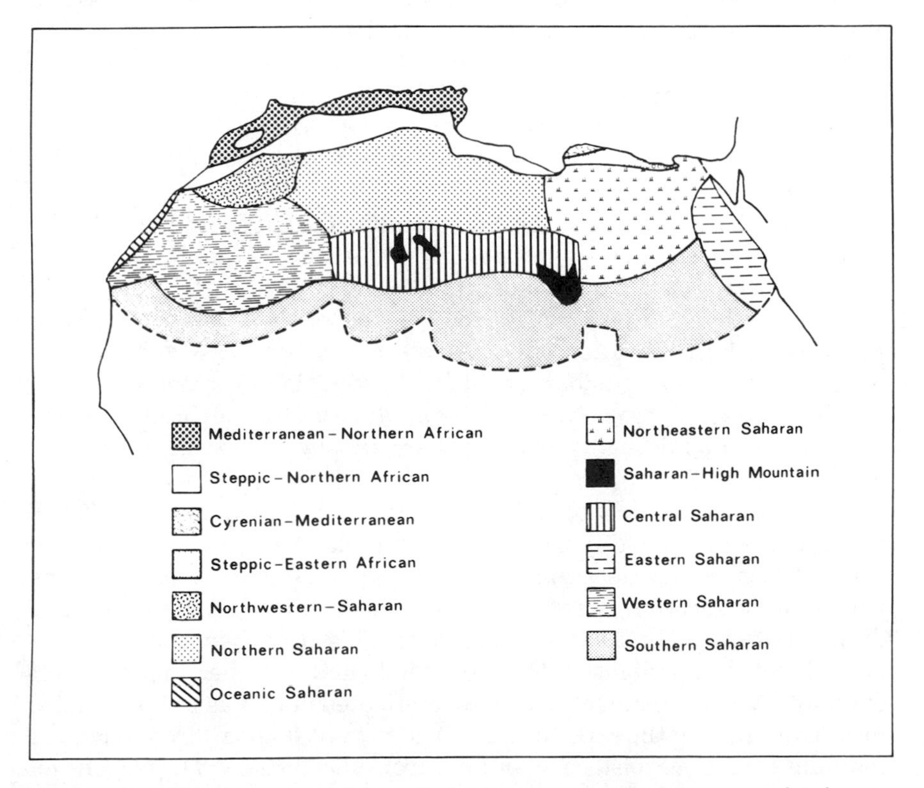

Figure 10.2. Phytogeographical subdivisions of Mediterranean and Saharan Africa.

distributions of these animals, as recently as the beginning of the Holocene epoch, would require a vastly different delineation of realms. Instead of being a center of radiation, then, Africa should be regarded as a refuge for mammalian taxa that have become regionally extinct in Eurasia and North America.

The living African mammalian fauna consists of 12 orders representing 52 families, 250 genera and 744 species (Bigalke, 1972), or about 25 percent each of the total world mammalian families and genera. Despite the large numbers, there is a suggestion that the African assemblage is somewhat less diversified than in other areas, especially when continental size is considered. The Tertiary fossil record is derived from fewer than 40 sites and consists of 150 genera and about 250 species. The Quaternary record is more complete. Several hundred species have been uncovered from 25 sites (Figure 10.3).

Table 10.2. summarizes the composition of contemporary African mammals, exclusive of those inhabiting Madagascar and the islands of the west Indian Ocean. All of the taxa are placental, the monotremes and marsupials being absent. Endemism, at least among the Insectivora, Rodentia, and Artiodactyla, is high. At the generic level endemism exceeds 50 percent among the insectivores, primates, hares, rodents, carnivores, elephants and aardvarks, dassies, and odd- and even-toed ungulates. Only the bats, anteaters, and manatees fail to meet this 50-percent criterion.

In general it appears that the African fauna is derived from the therapsids of the Permo-Triassic period, supplemented by immigrant waves from Eurasia. Insectivores and primates were an integral part of the early mammalian fauna. During the Oligocene, there was a large-scale exchange of forms into and out of Africa, at which time Africa acquired many perissodactyls, its basic suid stock, some bovines, and numerous carnivores, mostly from Europe (Cooke, 1972). Diversification was the hallmark of the Pliocene, particularly among the Proboscidea, Bovidae, and Suidae.

Based on fossil evidence from sites shown on Figure 10.3, few faunal changes seem to have taken place during the Pleistocene. There was certainly less extinction than in North America or Eurasia. The overwhelming suggestion is that the fauna arose from Tertiary stock and differentiated during the Plio-Pleistocene. This statement, however, does not rule out the possibility that the eutherian and metatherian progenitors of today's African mammals may have originated in Laurasia and migrated into Africa during the early Tertiary. The current fauna is, by and large, a remnant from the Pleistocene, but not from the Tertiary. The reasons for the persistence of this fauna are related to the stable, tropical wet and dry environment, and the diversity of habitats found there.

The Maghreb sites (1-8) in Figure 10.3 supported a varied mammalian fauna including giraffe, elephant, and hippopotamus, even in historic times. Hippo from as recently as the Holsteinian interglacial episode has also been recovered in southern England. In the Maghreb, these typical mammalian forms probably became extinct at the hand of *Homo sapiens*.

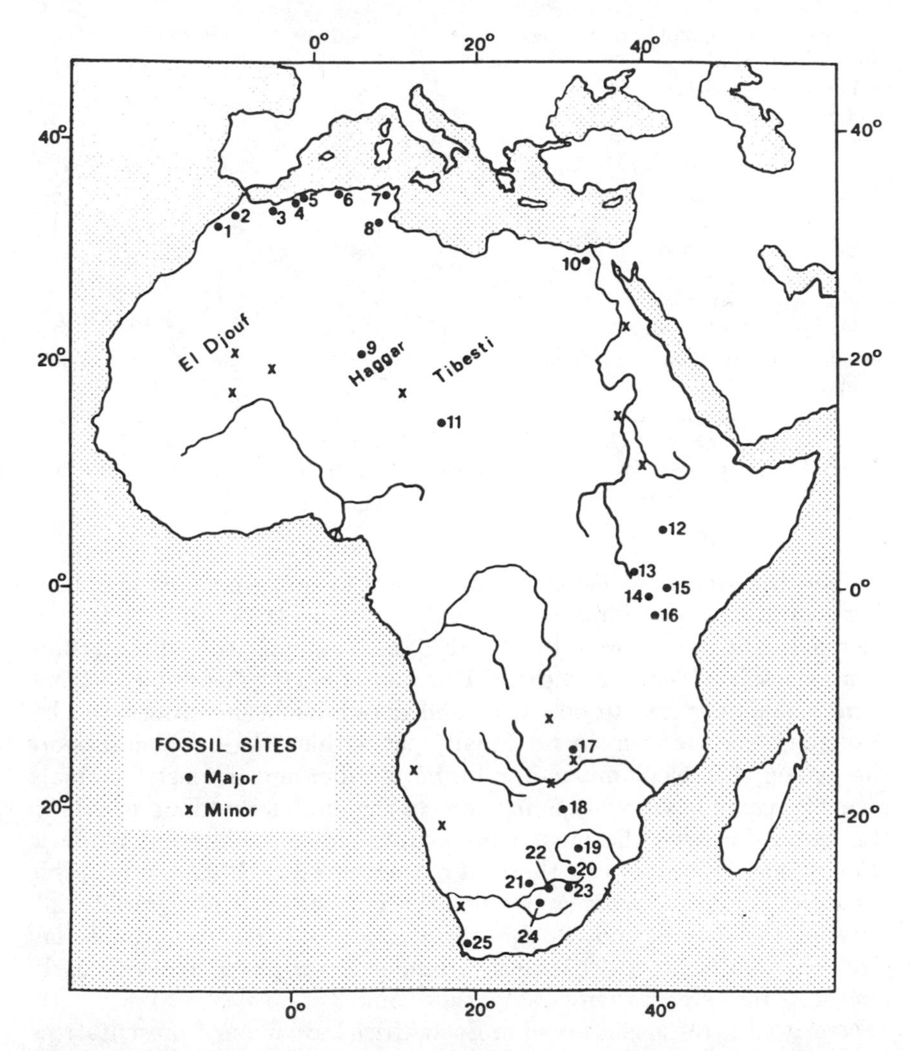

Figure 10.3. Major fossil localities of Africa. (1) Sidi Abderrahman; (2) Fouarat; (3) Lac Karar; (4) Ternifine; (5) Bel Hacel; (6) St. Arnaud; (7) Lac Ichkeul; (8) Ain Brimba; (9) Tihoudaine; (10) Wadi Natrun; (11) Koro Toro; (12) Omo; (13) Kaiso; (14) Kanam, Rawi, Kanjera; (15) Nakuru-Naivasha basin; (16) Laetolil, Olduvai; (17) Broken Hill; (18) Chelmer; (19) Makapansgat; (20) Sterkfontein area; (21) Taung; (22) Vaal River gravels; (23) Cornelia; (24) Florisbad, Vlakkraal; (25) Hopefield.

Table 10.2. Composition of Contemporary African Mammals Exclusive of Madagascar and the Indian Ocean Islands (% = Endemism)

Order	No. of families (%)	No. of genera (%)	No. of species
Insectivors (shrews, moles, hedgehogs)	5 (60%)	24 (79%)	94
Chiroptera (bats)	9 (0)	44 (41)	174
Primates (apes, monkeys)	4 (25)	11 (100)	58
Pholidota (anteaters)	1 (0)	1 (0)	4
Lagomorpha (hares)	1 (0)	4 (75)	10
Rodentia (squirrels, rats, mice, porcupines)	14 (43)	84 (81)	227
Carnivora (cats, jackals weasels, hyenas)	6 (17)	36 (66)	69
Tubulidentata (aardvark)	1 (100)	1 (100)	1
Proboscidea (elephants)	1 (0)	1 (100)	1
Hyracoidea (dassies)	1 (0)	3 (66)	11
Sirenia (manatees)	2 (0)	2 (0)	2
Perissodactyla (zebras, rhinos)	2 (0)	3 (66)	7
Artiodactyla (pigs, hippos, giraffe, antelopes)	5 (40)	36 (88)	86
Totals	52	250	744

Source: R. C. Bigalke, 1972, The Contemporary Mammal Fauna of Africa, in *Evolution, Mammals, and Southern Continents,* A. Keast, et al., eds. State University of New York Press, Albany, N.Y., p.148.

There has, of course, been a long controversy over whether the Sahara is natural or human induced. There is ample evidence of Saharan encroachment southward into the sub-Saharan steppe. In addition, there is much fossil evidence from the El Djouf and the Haggar to suggest that climatic changes had to be recent and may have been accentuated by short-term cyclical phenomena. Finally, paleolithic paintings also support the notion that elephant, rhino, hippo, giraffe, and domestic animals recently roamed over what is now one of the most forbidding deserts in the world. Perhaps the most convincing evidence for recent climatic change, however, has been the find of an australopithecine skull. This discovery suggests that there may have been a recent biogeographic link between the Sahara and Olduvai regions. Put another way, there may have been no ecological barriers to hinder the movement of our early ancestors into areas where, today, they could not have survived.

Nearly all of the fossils found in East Africa have living representatives in the same area. Furthermore, all of the forms recovered seem to have been adapted to savanna and steppe conditions. It is from the area of Bechuanaland and the Kalahari in southern Africa, however, that evidence has emerged for the pre-Pleistocene differentiation of large sectors of the African mammalian fauna. If this evidence is accurate, it might be supposed that these forms arose in the mid-Tertiary and radiated throughout Africa. Eventually, some forms migrated to Eurasia. In this scenario Africa can be viewed as a source rather than a refuge for Eurasian fauna.

Hominid Evolution

Certainly no comment on the African mammalian fauna would be complete without at least brief reference to the origins of *Homo sapiens.* In view of the abundance of evidence for massive differentiation of mammals in Africa, and the abundance of hominid fossil remains, particularly in East Africa, there is good reason to believe that *Homo sapiens* also evolved out of stock originating in this area. It is believed that *Ramapithecus* arose in Miocene or Pliocene time in South Africa and that the hominid line became gradually erect by the early Pleistocene (see also Figure 3.9). At this point our ancestors were dominantly vegetarian but assumed omnivorous habits as the climate became slightly drier.

The phylogeny of the hominid line, beginning with *Ramapithecus,* has recently come under question by Zihlman and Lowenstein (1979). Through molecular anthropology, it now appears improbable that there are many ancestral humans older than six million years. Molecular anthropology is the study of the differences and similarities between living proteins to deduce how long ago they diverged from a common ancestor. Such studies indicate that *Homo's* ancestors, chimpanzees, and gorillas diverged on the order of four to six million years ago. Thus, Homo sapiens is as closely related to these primates as the horse is to the zebra, and scientists should consequently not be looking for hominid lineages older than about six million years. Since *Ramapithecus* is Miocene in age, and since there was a rich cosmopolitan primate fauna at this time, it is possible that this ancient form is ancestral to both modern apes and *Homo sapiens,* but not to *Homo sapiens* alone.

Whatever the precise origins of the hominid stream, it is clear from the fossil record that as long ago as two million years, representatives from *Australopithecus* and *Homo* were present in Africa. *A. robustus* and *A. africanus* were two noninterbreeding allopatric lines, the former being more herbivorous and the latter more carnivorous. It is believed that *A. robustus* had a competitive advantage in areas of denser vegetation and higher rainfall, while *A. africanus* had the edge in open savannas. It is further believed that *A. robustus* and *Homo habilis* were sympatric but had developed substantially different adaptations to the same environment. Thus, a gradual broadening of resource use, as much as predation, may have left *A. robustus* without a viable niche and led to that species' gradual extinction. The subsequent progression of *H. habilis* to *H. erectus* and finally to *H. sapiens* is essentially a story of both continuing adaptations to a changing physical milieu and the opening of new adaptive zones.

THE ATLANTIC ISLANDS

The Atlantic Islands include, among others, the Azores, Canaries, Bermuda, and Cape Verde groups in the North Atlantic, and the Tristan da Cunha, South Georgia, and South Orkney groups in the South Atlantic. The biogeography of most of these islands is fairly straightforward, in the sense that their affinities are to the nearest continental land mass. Of all the islands, those most distant from the nearest land mass are the Azores, Bermuda, and Tristan groups. The biotic affinities of the Azores are to southern Europe and Africa, while those of Bermuda are to the West Indies and southeastern United States. Both Bermuda and the Azores lie along the route of the Gulf Stream, which means that their littoral biotas are very closely tied to littoral forms of the West Indies. In the South Atlantic, the Tristan de Cunha Islands can truly be described as a far, small group in terms of the theory of island biogeography. Of all the Atlantic islands, this group has, in some respects, the most interesting biota.

The islands of the Tristan-Gough group are Miocene to early Pliocene in age. The island of Nightingale is estimated to be 18 million years, while Tristan, itself, is believed to be about 1 million years old. Since continental drift is supposed to have placed the continents in their present position by mid-Tertiary time, it seems unlikely that dispersal distances have ever been much shorter than at present. Moreover, there is no evidence that island stepping stones have ever been present.

Four elements have been identified in the Tristan flora: the southern temperate circumpolar; South American (mainly Fuegian and Falkland); the South African/Indian Ocean; and, the Cosmopolitan. Most of the Fuegian species are coastal or montane types, whereas those from the Falkland Islands occupy mixed habitats. The South African/Indian Ocean element comprises less than 8 percent of the vascular flora, and most of the species are characteristic of lower ground. They may represent a relict element that has not had time to adapt to noncoastal habitats.

Most bryophytes and vascular plants of Tristan and Gough have affinities with South America, even though they lie closer to Africa. All families are widespread. Of the 31 genera, 20 are bihemispheric in distribution and widespread, 10 are confined to the Southern Hemisphere, and one has a pantropical range. For the 33 pteridophytes, 30 have relatives in America and about 15 are known from the sub-Antarctic. Two-thirds of the pteridophytes are also related to African species and in general have much wider ranges. This fact may reflect the greater age of the pteridophyte flora. Their evolution and distribution predates continental drift.

In terms of dispersal, few species are adapted to anemochory or hydrochory. Biochory through birds, both as internal and external carriers, may be common, but the details still need to be worked out in terms of flying distances and feeding habits. All of the evidence suggests that the flora is, in fact, only Quaternary in age. That is, endemism is at the species level only; the flora is notably impoverished and the communities are not well developed; the diaspores of most of the native flora are capable of long-range dispersal; and the affinity with South America is consistent with the directions of ocean currents. What happened to the pre-Quaternary flora is unknown. It may have been completely destroyed by vulcanism.

Biogeographic relations in the Tristan fauna are obscured by high endemism. Of the 132 terrestrial forms found on the island group, 61 are endemic and 43 were probably introduced. When the cosmopolitans are subtracted, only 8 species are left that are useful as indicators of faunistic affinity. Of these 8, 3 are butterflies (Lepidoptera) and 3 are flies (Diptera). One must therefore consider the ancestral affinities of the endemic species to ascertain source regions, and when this consideration is made, an overwhelming link with South America emerges. Of the 37 species whose affinities can be traced outside Tristan, 31 are closest to South American forms. Most of them can be explained as hydrochores, considering the location of the islands with respect to ocean currents. The cosmopolitan nature of the littoral fauna is consistent with the directions of currents and gives testimony that despite isolation, there is a constant influx of immigrant species.

Another large segment of the terrestrial fauna arrived as anemochores from upwind in South America. It is interesting to note that the nonendemic terrestrial forms are small airborne insects. There are, nevertheless, forms like the pseudoscorpion (*Chelanops atlanticus*) that disperse without flying. These forms are hard to explain because the islands have never been connected either to South America or Africa. Holdgate (1960) believes that these immigrants may have arrived by attachment to migrating birds or flies—in other words, that they are biochores.

The faunas of the South Atlantic islands are considered to be young because of the relatively low number of species. It has been calculated, for example, that in the tropical Pacific, 3,722 known endemics could have differentiated from as few as 250 initial immigrant insects in Hawaii. In the South Atlantic groups only 47 endemic insects have arisen from 20 immigrants. It is true that the South Atlantic islands are not tropical, but neither have they ever been glaciated. The only reasonable explanation for the low rate of endemism is that the islands are very young and the biota only poorly developed.

MADAGASCAR AND THE INDIAN ISLANDS

The main island groups of the Indian Ocean west of the Carlsberg and mid-Indian ridges include Madagascar, the Mascarenes (including Mauritius), the Seychelles, and the Aldabra group. In the eastern part, between the above named ridges and the East Indian Ridge, can be found Sri Lanka, the Laccadive and Maldive chains and their southern isolated extension, Diego Garcia.

Madagascar, lying 400 kilometers off the southeast coast of Africa, is by far the most interesting from a biogeographic point of view. It has a unique assortment of plants and animals confined on a land area of some 590,000 square kilometers. It is an ancient land mass that served as a link between Africa and India until the Late Cretaceous. After India began its drift northeastward, island groups such as the Comoros (2,072 square kilometers), the Mascarenes (1,840 square kilometers), and the Seychelles (256 square kilometers) emerged. They are of varying heights, compositions, and ages and consequently display differing affinities for the Africotropical and Oriental realms.

Based on current evidence, there is little doubt that Africa has served as the source region for much of the biotic stock of Madagascar, even though the latter has been separated from Africa since Permian times. The Mozambique Channel has served as a water barrier ever since, and as a sweepstakes route for distributions out of Africa. Over a long period of time, founder populations from Africa have successfully migrated to Madagascar, which helps to explain the similarity of biotic forms between the areas. Nevertheless, the striking differences that are also apparent can only be explained as the failure of many forms to make a successful crossing.

The birds of Madagascar exhibit high endemism at the species level, as do many of the other groups. The birds are mainly of African origin and show a distinct impoverishment over the forms present on the mainland. A small but important element of forms represents the Indo-Oceanic region and points east. These residents do not necessarily indicate that there has ever been a land connection between Madagascar and, say, India. Rather, they suggest a second sweepstakes route of forms originating from the east.

Flora of Madagascar

The history of Madagascar can be divided roughly into three periods: (1) the ancient period through the Middle Cretaceous, when there were links to other southern continents and the basic floral elements became established; (2) the Tertiary period, during which time only the African connection was maintained; and (3) Pliocene to Recent isolation, which fostered endemism and reduced immigrants to a few birds. A very thorough account of the more than 160 vascular families is given by Leroy (1978), who confirms this generalized three-stage outline.

In Madagascar, 100,000 species of angiosperms give the island a very rich flora indeed, for its size. In some of the groups, endemism exceeds 90 percent (Koechlin, 1972). The eastern region, up to about 800 meters in elevation, contains some 500 genera and 5,500 species; the western zone contains only 200 genera and 1,800 species; and some 600 species are common throughout the island.

A pantropical element accounts for more than 40 percent of the families present. About 27 percent of the species have affinities with Africa and they seem to have relatively recent origins. The eastern, Indo-Malaysian element accounts for only about 7 percent of the flora. Affinities with South Africa, South America, and Oceania are even less important. As a whole, the flora is considered to be autochthonous and to have differentiated from ancient Gondwanaland stock. Time and isolation have both stimulated and protected the evolution of its members, so that today the area survives as a huge natural laboratory for tracing the developmental history of angiosperms.

The Fauna

Next to that of Australia, the mammalian fauna of Madagascar is the oldest in the world. Millot (1972) indicates that most of the endemism (which, excluding the civet cat, is 100 percent) is ancient. The tenrecs (family Tenrecidae of the Insectivores) are probably the most ancient of the placentals, and roughly the same statement can be made of the present rodents and carnivores. The Carnivora of Madagascar consist of one family, the Viverridae consisting of three subfamilies and eight genera (Albignac, 1972). The forms appear to be ancient ones with links mainly to Europe. Except for the genus *Viverricula,* which has only a single species (the civet cat—*Viverricula rasse*), all of the remaining genera of carnivores are endemic to Madagascar.

Far more interesting in a global sense is the diversity of lemurs. Although once very widespread, the lemurs are today more or less confined to Madagascar. Four Lemur families exist: Lorisidae, with several African species; Lemuridae; Indridae; and Daubentonidae. Altogether there are 3 families, 12 genera, 19 species, and 28 subspecies occupying Madagascar (Table 10.3). They are of Eocene age and roamed in Europe, North America, and to a lesser extent, in Asia. They suddenly disappeared in the Oligocene, and no fossil records exist until they

Table 10.3. Diversity among the Lemurs of Madagascar

Family	*genera*	*No. of species*	*No. of subspecies*
Lemuridae	Microcebus	2	—
	Cherrogaleus	2	—
	Phaner	1	—
	Allocebus	1	—
	Lepilemur	1	5
	Hapalemur	2	—
	Lemur	4	9
	Varecia	1	2
Indridae	Avahi	1	2
	Propithecus	2	10
	Indri	1	—
Daubentonidae	Daubentonia	1	—

reappear in the Quaternary of Madagascar. One hypothesis is that primitive lemurs rafted to Madagascar from Africa and radiated into current habitats. Unfortunately, no traces have been found of these ancient lemurs in Africa.

The Aldabra Group

The flora of the Aldabra group (Aldabra, Cosmoledo, Astove, and Assumption) consists of 206 species, 40 of which are introduced. The 166 indigenous species can be grouped into 8 affinity groupings: (1) pantropical; (2) Indo-Pacific; (3) African; (4) East African; (5) Madagascan; (6) Mascarene; (7) Seychellian; and (8) endemic. Fifty-five percent are widespread; 45 percent are restricted to East Africa and Madagascar. The widespread species are largely those occupying coastal environments, whereas the more localized species occupy the inland habitats.

Aldabra is unique, or at least significant, in that 31 species (19 percent) are endemic. This figure is exceptionally high for a coral island, and the only immediate explanation is that the island is large and has high habitat diversity. Of the endemic species, 12 are closely related to East African forms, 15 to Madagascan types, and 3 to Mascarene forms; 1 is most closely related to Sri Lanka. It is curious that some Aldabran species originating in Africa, Madagascar, the Seychelles, and the Mascarenes are not widespread throughout the western Indian Ocean. The reason may be that Aldabra is younger than the others and therefore has received only those forms that could survive bare limestone with little soil. Such forms may have been excluded from older islands in the area because of species competition and low habitat diversity. Table 10.4 gives percentages of coastal and inland species as a function of geographic affinities.

Table 10.4. Floral Affinities of Aldabra

Affinity (element)	*No. of species*	*% of total*	*No. coastal (%)*	*No. inland (%)*
Pantropical	36	21.7%	25 (69%)	11 (31%)
Indo-Pacific	39	23.5	26 (67)	13 (33)
Widespread African	14	8.8	5 (36)	9 (64)
East African	15	9.0	2 (13)	13 (87)
Madagascan	24	14.2	5 (21)	19 (79)
Mascarene	2	1.2	0 (0)	2 (100)
Seychellian	5	1.3	0 (0)	5 (100)
Aldabran	31	18.6	1 (3)	30 (97)

SUMMARY

The processes in Africotropical development that distinguish this realm from the Western and Northern Hemispheres relate in part to its Tertiary northward migration, its long period of isolation, changing oceanic circulation patterns, and broad-scale continental rejuvenation. The combination of these factors created a constant stimulus for adaptations from tropical wet to tropical savanna, and finally, to tropical desert environments. Unlike the Nearctic and Neotropical realms, which have migrated essentially westward, the northward migration of Africa means that the vast bulk of the modern day Saharan region once passed under the equator. A line connecting Dakar with Khartoum would roughly delineate that ancestral equatorial zone. This territory is influenced today by the drying effects of atmospheric high-pressure cells. Concomitant changes were promoted in the angiosperm flora as it adjusted, first to mild dryness, then to definite winter drought. Savanna, thorn forest, and shrublands spread over a large real estate. Areas subjected to prolonged summer drought, on the other hand, developed their own distinctive sclerophyllous adaptations.

Alterations in oceanic circulations greatly affected the directions of change initiated by continental migration. In the north, Miocene closing of the Tethys Sea at Iraq terminated the east-to-west flow of ocean current that had enhanced the trend toward aridity. As the gap narrowed, there was tremendous increase in mammalian diversity by virtue of species moving southward through eastern Europe and possibly Gibraltar. These late Oligocene-Miocene migrations can be roughly equated in scope to Pleistocene Nearctic migrations across the Panamanian Bridge, but in the case of Africa, gradual development of the Sahara Desert precluded any continuing interchange, further isolating mammalian stock reaching or residing in southern Africa. The opening of the Straits of Magellan in the south had an opposite effect to the Tethyan experience. Separation of South America and Australia from Antarctica at a time commensurate with glacial developments in the far south provided an opportunity for cold Antarctic water to bathe southwest Africa. Floral adaptations to this increased aridity provide our first glimpse of true desert vegetation in the south, while similar developments in the north owe their stimulus to Pleistocene glacial surges, and perhaps to the impact of humankind.

With the closing of the Tethys Sea, broad-scale continental warping, combined with vulcanism and rifting, brought on a period of topographic rejuvenation. The East African landscape, familiar to all of us, was initiated in the Miocene. It sparked the evolution of new habitats and in turn the diversification of species. Origins of the Anthropoidea can be traced to this time and place, but it is only in the early Pleistocene that one can discern the early forms of *Homo*. This later expansion of forms coincides with speciation induced by the mild cooling that attended oscillations of forest and savanna similar to those hypothesized for Amazonia.

Finally, it should be noted that the Africotropical Realm, though virtually isolated for some 60 million years, still reveals a distinct tie through Madagascar to the Oriental Realm. India was a Noah's Ark from Africa that carried many members of the ancient stock eastward.

REFERENCES

Albignac, R., 1972, The Carnivore of Madagascar, In *Biogeography and Ecology in Madagascar,* R. Battistini and G.R. Vindard, eds., W. Junk, The Hague, pp. 667-682.

Axelrod, D. I., and P. H. Raven, 1978, Late Cretaceous and Tertiary Vegetation History of Africa, in *Biogeography and Ecology of Southern Africa,* vol. 1, M. J. A. Werger, ed., W. Junk, The Hague, pp. 77-130.

Bigalke, R. C., 1972, The Contemporary Mammal Fauna of Africa, in *Evolution, Mammals, and Southern Continents,* A. Keast, et al., eds., State University of New York Press, Albany, N. Y., pp.141-194.

Brenan, J. P. M., 1978, Some Aspects of the Phytogeography of Tropical Africa, *Missouri Bot. Gard. Ann.* **65**:479-534.

Coe, M. J., 1967, Biogeography and the Equatorial Mountains, *Palaeoecol. Afr.* **2**:68-73.

Cooke, H. B. S., 1972, The Fossil Mammal Fauna of Africa, in *Evolution, Mammals, and Southern Continents,* A. Keast, et al., eds., State University of New York Press, Albany, N. Y., pp. 89-140.

Darlington, P. J., 1957, *Zoogeography: The Geographical Distribution of Animals,* Wiley, New York, 675p.

Goldblatt, P., 1978, An Analysis of the Flora of Southern Africa: Its Characteristics, Relationships, and Origins, *Missouri Bot. Gard. Ann.* **65**:369-436.

Holdgate, M. W., 1960, The Fauna of the Mid Atlantic Islands, *R. Soc. London, Proc.* **B152**:550-567.

Killick, D. J. B., 1978, The Afro-Alpine Region, in *Biogeography and Ecology of Southern Africa,* vol. 1, M. J. A. Werger, ed., W. Junk, The Hague, pp. 515-560.

Koechlin, J., 1972, Flora and Vegetation of Madagascar, in *Biogeography and Ecology in Madagascar,* R. Battistini and G. R. Vindard, eds., W. Junk, The Hague, pp. 145-190.

Leroy, J. F., 1978, Composition, Origin, and Affinities of the Madagascan Vascular Flora, *Missouri Bot. Gard. Ann.* **65**:535-589.

Millot, J., 1972, In Conclusion, in *Biogeography and Ecology in Madagascar,* R. Battistini and G.R. Vindard, eds., W. Junk, The Hague, pp. 741-756.

Quézel, P., 1978, Analysis of the Flora of Mediterranean and Saharan Africa, *Missouri Bot. Gard. Ann.* **65**:479-534.

Raven, P. H., and D. I. Axelrod, 1974, Angiosperm Biogeography and Past Continental Movements, *Missouri Bot. Gard. Ann.* **61**:539-673.

Richards, P.W., 1973, Africa, the "Odd Man Out", in *Tropical Forest Ecosystems in Africa and South America: A Comparative Review,* B. J. Meggers, E. S. Ayensu, and W. D. Duckworth, eds., Smithsonian Institution Press, Washington, pp. 21-26.

Werger, M. J. A., 1978, Biogeographical Division of Southern Africa, in *Biogeography and Ecology of Southern Africa,* vol. 1, M. J. A. Werger, ed., W. Junk, The Hague, pp. 145-170.

Zihlman, A. L., and J. M. Lowenstein, 1979, False Start of the Human Parade, *Natural History* **88**(7):86-91.

FURTHER READING

Bakker, E. M. van, Sr., 1978, Quaternary Vegetation Changes in Southern Africa, in *Biogeography and Ecology of Southern Africa,* vol. 1, M. J. A. Werger, ed., W. Junk, The Hague, pp. 131-143.

Balsac, H. H. de, 1972, Insectivores, in *Biogeography and Ecology in Madagascar,* R. Battistini and G.R. Vindard, eds., W. Junk, The Hague, pp. 629-660.

Boaz, N. T., 1979, Early Hominid Population Densities: New Estimates, *Science* **206**:592-595.

Brown, W. L., Jr., 1958, Some Zoological Concepts Applied to Problems in Evolution of the Hominid Lineage, *American Scientist* **46**(2):151-158.

Cartmill, A., 1967, The Early Pleistocene Mammalian Microfaunas of Sub-Saharan Africa and Their Ecological Significance, *Quaternaria* **9**:169-198.

Clark, W. E. Le G., 1959, The Crucial Evidence for Human Evolution, *American Scientist* **47**(3):299-313.

Clark, W. E. Le G., and L. S. B. Leakey, 1951, *The Miocene Hominoidea of East Africa,* London: British Museum of Natural History, 117p.

Dixson, A. F., 1981, *The Natural History of the Gorilla,* Columbia University Press, New York, 202p.

Dobzhansky, Th., 1961, Man and Natural Selection, *American Scientist* **49**(3):285-299.

Douglas-Hamilton, O., 1980, Africa's Elephants: Can they Survive? *National Geographic* **158**(5):568-603.

Johanson, D. C., and M. Edey, 1981, *Lucy: The Beginnings of Humankind,* Simon and Schuster, New York, 409p.

Johanson, D. C., and T. D. White, 1979, A Systematic Assessment of Early African Hominids, *Science* **203**:321-330.

King, M. C., and A. C. Wilson, 1975, Evolution at Two Levels in Humans and Chimpanzees, *Science* **188**:107-116.

Klein, R. G., 1977, The Ecology of Early Man in Southern Africa, *Science* **197**:115-126.

McHenry, H. M., 1975, Fossils and the Mosaic Nature of Human Evolution, *Science* **190**:425-431.

McIntosh, S. K., and R. J. McIntosh, 1981, West African Prehistory, *American Scientist* **69**(6):602-613.

Miller, D. A., 1977, Evolution of Primate Chromosomes, *Science* **198**:1116-1124.

Moreau, R. E., 1963, Vicissitudes of the African Biomes in the Late Pleistocene, *Zool. Soc. London Proc.* **141**(2):395-421.

Moreau, R. E., 1966, *The Bird Fauna of Africa and Its Islands,* Academic Press, New York, 424p.

Peake, J. F., 1971, The Evolution of Terrestrial Faunas in the Western Indian Ocean, *R. Soc. London Philos. Trans.* **B260**:581-610.

Petter, J. J., 1972, Order Primates: Sub-Order of Lemurs, in *Biogeography and Ecology in Madagascar,* R. Battistini and G.R. Vindard, eds., W. Junk, The Hague, pp. 683-702.

Renvoize, S.A., 1971, The Origin and Distribution of the Flora of Aldabra, *R. Soc. London Philos. Trans.* **B260**:227-336.

Tattersall, I., 1982, *The Primates of Madagascar,* Columbia University Press, New York, 382p.

Terrell, J., 1977, Biology, Biogeography and Man, *World Archeology* **8**(3):237-248.

Wace, N. M., 1961, The Vegetation of Gough Island, *Ecol. Monogr.* **31**:337-367.

Wace, N. M., and J. H. Dickson, 1965, The Terrestrial Biota of the Tristan da Cunha Island, *R. Soc. London Philos. Trans.* **B249**:273-360.

Watanabe, H., 1972, Periglacial Ecology and the Emergence of Homo sapiens, in *The Origin of Homo sapiens,* F. Bordes, ed., Presses Universitaires de France, Vendome, France, pp. 271-283.

Chapter 11

The Oriental Realm

The Oriental Realm has a land-sea area about the size of North America, but a land area approximately the size of the conterminous United States. This landed part is some 8,000,000 square kilometers, consisting of the subcontinent of India, the Himalaya, Burma, all of Indochina and Malaysia, Indonesia, the Philippines, and the Pacific islands. India is the largest component, accounting for about 40 percent of the land. Only two-thirds of the realm, that part from Lahore to Singapore, has direct land connections to the Palearctic Realm. The remaining one-third of the realm consists of island archipelagos traversing a longitudinal range from 95°E to approximately 130°E. In a geologic sense the realm grows progressively younger from west to east. India is an ancient body of land recently united with Burma and Indochina. Toward the east, the island archipelagos are mostly early to mid-Tertiary or Recent in age. The Himalayan environment is the most recent addition to the range of environments occupying the realm.

THE WESTERN ORIENTAL REALM

The Indian Subcontinent

The Indian peninsula, which in the broadest geographical sense includes the Seychelles, Laccadives, Maldives, and Sri Lanka, is an ancient topographic surface similar in history to Madagascar and Australia, except that the land mass has rafted under the equatorial zone into the Northern Hemisphere. Figure 11.1 shows the hypothetical position of this land mass prior to its contact with Asia in the Assam-Burma area. Except for the rather recent lava flows that form the Deccan Plateau, India experienced long-term stability throughout its journey northeastward from Gondwanaland. However, the fauna and flora have undergone complex changes since their contact with the Asian Plate and the consequent lifting of the Himalayan Mountains. These changes have culminated in the past 200 years with wholesale human destruction of entire ecosystems.

In general, the ecology of the subcontinent is most strongly influenced by (1) ancient topography; (2) the physical proximity of the Himalayas;

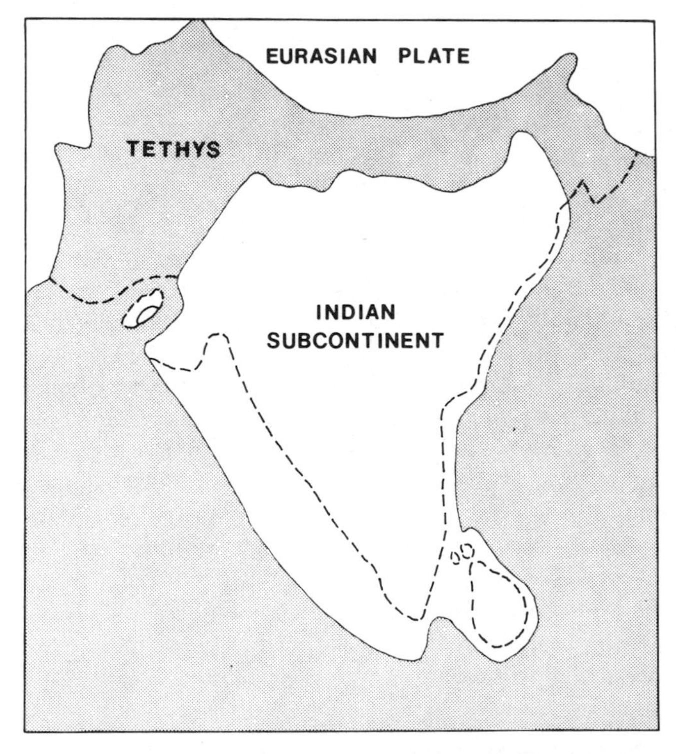

Figure 11.1. Inferred position of the Indian subcontinent prior to its contact with Asia. *(From M. S. Mani, 1974, Biogeographical Evolution in India, in Ecology and Biogeography in India, M. S. Mani, ed., W. Junk, The Hague, p. 706; copyright © 1974 by W. Junk. Reprinted with permission of the publisher.)*

(3) the monsoon; and (4) human occupance. Modern distributional patterns can be almost wholly explained as the result of intense human disruption. Nevertheless, there are sufficient refugia to piece together crude historical outlines for the development of floral and faunal assemblages.

India south of the Ganges Plain is distinct biogeographically from the plain itself, from the Punjab and region of the Indus tributaries, and from Assam. South of the Ganges Plain, faunistic affinities, for example, are closer to Madagascar than to the Oriental or Malaysian regions. The subcontinent may be considered to consist of (1) the peninsula *sensu strictu;* (2) the eastern and western borderlands; and (3) the Ganges Plain. Of these various parts, only the peninsula, with its insular appendages, has a distinct ancient floral and faunal component derived from Gondwanaland. Even this component must have been altered by the passage of time as the subcontinent rafted under various climatic regimes. The eastern areas contain mixtures of Oriental and Malaysian forms; the western region contains admixtures of Mediterranean and Africotropical forms; and the higher Himalaya contains a new and rapidly evolving postglacial Turkmenian and central Asian component. The general outlines of these relationships are given in Figure 11.2, modified from Mani's (1974) "Biogeographical Evolution in India."

As ancient India rafted into its present position in south Asia, the Tethys Sea narrowed and was gradually erased by sediments pouring out of the rising Himalaya. In time, these sediments established land routes for the migration of younger biotic elements extending their ranges from east, west, and north. They penetrated and commingled with the older biotic elements of the peninsula. In these Pliocene days, both the peninsular and eastern floras were humid tropical types, but the eastern assemblage was, and is, phylogenetically younger and taxonomically higher. It is derived almost exclusively from Asia and has spread westward along the forested zones on the southern flank of the Himalaya.

The fauna of the subcontinent, like the flora, consists of an autochthonous older assemblage derived from Gondwanaland, and a more recent assemblage. This second group consists largely of Mio-Pliocene migrants from eastern humid tropical forests located in China and Southeast Asia, but there are also large numbers of species from Eurosiberia, Turkmenia, and the Mediterranean-Ethiopian regions. Figure 11.3 mirrors, in many ways, the regions delineated in Figure 11.2, but shows the importance of Tertiary mountain centers in contributing to the fauna of the western Oriental Realm. The region as a whole can be regarded as a vast insular mass separated by lowland valleys.

Due to the destruction of forests by people, the present range of most species is restricted to small, isolated refugia. In effect, the existing centers are islands of favorable conditions to which the fauna has retreated.

Basically, the pattern shows the impoverished remnants of a vanished fauna. "Not only have the Pliocene and nearly all the Pleistocene vertebrates completely disappeared from the Indian fauna within the past ten or eight thousand years, but the greatest majority of those that differentiated in recent times, and were abundant until the middle of the last century, are either rapidly disappearing or have already vanished" (Mukherjee, 1974, 331).

In summary, the fauna of India is comprised of ancient endemics and relict fauna on the peninsula; a young and highly differentiated humid-tropical-forest fauna encroaching from the east; and a youthful, ecologically specialized, endemic, autochthonous fauna in the higher Himalaya. As the peninsula migrated northeastward, the ancient Tethys Sea was obliterated, becoming gradually filled with the sediments from the

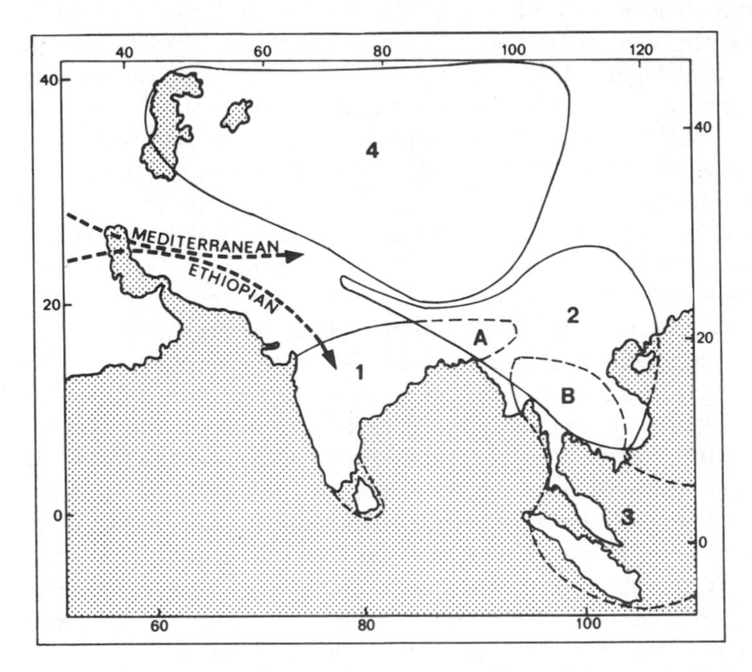

Figure 11.2. Major centers of floral and faunal differentiation: (1) peninsular India; (2) Indo-Chinese center with a westward extending tongue in the forest zone of the southern Himalaya; (3) Malasian Center; (4) Turkmenian Center. Immigrant floras and faunas have also penetrated the region from both the Mediterranean and Ethiopian areas. *(From M. S. Mani, 1974, Biogeographical Evolution in India, in Ecology and Biogeography in India, M. S. Mani, ed., W. Junk, The Hague, p. 701; copyright © 1974 by W. Junk. Reprinted with permission of the publisher.)*

Himalaya, and forming what is now the Ganges Plain. As this so-called Assam Gate formed, life forms infiltrated from the younger Oriental and Malaysian areas. Later episodes of glacial and interglacial fluctuations permitted southward and northward migrations of Turkmenian and Indian life forms across this Gangetic transition. Figure 11.4 illustrates the main routes of immigration of the mammalian faunal assemblages.

Since the strong monsoon climate has originated rather recently (perhaps in the post-Pleistocene), it is not considered to be an important factor in the evolution of the present-day biotic character of India. The modern

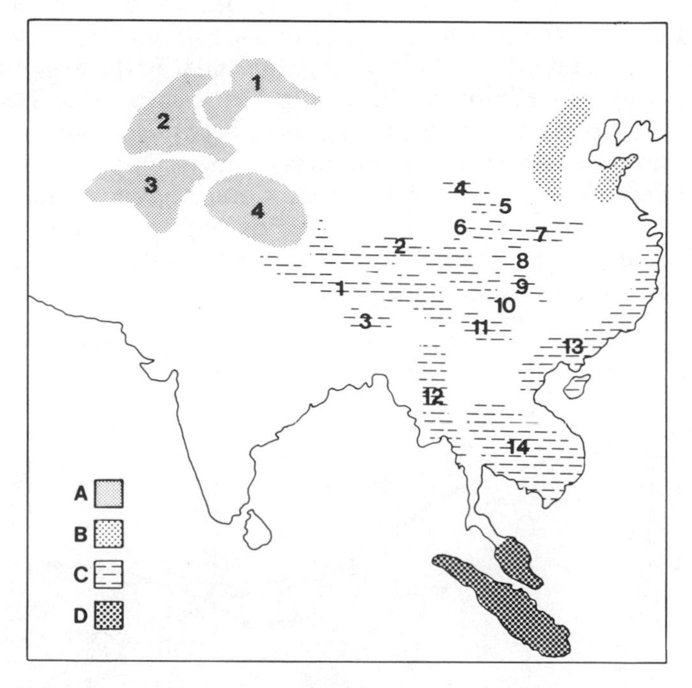

Figure 11.3. Centers of origin for the Oriental faunas of the Tertiary mountains. *(A)* Turkmenia; *(B)* Manchuria; *(C)* Indo-China; *(D)* Malaya. *(A)* and *(B)* belong to the Palearctic Realm; *(C)* and *(D)* to the Oriental Realm. The subcenters in *(A)* are (1) Ala-Tau-Tien-Shan Mountains; (2) South Turkestan; (3) Afghanistan; and (4) Northwest Himalaya. The subcenters in *(C)* are (1) the forest covered Himalaya to Assam; (2) Tibet; (3) Khasi-Jaintia Hills of Assam; (4-11) Eastern Tibet and Yunnan; (12) Burma; (13) Indochina; (14) Thailand. (*From M. S. Mani, 1974, Biogeography of the Himalaya, in Ecology and Biogeography in India, M.S. Mani, ed., W. Junk, The Hague, p. 680; copyright © 1974 by W. Junk. Reprinted with permission of the publisher.*)

pattern is very much a reflection of intense human impact on all natural habitats. Extensive areas of former humid tropical forests have been transformed into semiarid deciduous forest, scrublands, and savanna. One might speculate that human impact on the Indian land mass has virtually erased the tropical humid character that the vegetation acquired through passage under the equator, imparting a physiognomy reminiscent of Madagascar and Australia.

The Himalaya

The region described as the Himalaya includes the area between the Ganges Plain on the south and the drainage divide separating the Himalayan Mountains from the Tibetan Highlands. To the west, the area is bounded by the Hindu Kush, and to the east by the Szechwan Highlands. There is a sharp climatic gradient between the far northwest and far southeast along the mountain system, such that within the region, one can recognize three subcenters: the northwest, dominated by *Artemisia, Eurotia,* and *Kochia* shrubs; the higher elevations of the eastern Himalaya; and, the forest zone of lower elevations connected to the northern Burmese mountains.

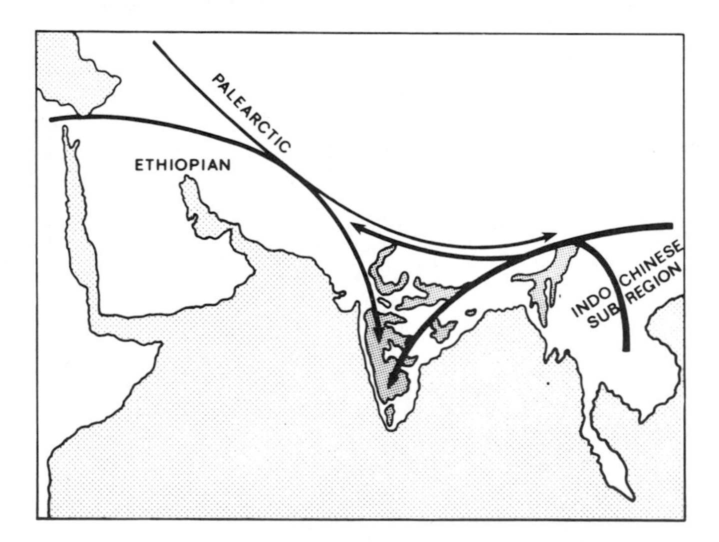

Figure 11.4. Mainstreams of faunal immigration into India. *(From G. U. Kurup, 1974, Mammals of Assam and the Mammal Geography of India, in Ecology and Biogeography in India, M.S. Mani, ed., W. Junk, The Hague, p. 611; copyright © 1974 by W. Junk. Reprinted by permission of the publisher.)*

Compared to the peninsula, the Himalaya is rich in relatively young and phylogenetically plastic forms of more recent and highly evolved Asiatic groups, but it is poor in Gondwanaland elements. The autochthonous and endemic elements are concentrated in the higher elevations above the forest line. Thus, in addition to a longitudinal segregation of assemblages, there are two biotic assemblages that can be differentiated on the basis of elevation.

It appears from the botanical pattern that the present-day Himalayan flora is related in its eastern sector to western and northwestern China, and on the western side to western, middle, and northern Asia. Basically, the mountain system has served as a route for emigration and colonization from the east and northwest. As uplift forced species to adapt to colder and drier conditions, consequent endemism occurred. The general area of 83° east longitude at an elevation of 3,500 meters must be one of the world's most interesting biogeographic pivotal points. The west Himalayan botanical province extends to about 83° and is defined by the eastern limit of *Cedrus deodara.* At this longitude, the elevational transition at 3,500 meters is from the intrusive elements of *Betula/Juniperus* forest to the highly endemic elements of the alpine zone.

The forest line and the snow line are two very important biological and biogeographic transition zones. The forested zones up to about 1,000 meters are covered by broadleaf wet forest with species derived primarily from the east. Sclerophyllous forests of *Quercus* and *Rhododendron* extend up to 3,000 meters on the west. Above these elevations, and reaching as high as 3,500 meters, is the alpine zone. As elevation increases, there is a corresponding increase in endemism. At the snow line, environmental conditions are such that life forms are assembled into essentially pioneer communities comprised of species still adapting to their extreme environments.

The forest fauna consists of (1) derivatives of the Oriental fauna that have attenuated westward; (2) Palearctic forms from the Mediterranean and Turkmenian areas; and, (3) Ethiopian elements. The biogeographical transition between eastward- and westward-spreading forms occurs at the defile of the River Sutlej, east of Lahore, Pakistan. Compared to the western sector, the eastern forest fauna is particularly rich in Lepidoptera of Indochinese and Malaysian origin and in endemic fish, almost 40 species of which do not extend westward beyond Nepal.

The alpine fauna consists of cold-adapted Palearctic forms, the bulk of which are endemics that arose passively and *in situ* with the Himalayan uplift. There is a general poverty of fish, amphibians, and reptiles, but birds and mammals are fairly abundant. Among the invertebrates, endemism increases westward, particularly in isolated glacial ponds, lakes, and streams. With increasing elevation, the communities become sparser and more pioneer. The Himalayan elements are characterized by the youngest and ecologically most specialized life forms. There is a remarkable absence of freshwater fish, amphibians, and reptiles, all of which are phylogenetically older forms. Birds and mammals are most common, as are the Arthropoda. Niches are highly unsaturated and ecologically unstable, as is typical of pioneer communities. A high degree of isolation and speciation exists in nearly all groups of animals, especially the insects. Differentiation is partly a result of adaptation to rapid topographic uplift into colder and drier climatic zones, and virtually all forms are related to the Turkmenian subregion of the Palearctic Realm.

Assam-Burma

Assam, Bhutan, and the northern portion of Burma are dominated by Tertiary mountains over which there has been extensive blending of Asiatic and Indian peninsular floras. The gradual narrowing of the Tethys Sea, leading to formation of the Assam Gate, permitted two-way passage of plants and animals between the eastern Ghats of India, the Khasi and Naga Hills of Assam, and the then juvenile Himalayas. The phytogeography of Assam-Burma shows the combined influences of a wide range of altitudes, the meridional orientation of the mountains (except the Khasi Hills), and the recent geologic past. The diversity of slopes, aspects, latitudinal and altitudinal ranges, microclimates, and soil and drainage conditions have all stimulated a high rate of endemism. There are an estimated 1,100 species of dicots listed as endemic to Burma alone. These conditions are not unlike those of California, except that the area is tropical. During the Tertiary, a common flora must have covered the whole of east Asia including the Himalayan countries, China, and Japan. With uplift, the flora became separated and highly modified during glacial and interglacial episodes. Judging from the diversity of elements currently comprising the forested zones of eastern Himalaya, large-scale migrations and exchanges of elements must have occurred. The two main migration avenues were from China on the east and from Malaysia in the south. The Chinese element consisted predominantly of temperate and alpine plants occupying higher elevations.

The modern flora of Assam is largely a relict of the Oligocene flora, which is similar to the fossils found in the London Clay of Europe. The assemblage probably occupied all paleotropical lands. Vacciniaceae and Ericaceae are well represented by about 70 species (64 endemic). Among the Ericaceae, *Rhododendron* is the most prominent, with almost 90-percent endemism. There are 34 families of monocotyledons in Assam, the largest of which is Orchidaceae. In the Khasi Hills, alone, there are 75 genera and 265 species—rivalling the diversity of forms found in the Philippines.

Gymnosperms are also well represented in Assam and Burma, especially by the conifers. Among the dominants are *Abies, Cephalotaxus, Cupressus, Juniperus, Larix, Picea, Pinus, Podocarpus, Taxus,* and *Tsuga. Pinus insularis* is the characteristic pine of the Khasi, Naga, and Lushai Hills in north Burma, and this species extends to northern Luzon in the Philippines. If species diversity accounts for anything, then by all odds the paleotropics of Southeast Asia, Malaysia, and Indochina must be close to the center of origin of the angiosperms.

THE EASTERN ORIENTAL REALM

Biogeographic thinking regarding the eastern reaches of the Oriental Realm is dominated by observations relating to the zone of contact with the Australian Plate. This insular region is known as *Wallacia* and has long been regarded as a sweepstakes route that effectively separated the mixture of these two worlds. Wallacia includes the Lesser Sunda Islands from Lombok, on the west, to Timor, then runs northward through Celebes and the Sula Islands to Mindinao, the Visayan group, Mindoro, and Luzon. *Sundaland,* on the west, consists of Palawan, Borneo, Java, and Bali; while *Papualand,* on the east, contains Halmahera, Buru, Ceram, New Guinea, and Australia. Figure 11.5 shows the locations of these areas and the placement of the biogeographic lines that separate them.

Wallace's Line was drawn between two distinct faunal realms—the Oriental and the Australian. Weber's Line is a geologic boundary, but one that has distinct zoogeographical significance—especially among the affinities of freshwater fish. Each of these lines represents a barrier across which only the more dispersive forms have migrated. Biologically, Weber's Line is more recent and pronounced, perhaps because of the recency of contact between the Australian and Asian plates.

Wallacia, as defined by Dickerson et al. (1928), is regarded as a meeting ground of three migration routes: a poorly understood pre- to early Tertiary trail from Taiwan and the Eurasian mainland; a primary, and more or less continuous path through Borneo and Palawan; and finally, an effective, but secondary and rather recent road from the Australian region through Celebes.

The climate of Wallacia is uniform throughout its latitudinal range (10°S to 18°N). Aside from the insular nature of the region, elevations seem to be the most profound barrier to biotic distribution. They range from sea level to several thousand meters. In the Philippine Archipelago alone there are over 7,000 islands, only a third of which have been named and fewer than 500 of which have areas larger than 2.5 square kilometers.

Luzon (ca. 106,600 square kilometers) and Mindinao (ca. 104,000 square kilometers) comprise more than half the total land area of the group.

The distribution of flora and fauna in Wallacia indicates that biotic elements are composed of relict species and their descendants, with

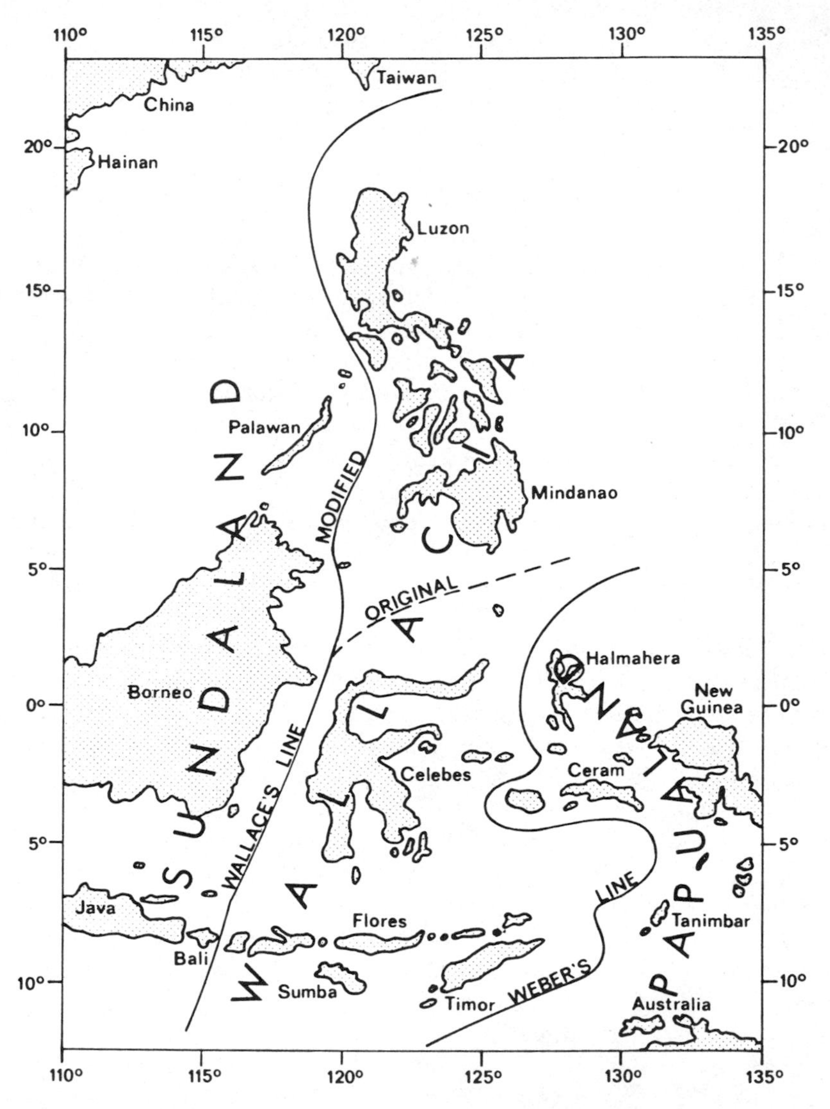

Figure 11.5. Wallacia, the unstable area between Asia and Australia. *(From R. E. Dickerson et al., 1928, Distribution of Life in the Philippines, The Bureau of Science, Manila, Monograph 21, Bureau of Printing, Manila, p. 27.)*

infiltrations from both Sundaland and Papualand. Within Wallacia, Asiatic types decrease as one proceeds from northwest to southeast. The Australian types decrease in number and importance in the reverse direction.

Wallacian Flora

There has been no general synthesis of the floristic history of Wallacia since Dickerson and his colleagues wrote in 1928. Although outdated in detail, the broad lines of Dickerson's evidence and reasoning are still valid, except that modern concepts of continental drift have altered some of the conclusions. Within Wallacia, the Philippine Islands are the area about which most is known. The flora of this archipelago contains more than 10,000 species of flowering plants and ferns. They can be subdivided into two groups: the introduced and the native. The native flora can in turn be divided according to its basic elements, and here one can appreciate the complicated aspects of its historical development.

The introduced flora now constitutes a high percentage of the total species. Prehistoric human immigration, the period of early European exploration (fifteenth to eighteenth centuries), and twentieth century practices of horticulture and mass tourism have all impacted on the native vegetation. In the hinterland around Manila, for example, there are about 1,000 species, of which 550 are indigenous, and about 450 (45 percent) are introduced. Of the introduced species, 180 (40 percent) were introduced from tropical America by the Spaniards, beginning in 1521. While there are no reliable records, it can safely be said that the rest were introduced in a similar manner over a period of some 1,500 years since the beginning of trade in the archipelago. In areas cleared of their original forest, endemism may go as low as 10 percent, whereas in the virgin forests, the rate is 80 percent. It is, indeed, the virgin forests and low-lying coastal vegetation that remain more or less free from human impact.

Since most of the coastal vegetation of the Philippines has pantropical or paleotropical affinities, the few remaining virgin forests of the middle and higher elevations provide the best clues of biotic history. During Cretaceous and early Tertiary time, Taiwan and the northern two-thirds of Luzon must have been a single land area. At that time a rather uniform temperate flora, spreading outward from the Chinese mainland, entered Luzon via Taiwan (Figure 11.6). This flora, dominated by *Pinus insularis* in Luzon and *P. mercusii* in Mindoro has persisted above the 1,500-meter contour. Other dominants include oak, rhododendron, and other temperate types. Undoubtedly, many of the members of that early invasion have subsequently become extinct. Today over 70% of the species descendent from this element are endemic in northern Luzon, while the remainder are more widely distributed over Taiwan, Japan, and the mainland. Confined, as they are to higher elevations, these pine forests have undergone prolonged development in isolation. Pleistocene glacial episodes, which led to oscillating temperature regimes and sea-level adjustments, seem not to have affected this upland flora.

Indeed, the alliance between the Philippines and Taiwan is remarkable

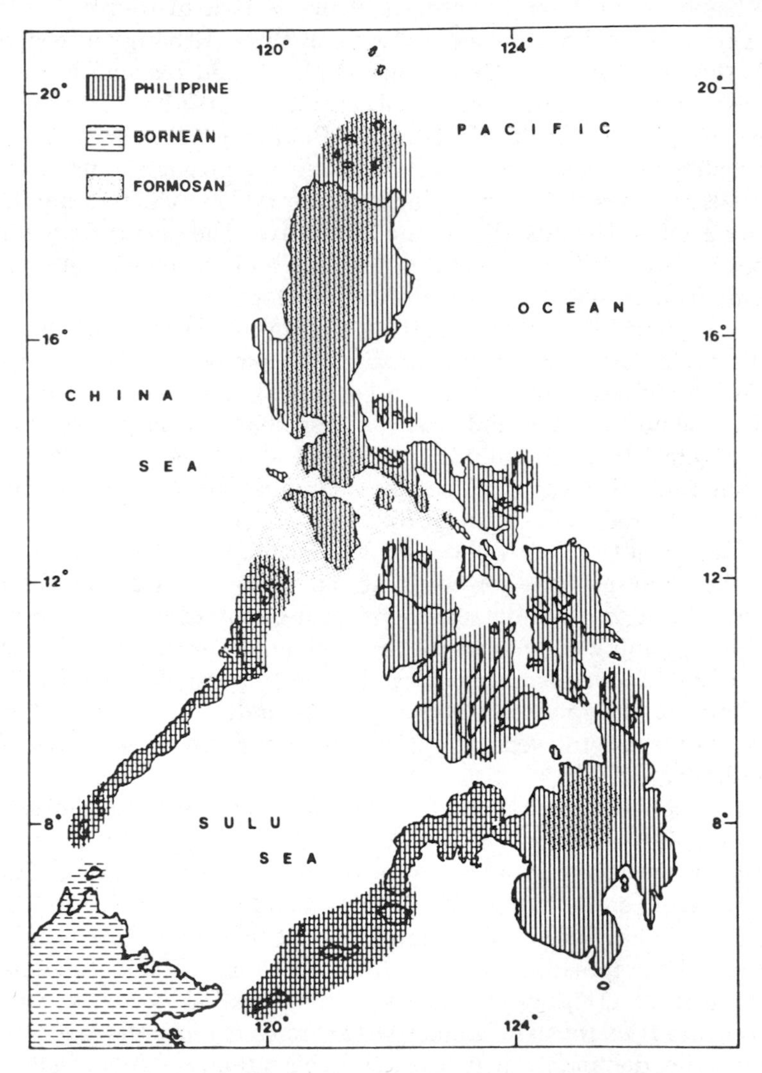

Figure 11.6. Philippine floral subprovinces. *(From R. E. Dickerson et al., 1928, Distribution of Life in the Philippines, The Bureau of Science, Manila, Monograph 21, Bureau of Printing, Manila, p. 290.)*

in its meagerness. Considering the proximity of northern Luzon to southern Taiwan, and their similar climatic environments, it is curious that there is not a single genus common to both islands. Even at the species level, at last report there were only 38 species known to be restricted to the area, and most of them have been disseminated either by birds or by wind. These facts point to the strength of the Bashi Channel as a dispersal barrier.

The seashore and low-altitude forests are predominantly Indo-Malaysian in character. Between 2,500 and 3,000 tree species are known from the primary dipterocarp forest. Prior to human occupancy, these forests covered more than 75 percent of the Philippine land area between the 600- and 1,000-meter contours. The family Dipterocarpaceae, which ranges from Sri Lanka to New Guinea and northward through Southeast Asia, is represented in the Philippines by some 50 species and 9 genera. There are only a few representatives in tropical Africa and Madagascar, and one in the Seychelles. The family is not native to tropical America or Australia. As presently known, Dipterocarpaceae has 16 genera and 375 species. Borneo supports 13 genera and over 100 species, while the Malay Peninsula has numbers roughly equal to those of the Philippines. As one progresses eastward into Celebes, the Moluccas and New Guinea, the representation dwindles rapidly.

In contrast to the Malaysian focus of the dipterocarps, the family Myrtaceae of the Philippines has a striking alliance with Papualand. This family includes such well-known genera as *Eucalyptus, Leptospermum, Tristania,* and *Melaleuca.* The family is worldwide in distribution and equally rich in both genera and species in Sundaland as in Papualand, although the genera just listed are of Australian origin. The widespread distribution of the family, with its peculiar overprint of Australian types confined to the eastern portion, points to its early development in Pangaea, followed by prolonged isolation of segments that later reunited and spread from Australia. The dipterocarps, on the other hand, must have differentiated after the breakup of both Pangaea and Gondwanaland (possibly originating in Borneo), so that the Australian region failed to acquire an early component for development in isolation. Celebes, which forms the base of Wallacia, contains floral representatives from both Sundaland and Papualand. Dickerson et al. (1928), report some 63 species common to Celebes and Sundaland, and about 45 common to Celebes and Papualand.

There is good evidence that vegetation in the Malay Archipelago has remained fairly constant for the past several thousand years. The dipterocarp forests have spread slowly via Palawan into the low and middle elevations throughout Wallacia. Palawan and the Clamian Islands are Bornean in both flora and fauna. Mindinao, farther east, is decidedly less Bornean and shows higher rates of endemism—a fact that suggests a somewhat longer separation from Borneo than is true for Palawan.

Wallacian Fauna

In broad terms the fauna of Wallacia shows distributional patterns similar to those found in the flora. This basic pattern is shown in Figure 11.7. The mammalian fauna is dominantly western Malaysian in character. The latest invasion came from Borneo when Palawan and the Sulu Archipelago were connected. A second avenue of migration apparently existed on the east coast of the Philippines from Celebes through Mindinao to northern Luzon along the mountain chain. A small element from Australia may have arrived by rafting. All of the mammals are small—shrews, lemurs, bats, rats, squirrels, monkeys, tarsiers, pigs, small deer, cats, and civets being the most numerous. The largest native mammal is the tamarou (*Anoa mindorensis*), a water buffalo of Mindoro, which probably arrived via Palawan during the Pleistocene.

The avian fauna of the Philippines consists of more than 750 species from 293 genera. Among the native land birds, the Indo-Malaysian element is represented by some 70 genera, while the Austro-Malaysian element has only about 20 genera. A few genera are shared with the Oriental Realm in the north, and about 25 genera are endemic. Specific endemism is very high among all the genera, and most species have restricted ranges within the archipelago. Luzon has the highest rate of endemism.

In summary, it may be stated that the biogeographical history of the Philippines begins in the early Tertiary. The lizard and insect (Lepidoptera) fauna, together with the upland flora of Mountain Province in Luzon, suggest an early direct connection with Taiwan and the Asian continent. During Miocene times, the present eastern Philippines coast may have been a continuous, elongated island connecting Luzon, the Visayans, and Mindinao with Celebes. Much of the mammalian fauna and some of the tropical Australian flora appears to have dispersed along this bridge. Since the forms are not coastal forms, it is unlikely that they arrived by rafting. A contemporary bridge through the Sulu Archipelago permitted the arrival of the dipterocarp flora and additional mammalian forms from Borneo.

This Miocene pattern persisted during at least part of the Pliocene. Mindinao probably consisted of several islands. The Palawan Bridge probably existed, but little is known of that chain. Throughout Mio-Pliocene times, the movement of life forms along island stepping stones and continuous elongated bridges established the basic patterns observed today. Complications and disruptions of the pattern have occurred from the Pleistocene to the present through glacial changes in sea level, vulcanism, and other tectonic adjustments. These adjustments stimulated speciation and further differentiation of types through forced isolation and have resulted in locally high rates of endemism in the flora. Most angiosperm genera contain a few widely distributed species, many of

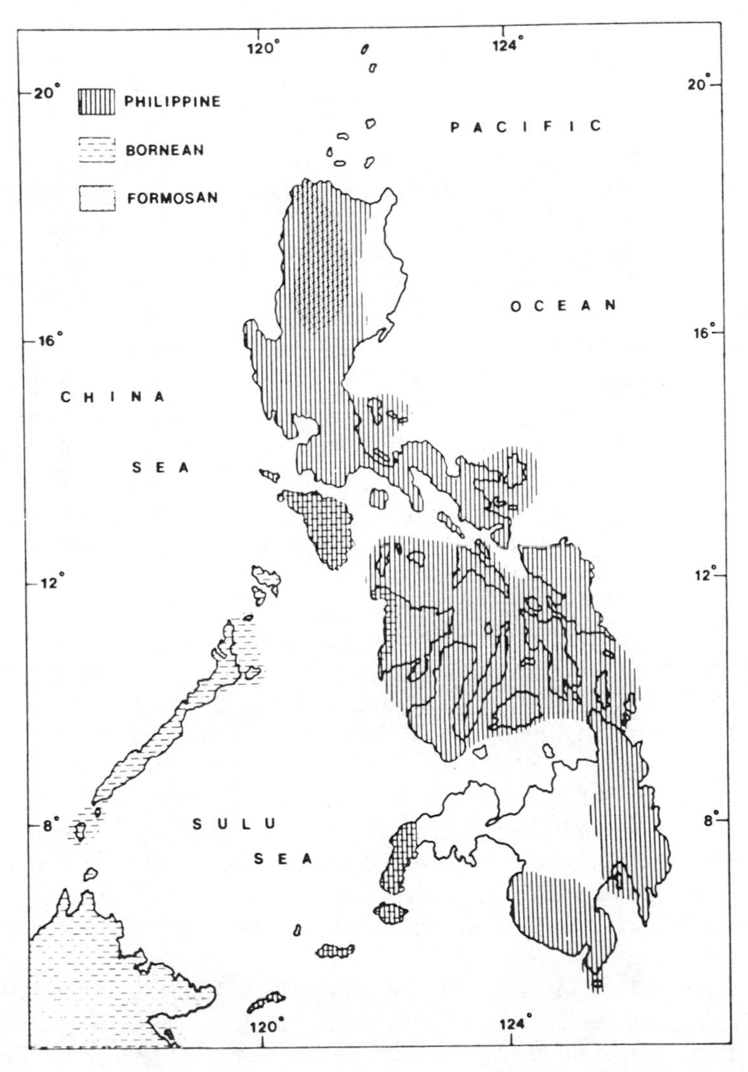

Figure 11.7. Philippine faunal subprovinces. *(From R. E. Dickerson et al., 1928, Distribution of Life in the Philippines, The Bureau of Science, Manila, Monograph 21, Bureau of Printing, Manila, p. 295.)*

which are endemic to the Philippines. Others range over the whole of the Paleotropical Realm. Thus, there is a pattern of paleo- and neoendemism similar in magnitude and importance to that found in California.

Wallacia's botany and zoology may be described as essentially Asian with Papuan and Australian overtones. To both the east (Weber's Line) and west (Wallace's Line) the neighboring regions display great differences in geologic history.

The Pacific Islands

In general, the Pacific islands include all those listed in Table 11.1. They are divided into a large Oriental group and smaller Australian, Antarctic, Neotropical, and Holarctic groups. Emphasis here is on the Oriental islands comprising the Bismarckian, Fijian, Polynesian, and Hawaiian provinces. Generally, within this geographic area (see Figure 11.8), one observes an attenuation eastward of Indo-Malayan and Papuan forms. For example, cassowaries, rodents, and a few marsupials extend eastward to the Bismarcks, but farther east, in the Solomon Islands, all terrestrial mammals except rats and mice, all freshwater mussels, most of the Paleo-Oriental snail fauna, and 97 genera of angiosperms reach their eastern distributional limits. No terrestrial vertebrates except for a few geckos and small rodents extend eastward from Tonga and Samoa. Some snakes get as far east as Tonga, then stop; and a few bats extend as far east as Samoa. Despite this attenuation, the Malaysian element generally dominates the biota of all islands in this region, with secondary contributions from the Australian Realm.

The earliest geologically dated events from the Pacific Basin are from truncated mid-Pacific mountains called *guyots*. They represent the Mid-Cretaceous of 100 to 110 million years ago. During the Cretaceous and early Tertiary, much of the Pacific Basin had a tropical or near tropical climate, giving the plants and animals a huge area over which to move. The land-bridge concept, as well as the "foundered continent" concept, are generally rejected in favor of island stepping stones as a dispersal avenue. Some authors believe that dispersal has taken a largely random course because there have been so many islands emerging and submerging and because such a large, uniform climatic area for geographic spread exists.

Durham (1963), for one, believes that occupation of the central and western Pacific took place either during the Late Cretaceous or early Tertiary, rather than during the later Tertiary or the Pleistocene. Part of the evidence for his view rests on the distribution of marine mollusks. The Pacific islands display mollusks having close affinities with Indonesian forms, despite prevailing winds and currents that run from east to west across the region, as shown in Figure 11.9. For these shallow marine forms, dispersal can take place only during the free-swimming larval stages; thus, widely separated islands with intervening deep water form effective barriers to these species. The oldest fossils found are from the Middle Cretaceous. Geologic dredging indicates that at that time, there were many more shallow banks, reefs, and high islands than exist today.

In fact, slow subsidence has been taking place over much of the

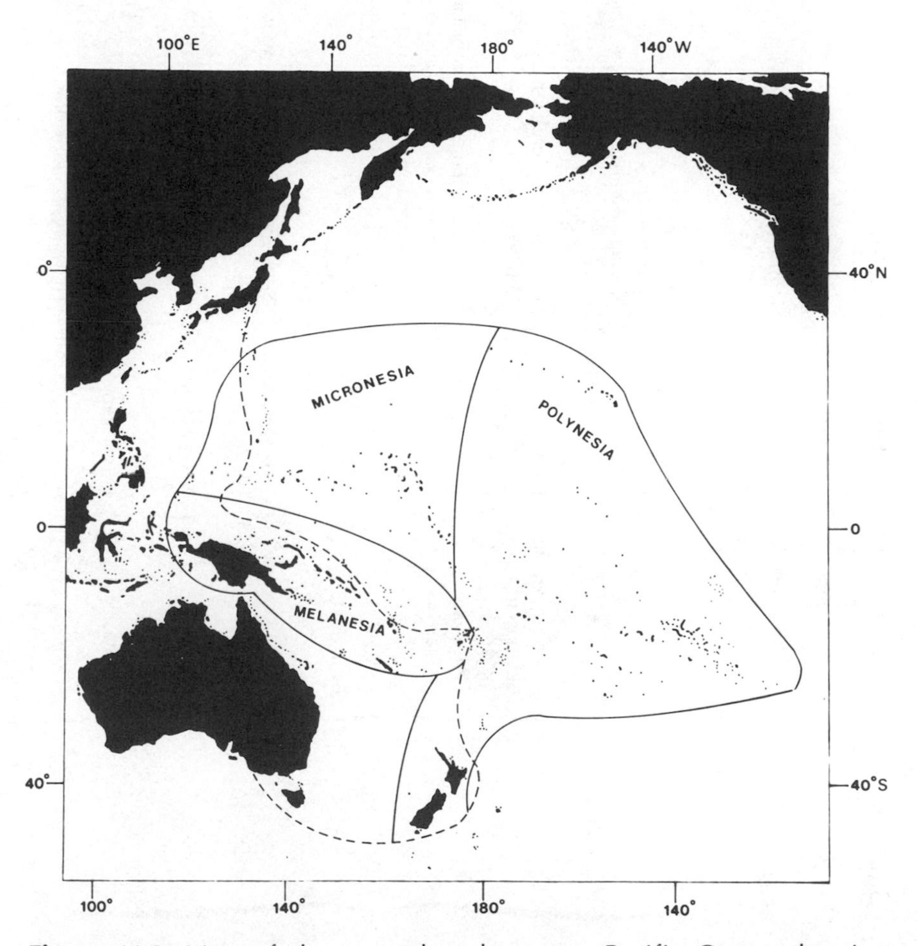

Figure 11.8. Map of the central and western Pacific Ocean showing boundaries of Polynesia, Micronesia, and Melanesia. The andesite line is shown by dashes. *(From H. S. Ladd, 1960, Origin of the Pacific Island Molluscan Fauna, Am. J. Sci.* **258A:***138; copyright © 1960 by the American Journal of Science. Reprinted by permission of the publisher.)*

Table 11.1. Biogeographical Subdivision of the Islands of the Pacific Ocean

Regions	*Subregions*	*Provinces*	*Districts*	*Islands of the Pacific Ocean*
Oriental	Indo-Chinese	South Chinese	Formosan	Taiwan (Formosa), Ryukyu
		Indo-Chinese		Hainan
	Indo-Malayan	Malayan	Malayan	Malaya
			Sumatran	Sumatra
			Javan	Java, Bali
			Bornean	Borneo, Balabac, Palawan, Calamianes
		Lesser Sundan		Lesser Sunda (Lombok to Timor and Babar)
		Philippinean		Philippine to Botel Tobago
		Celebesian		Celebes
	Papuan	Moluccan		Mulucca, Banda, Kei, Timor Laut
		Papuan		Aroc, Misol, Salawati, Waigeo, New Guinea, D'Entrecasteaux, Trobriand, Woodlark, Louisiade
		Torresian		Torres Strait, coastal Queensland, northeastern New South Wales
		Bismarckian	Bismarckian	Bismark, Admiralty
			Solomonian	Solomon
	Polynesian	Fijian	New Hebridean	Santa Cruz, Banks, New Hebrides
			Fijian	Fiji, Tonga, Somoa
		Polynesian	Micronesian	Bonin, Volcano, Mariana, Palau, Caroline, Marcus, Wake, Marshall, Nauru, Ocean, Gilbert, Ellice, Wallis, Horn, Howland, Baker, Phoenix, Tokelau
			Polynesian	Line, Niue, Cook, Society, Austral, Rapa, Tuamotu, Marquesas, Gambier, Pitcairn, Henderson, Easter, Sala-y-Gomez
		Hawaiian		Hawaii to Midway, Johnston
	Neocaledonian	Neocaledonian		New Caledonia, Loyalty

Australian	Australian	Australian		Australia (except coastal Queensland and northeastern New South Wales), Tasmania
	Neozeylandic	Kermadecian	Lord Howean	Lord Howe
			Norfolkian	Norfolk
			Kermadecian	Kermadec
		Neozeylandic	Neozeylandic	New Zealand, Stewart
			Chathamian	Chatham
			Antipodean	Bounty, Antipodes, Snares, Auckland, Campbell
Antarctic	Subantarctic	Kerguelenian	Macquarian	Macquarie
		Magellanian	Fuegian	Fuegian and other southern Chilean coastal islands to Gulf of Penas
	Antarctic	Antarctic		Antarctica and adjacent islands
Neotropical	Chilean	Fernandezian		Juan Fernandez, Desventuradas
	Peruvian	Galapagean		Galapagos, Cocos
Holarctic	Nearactic	Caribbean	Mexico	Revilla Gigedo, Tres Marias, Clipperton
		Sonoran	Sinaloan	Cedros
		Californian	Californian	San Benito, Guadalupe, California coastal islands
			Vancouverian	Vancouver, Queen Charlotte, Southern Alaskan, Aleutian, Commander
		Aleutian		
	Palearctic	Siberian	Kurilean	Kuril, Sakhalin
		Manchurian	Japanese	Japan

Source: R. F. Thorne, 1963, Biotic Distribution Patterns in the Tropical Pacific, in *Pacific Basin Biogeography,* J. L. Cressitt, ed., Bishop Museum Press, Honolulu, pp. 332–333.

southwest Pacific east of the Andesite Line, including the areas of Micronesia and Polynesia (see Figure 11.8). In this now sunken region, where some areas have descended more than 120 meters below the surface, many flat-topped sea mounts have been discovered. When dredged, many of them produce rounded cobbles of basalt with a variety of shallow-water fossils of Tertiary and Cretaceous age. In the area between Hawaii and Micronesia, 50 such guyots have been discovered, whereas only two have been found between Hawaii and the Americas. As a group, they could have served as stepping stones for shallow marine mollusks,

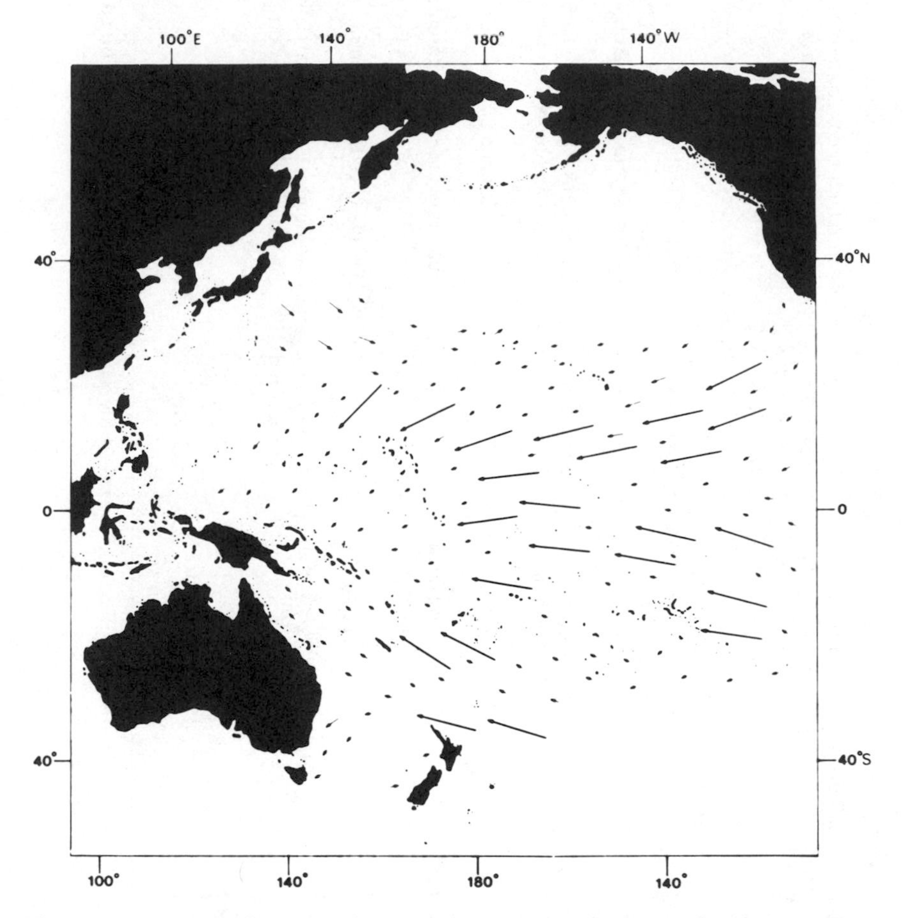

Figure 11.9. Prevailing winds in and near island areas. The figure shows the pattern in January; in July there is a shift to the north, but no essential changes are involved. *(From H. S. Ladd, 1960, Origin of the Pacific Island Molluscan Fauna, Am. J. Sci.* **258A:** *139; copyright © 1960 by the American Journal of Science. Reprinted by permission of the publisher.)*

and while above water, could also have permitted the spread of land snails and flora. In all, there are about 300 reefs, shallow banks, and atolls that could have served this function.

Dispersal could very well have been by way of free-swimming larvae, as well as by rafting, avian transport, or storms. In other words, migration may have been toward Indonesia rather than originating there. According to Ladd (1960) the present diversity of Indonesian forms may reflect the successful colonization of species finding suitable habitats in warm, tropical waters. The difficulty with this theory is that the original stock of mollusk forms is not clear. It is unlikely that Hawaii represents the original home of forms migrating westward, because these islands are too young; nor is there much affinity between their living or fossil forms and those of North and South America.

The land snails of the New Hebrides provide a second line of puzzling data. Excluding introduced species, these snails show almost 100-percent specific endemism. Within the southwest Pacific region, the snails of each of the archipelagos have affinities that can be described as paleo-Oriental (Malaysian), southern relict (Australian), and Pacific Ocean (Polynesian). The paleo-Oriental snails extend from the Malay Peninsula through New Guinea to the Solomon Islands and down the Queensland coast of Australia. The southern relict snails dominate southern Australia, Tasmania, New Zealand, and New Caledonia. The high islands of Polynesia and Micronesia have a few southern relict families but are otherwise dominated by Pacific Ocean forms. Melanesia, on the other hand, has all three elements. The Solomon Islands, for example, are dominated by the paleo-Oriental forms, Fiji by Pacific Ocean forms, New Caledonia and the Loyalty Islands by southern relict types, and the New Hebrides by a mixture of all three. This pattern is contrary to the plant and vertebrate patterns of New Guinea and the Pacific islands.

Solem (1958) has speculated that time may account for this basic difference. He thinks the mollusks represent relict forms that achieved their distribution in the late Paleozoic, whereas the vertebrate and vascular plants radiated later. When viewed as a unit, New Hebridean land snails show as much affinity to New Zealand as to New Caledonia. "The present contours of the ocean floor suggest that New Caledonia and the New Hebrides may have had separate land connections to New Guinea rather than any joint connection" (Solem, 1958, 1255). The only trouble is that for this theory to work, and for the snails to have achieved their distribution in the late Paleozoic, New Guinea would necessarily have been above water from the Paleozoic forward—a prospect that will be shown in Chapter 12 to be unlikely.

In the central Pacific area the Hawaiian Islands also display an interesting pattern. From 16 to 20 percent of the Hawaiian genera of flowering plants are considered to be endemic. The families with the greatest number of endemics are Compositae, having their origins in western North America, and Lobeliaceae. There are no endemic families, as such, but recent estimates place specific endemism at about 96 percent. Only 83 of the 418 families of angiosperms are present (20 percent of the world total). Hawaii and New Guinea share 41 percent of their families, while Hawaii and Australia share 44 percent. The Hawaiian fauna consists of highly endemic insects, terrestrial mollusks, and nonoceanic birds. There are no native mammals, except perhaps one species of bat; amphibia and reptiles are absent, or at least nonindigenous; and there are no freshwater fish.

It is generally agreed that the bulk of the indigenous Hawaiian genera are either related to, or derived from, stock from Austral and Malaysian regions with minor, but significant, additions from the Boreal and American regions. The implication is either that the disseminules arrived, adapted, and differentiated into endemic species, or that the present flora represents a contraction from a once greater diversity. The former idea fits the general model of island biogeography better.

SUMMARY

Although the Oriental Realm is second only to the Palearctic in longitudinal breadth, its overwhelming insular character reduces its land area to one of the smaller areas of biotic interaction. It is this same insular character that provides a laboratory for testing concepts and theories of island biogeography. The realm boasts not only the world's highest elevation—Mt. Everest, at 8,850 meters—but also ocean depths of 10,860 meters, south of the Mariana Islands. While most of the land area enjoys a tropical humid climate, the Himalayan biota, in particular, has been forced to adapt to extreme polar-like conditions, and some of the life forms of western India have adjusted to true desert.

A number of other unique attributes of the Oriental Realm have influenced the flavor of biotic development there since the early Tertiary. Aside from the obvious differences in climate, age, and topography between the western and eastern portions, one can cite the speed and distance of the Indian subcontinent's move northeastward from Africa, the abruptness and extent of Himalayan uplift, and the recent devastating impact of human occupancy in India.

Biological development in India has been conditioned by the subcontinent's rapid passage from south temperate latitudes through subtropical and equatorial climates and into the Northern Hemisphere. Bombay, at roughly 19° north latitude, for example, began its journey northward some 200 million years ago at about 50° south latitude; and depending upon whose evidence one accepts, it had its most recent contact with Africa and Madagascar some 60 million years ago at perhaps 30°S. For 40 million years, from the Paleocene to the Pliocene, the subcontinent drifted in isolation under a series of climatic belts while the resident flora and fauna differentiated and adjusted. Even crude calculations of time and distance seem to indicate a rate of progress ranging from 9 to 12 centimeters a year (cf. McKenzie and Sclater, 1973, 145). If one assumes that the land mass has decelerated considerably since making contact with Asia, its peak rate of movement must have been truly phenomenal.

The rapid rise of the Himalaya not only forced its passenger biota to adapt to boreal and polar conditions, but also initiated the monsoon over all of India. So recently did this uplift happen, however, that its effects are now largely masked by even more pronounced human alterations, magnified by a currently exploding human population. Protectionist and preservationist attitudes, so much a part of the Moghul Period, were replaced by European attitudes of resource utilization. Today, less than 200 years after European contact, the pristine flora and fauna have been not only decimated in terms of diversity, but also reduced in size to small and diminishing refugia.

Similar alterations have occurred in the insular portions of the realm. The Phillippines have perhaps the most documented history of plant introductions, but many of the islands of the smaller groups in the mid-Pacific have also been significantly modified to the detriment of their endemic biotas. The effect of humans in these areas has been to imprint a more cosmopolitan flora and fauna atop an already attenuated biotic series stretching eastward from the Southeast Asian mainland. The question of refugia throughout the island world is rhetorical. The islands are themselves refuges from which "escape" is often only successful through chance.

REFERENCES

Dickerson, R. E., E. D. Merrill, R. C. McGregor, W. Schultze, E. H. Taylor, and A. W. C. T. Herre, 1928, *Distribution of Life in the Philippines,* The Bureau of Science, Manila, Monograph 21, Bureau of Printing, Manila, 322p.

Durham, J. W., 1963, Paleogeographic Conclusions in Light of Biological Data, in *Pacific Basin Biogeography,* J. L. Gressit, ed., Bishop Museum Press, Honolulu, pp. 355-368.

Ladd, H. S., 1960, Origin of the Pacific Island Molluscan Fauna, *Am. J. Sci.* **258A**:137-150.

Mani, M. S., 1974, Biogeographical Evolution in India, in *Ecology and Biogeography in India,* M. S. Mani, ed., W. Junk, The Hague, pp. 698-724.

McKenzie, D. P., and J. G. Sclater, 1973, The Evolution of the Indian Ocean, in *Continents Adrift and Continents Aground,* Readings from Scientific American, W.H. Freeman, San Francisco, pp. 138-148.

Mukherjee, A. K., 1974, Some Examples of Recent Faunal Impoverishment and Regression, in *Ecology and Biogeography in India,* M. S. Mani, ed., W. Junk, The Hague, pp. 330-368.

Solem, A., 1958, Biogeography of the New Hebrides, *Nature* **181**(4618):1253-1255.

Thorne, R. F., 1963, Biotic Distribution Patterns in the Tropical Pacific, in *Pacific Basin Biogeography,* J. L. Gressit, ed., Bishop Museum Press, Honolulu, pp. 311-354.

FURTHER READING

Banerji, M. L., 1963, Outline of Nepal Phytogeography, *Vegetation* **11**:288-296.

Bhatt, D. D., 1977, *Natural History and Economic Botany of Nepal,* Orient Longman Ltd., New Delhi, 238p.

Gressitt, J. L., 1956, Some Distribution Patterns of Pacific Island Faunae, *Syst. Zool.* **5**:11-32.

Janaki Ammal, E. K., 1960, The Effect of the Himalayan Uplift on the Genetic Composition of the Flora of Asia, *Indian Bot. Soc. J.* **39**:327-334.

Keast, A., 1972, Australian Mammals: Zoogeography and Evolution, in *Evolution, Mammals and Southern Continents,* A. Keast, F. C. Erk, and B. Glass, eds., State University of New York Press, Albany, N.Y., pp. 195-239.

Kurup, G. U., 1974, Mammals of Assam and the Mammal-Geography of India, in *Ecology and Biogeography of India,* M. S. Mani, ed., W. Junk, The Hague, pp. 585-613.

Mani, M. S., 1974, Biogeography of the Peninsula, in *Ecology and Biogeography in India,* M. S. Mani, ed., W. Junk, The Hague, pp. 614-647.

Mani, M. S., 1974, Biogeography of the Himalaya, in *Ecology and Biogeography in India,* M. S. Mani, ed., W. Junk, The Hague, pp. 664-681.

Mani, M. S., 1978, *Ecology and Phytogeography of High-Altitude Plants of the Northwest Himalaya,* Methuen, New York, 199p.

Mayr, E., 1944, Wallace's Line in the Light of Recent Zoogeographic Studies, *Q. Rev. Biol.* **19**(1):1-14.

Merrill, E. D., 1945, *Plant Life of the Pacific World,* Macmillan, New York, 245p.

Mittre V., 1963, The Ice Ages and the Evolutionary History of the Indian Gymnosperms, *Indian Bot. Soc. J.* **42**:301-308.

Peake, J. F., 1971, The Evolution of Terrestrial Faunas in the Western Indian Ocean, *R. Soc. London Philos. Trans.* **B260**:581-610.

Rao, A. S., 1974, The Vegetation and Phytogeography of Assam-Burma, in *Ecology and Biogeography in India,* M. S. Mani, ed., W. Junk, The Hague, pp. 204-246.

Rau, M. A., 1974, Vegetation and Phytogeography of the Himalaya, in *Ecology and Biogeography in India,* M. S. Mani, ed., W. Junk, The Hague, pp. 247-280.

Raven, H. C., 1935, Wallace's Line and the Distribution of Indo-Australian Mammals, *Mus. Nat. Hist. Bull.* **68**:177-293.
Ridley, H. N., 1937, Origin of the Flora of the Malay Peninsula, *Blumea* (Suppl. 1), pp. 183-192.
Schaller, G. B., 1977, *Mountain Monarchs, Wild Sheep and Goats of the Himalaya,* University of Chicago Press, Chicago, 426p.
Stainton, J. D. A., 1972, *Forests of Nepal,* The Camelot Press, London, 280p.
Steenis, C. G. G. J. van, 1936, On the Origin of the Malaysian Mountain Flora, *Jard. Bot. Buitenzorg Bull.* **13**:135-262, 289-417.
Stone, B. C., 1967, A Review of the Endemic Genera of Hawaiian Plants, *Bot. Rec.* **33**(3):216-269.

Chapter 12

The Australian Realm

The Australian Realm consists of Australia and New Guinea, North Island and South Island of New Zealand, and New Caledonia. It is essentially the terrestrial portion of the Australian Plate, except for the outlying islands comprising the Fiji, New Hebrides, Bismarck, and Solomon groups. The region extends from 112° to 180° east longitude and from 46° south latitude to the equator. Altogether, the 8,450,000-square-kilometer area is larger than the conterminous United States, Australia, alone, being roughly equal in size. New Guinea is 461,700 square kilometers, and the combined islands of New Zealand add another 262,000 square kilometers. The entire area spans climatic belts from south temperate to humid tropical with a vast area of subtropical desert in the interior of Australia.

Raven and Axelrod (1972) have aptly described the Australian Realm as a set of former microcontinents and present-day archipelagos comprising a tectonic plate that experienced fusion with the Asian Plate during the Tertiary. Figure 12.1 shows the basic configuration of these features, and Figure 12.2 illustrates their possible predrift locations when they were part of Gondwanaland. From sea-floor magnetic records, it appears that creation of new sea-floor began to separate the Australian Plate from Antarctica in Late Cretaceous time, about 80 million years ago. The last contact with South America is dated around 45 to 50 million years ago. At the beginning of the northward migration, New Guinea was on the leading edge of the plate and was submerged. Its gradual emergence in Miocene time not only advanced the land edge of the plate some 10° of latitude into the tropics, but during glacial sea lowering it also may have provided lowland or swampy avenues for the transfer of biota across Torres Strait.

The shaded portions of Figure 12.1 show former "continental" land masses that are presently submerged except for their island fringes. The thesis advanced by Raven and Axelrod is that the history of the biota in these widely scattered lands has been determined by (1) the Late Cretaceous position of the Australian Plate; (2) its northward movement throughout the Tertiary, during most of which time it was isolated and provided good opportunity for speciation; (3) the disruption of its eastern flank to form a series of microcontinents and archipelagos; and, (4) its mid-Tertiary fusion with the Asian Plate, which built island stepping stones to Asia. Evidence from sea-floor spreading suggests that Australia

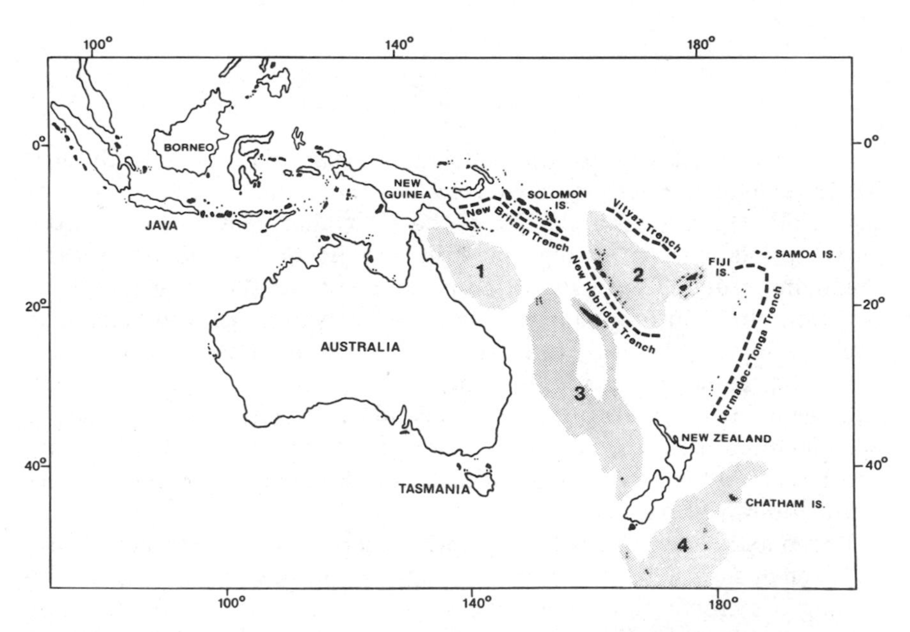

Figure 12.1. Major components of the Australian Realm showing submerged plateaus that constitute former microcontinents. Compare also with Figure 6.2. *(Redrawn from P. H. Raven and D. I. Axelrod, 1972, Plate Tectonics and Australasian Paleobiogeography, Science* **176:***1380; copyright © 1972 by the American Association for the Advancement of Science. Reprinted with permission.)*

only came into contact with the Asian region during Miocene time and that island stepping stones were all that ever connected the two realms. The persistence of Wallacia and the sharpness of its eastern and western limits confirm this widely held view.

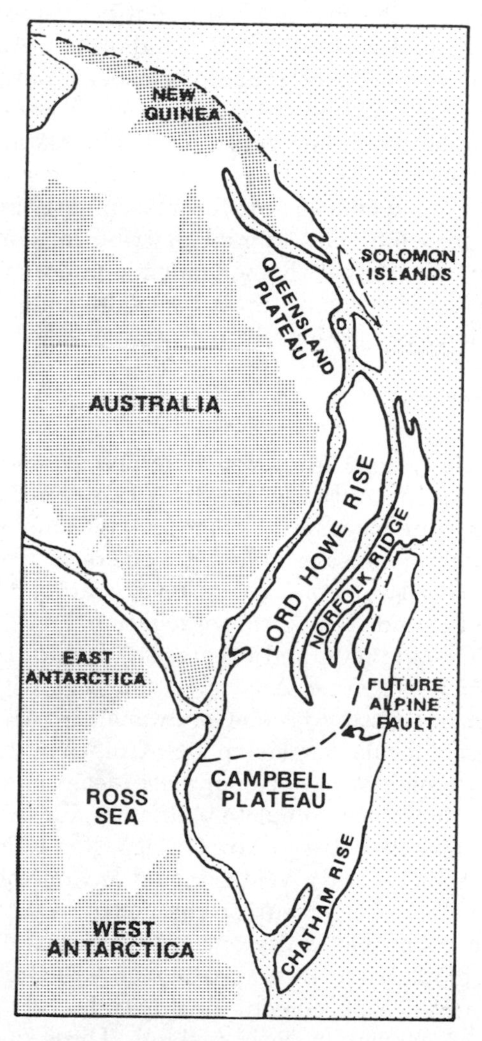

Figure 12.2. Reconstruction of Australian microcontinents prior to their separation from Antarctica. *(Redrawn from P. H. Raven and D. I. Axelrod, 1972, Plate Tectonics and Australasian Paleobiogeography, Science* **176:***1381; copyright © 1972 by the American Association for the Advancement of Science. Reprinted with permission.)*

AUSTRALIA

Considering the uniqueness of the Australian flora and fauna, it is remarkable that the biota conforms to the higher levels of taxonomic abstraction. This conformity, in itself, is evidence that angiosperm and mammalian differentiation on this "island continent" took place after the development of the major world families and orders. As early as 1867, Hooker realized that most Australian angiosperm families were represented in other realms, and that the largest families were also the largest world families. On the other hand, it is remarkable that at the generic and specific levels, the Australian fauna is more than 95-percent endemic and its flora is 85-percent endemic. These figures suggest two ideas: (1) that the explanation for plant and animal distributions must be sought within the region and (2) that the nonendemic minority of life forms is the most interesting element in terms of establishing both the earlier and more recent land connections.

General Development

There are essentially four intertwined explanations for Australia's biological development. Actually, the first two (continental drift and sea-floor spreading) are inseparable in modern terminology, but they are discussed separately because each handles the question of New Guinea differently. The other two explanations revolve around the concepts of "isolated proximity" and "ecological segregation."

According to Good (1960), and others, the idea of continental drift holds two possibilities: either Australia, together with New Guinea, drifted northward and eastward from Gondwanaland; or Australia drifted alone into its present locale, displacing New Guinea in the process. There is at least some evidence for the latter of these ideas in that the geology of New Guinea and Australia is completely different. Floristically, moreover, there is greater similarity between New Guinea, New Caledonia, and New Zealand than there is between Australia and any one of these areas. In the New Guinea flora, 37 percent of the genera are Paleotropical, 37 percent are Indo-Malaysian, 9 percent are Pacific, and 4.5 percent are Australian. By comparison, the percentage of New Guinean flora found in other areas includes 50 percent in Australia, 28 percent in New Caledonia, and 13 percent in New Zealand. These data suggest that Australia wedged its way into a previously existing New Guinean archipelago. Although vague in details, this concept does provide an explanation for the uniqueness of Australia's biota in view of its proximity to the Oriental Realm—only 120 kilometers across the Torres Strait.

A revision of Good's 1960 concept of continental drift is presented by

Raven and Axelrod (1972). In their view, Australia split from Antarctica about 80 million years ago and drifted slowly northward as the sea-floor spread laterally. New Guinea was part of the leading edge of the mass but was submerged until the Oligocene, when contact with the Asian Plate caused it to emerge. The disjunct temperate forest species of Australia and South America testify that the two were connected as recently as 50 million years ago. In contrast, Africa is believed to have separated some 110 million years ago, which is why that realm has few biotic connections with either South America or Australia, except in its very oldest flora.

After Australia's contact with the Asian Plate in Miocene times, a movement of life forms into and out of the island continent began. Many of these forms apparently evolved endemic species while Australia was an isolated "raft." The Indo-Malaysian element utilized stepping stones into Australia, while those same stones created a sweepstakes route for biota trying to leave Australia. The story of speciation in Australia is fascinating in that for several million years, the initial Cretaceous stock and early Tertiary immigrants were virtually isolated. From the Cretaceous onward, Australia was entering and leaving latitudinal climatic belts that drastically altered its environments and stimulated adaptive radiation. Like India, Australia has literally moved beneath a series of climatic zones.

Projecting the sea-floor-spreading concept into the future, one can imagine that the Mediterranean-like flora of southwest and southern Australia will eventually become extinct. The present desert of central Australia will shift southward as the land mass proceeds northward. In similar fashion, the temperate forest of Tasmania will shift, first to a subtropical form, then possibly to desert, and eventually into a humid tropical rain forest.

In the concept of isolated proximity, Papualand is believed to have come into contact with Sundaland through island stepping stones. To summarize the evidence for this possibility, the majority of nonendemic angiosperms and all of the marsupial fauna of Australia can be traced to progenitors in Asia. Of the nonendemic angiosperms, 25 to 35 percent are from New Guinea, 10 to 15 percent are from Africa, 10 percent are from the Pacific region, and 8 percent are from South America. Some families, like the Proteaceae, are believed to have evolved in Papualand. Cytologically, there is good evidence for a polyploid attenuation series along a Sundaland migration route. And, finally, there is fossil evidence to suggest an "angiosperm invasion" of Australia during the Cretaceous. One of the main problems with the isolated proximity concept lies in the marsupials. If these forms came to Australia across stepping stones from Asia, how can one reconcile the notion that Australia has only resided in the area for, say, the past 20 million years? The time required to achieve the diversity of marsupial species is far longer than this period of residence would permit.

The concept of ecological segregation attempts to explain the well-developed phytogeographic zones of Australia. These zones are shown in Figure 12.3. They consist of a tropical rain forest zone in the northeast, temperate zones in the southeast and west, and a vast desert and semidesert zone in the interior. Herbert (1967) first recognized that the floras of these regions were taxonomically as well as ecologically distinct, and he suggested that they probably arose out of a common paleotropical stock during a very long period of isolation. The tropical zone is rich in Malaysian forms and is only moderately endemic. The temperate zone is highly endemic and has disjunct distributions in the east and west. Lastly, the dry interior is almost wholly endemic and contains refugia for

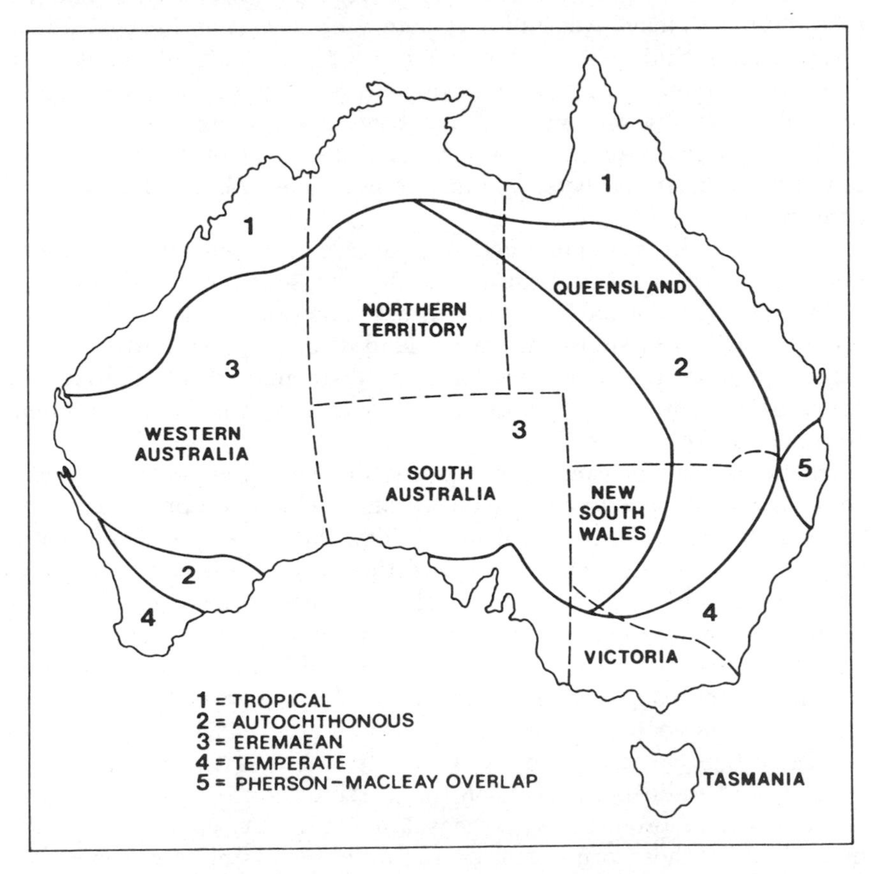

Figure 12.3. Phytogeographic regions in Australia. *(From N. T. Burbidge, 1960, The Phytogeography of the Australian Region, Aust. J. Bot.* **8***(2):79.)*

other elements. These patterns suggest that there is a continual shifting (and sifting) of the angiosperm flora to adjust to local internal environmental changes. Speciation on the continent seems to have been an explosive evolutionary process stimulated by Miocene climatic changes. The Miocene marked the end of the period of uniform climate, presumably because the continent had rafted into a subtropical arid zone.

The Flora

Based upon the concepts of Australia's continental history, several chronologies have been proposed for the Cenozoic development of it's flora. Specht and Mountford (1958), Smith-White (1959), Burbidge (1960), and Herbert (1967) have all discussed the main stages of floristic differentiation amidst climatic alterations. These events are charted in Table 12.1. Although the question of continental drift is muted by these authors, there is a definite suggestion that Australia was first visited by angiosperms during the Mid-Cretaceous. Myrtaceae, Proteaceae, Leguminosae, Casuarinaceae, and other typical families gained their foothold prior to separation from Gondwanaland, then proceeded to evolve in isolation throughout the Tertiary. Any inquiry into the present relationships of the flora should recognize that the continent has been inhabited by a diversity of both rain forest and savanna types since at least the early Tertiary, and that their present geographical ranges have been molded by post-Miocene climatic and geologic changes.

Basically, the flora of Australia can be partitioned into a large autochthonous element that has evolved out of the original Cretaceous stock; an Indo-Malaysian, or Oriental, element that represents a second wave of angiosperm invasion; and an alpine element that has recently arrived from New Zealand. The Indo-Malaysian species are largely restricted to the humid tropical coast of Queensland, but they extend all along the north coast of the continent. Nearly all of the rain forest species have apparently arrived via a New Guinean gate across the Torres Strait and the Coral Sea. Both montane and lowland types are represented, including species from 50 different families (Francis, 1951). There are two families of gymnosperms (Podocarpaceae and Araucariaceae); one of monocotyledons (Palmae); and 47 of dicotyledons. Among the more notable of the latter are Fagaceae, Proteaceae, Lauraceae, Capparidaceae, Pittosporaceae, Cunoniaceae, Leguminosae, Rutaceae, Burseraceae, Euphorbiaceae, Anacardiaceae, Celastraceae, Sterculiaceae, Myrtaceae, Sapotaceae, Apocynaceae, Boraginaceae, Solanaceae, and Rubiaceae.

The autochthonous genera of the Australian element include *Eucalyptus,* with over 500 species, *Acacia, Grevillea, Styphelia, Melaleuca, Goodenia, Hakea, Hibbertia, Eremophila, Banksia,* and *Schoenus.* With

Table 12.1. Comparative Chronology for the Cenozoic Development of Australian Vegetation

Time	*Smith-White, 1959*	*Burbidge, 1960*		*Specht, 1958*	*Herbert, 1950; 1967*
Recent	Aridity led to contraction of rain forest & sclerophyll forest with expansion of the eremaean element	Arid		Continued fluctuations in climate stimulating migration of rain forest, open woodlands, etc., in tropical Australia	Eremaean expansion w/contraction of rain forest and sclerophyll forest—aridity
Pleistocene	Uplift plus colder conditions favored adaptation to alpine conditions for some Paleo-Aust. plants but also favored colonization by "long distance dispersal" of Antarctic and other Southern Hemisphere plants	Arid Wet with minor glaciation Arid	migration of plants sometime during Pleistocene	Lowering sea level permitted migration of species between northern Australia and New Guinea; fluctuations in climate stimulated N-S and E-W migrations in Australia and development of present-day plant communities	Arrival of alpine types permitted by colder conditions
Pliocene		Subtropical conditions		Miocene-Pliocene peneplanation with monotonous soil and climate gave rise to plant genera which were extremely widespread	[Tertiary yielding its present character] Retreat of seas in early Pliocene followed by Koscinsko uplift; creation of new habitats (climatic and edaphic)
Miocene	[yields flux in vegetation] Widespread tropical conditions in Australia; contact between east and west Australia	Change to semiarid conditions			Separation of east from west Australia by encroaching seas in mid-Miocene
Oligocene	Second invasions of angiosperms in early- to mid	Subtropical climate from Eocene throughout most of		(Isolation of Australia through breakdown of land	Nearly pan-Australian flora from east to west coinciding

Eocene	Tertiary changes in climate and topography: Tertiary; dominantly the Indo-Malayan element but restricted to habitats to which already adapted	Miocene; 2-way migration of plants from early- to mid-Tertiary with interchange to eastern Malaysia and possibly New Caledonia and New Zealand via New Guinea	bridges effectively stranded the three main groups of plants [endemics, Indo-Malayan, and southern] and led to intermixtures of the three) Brackets mine	Evolution and differentiation of flora throughout: with uniform climate and topography during Oligocene. Indo-Malayan element arrived in early or middle Tertiary and mixed with already developed paleo-Australian element
Paleocene	Adaptive radiation; evolution at generic and species levels of the genetically more flexible portions of the flora		Outward migration of *Eucalyptus, Melaleuca, Banksia, Grevillea,* and other typical Australian genera to Indo-Malaya from Upper Cretaceous to Eocene along land bridges and archipelagos. Similar migration of southern elements along land bridges to and from Antarctica and South America	[Australian-wide cycle of evolution resulting in distinctive paleo-Australian element] Brackets mine
Upper Cretaceous		2-way migration		Dominance achieved in Angiosperm flora following their appearance in early Cretaceous
Mid-Cretaceous	Massive angiosperm invasion of Australia; breakdown of isolating barriers			Arrival of Angiosperm flora in Australia

Source: S. A. Morain, 1970, Ecological Segregation of Phytogeographic Elements, Upper Burdekin Valley, North Queensland, Ph.D. dissertation, Kansas University, Lawrence, Kans., p. 164.

the exception of *Acacia* and *Eremophila,* which are common to the more xeric communities of the interior, all of these genera are forest dwellers with eucalyptus species as dominants. Most are sclerophyllous shrubs or undershrubs that indicate a post-Miocene origin in response to drying trends. An interesting feature of this assemblage is that identical species occur in both the southwest and southeast, although they are separated by a vast and effective arid barrier known as the Nullarbor Plain. It is probable that they once enjoyed a more continuous distribution across the Great Australian Bight but have been "pushed off" the continent as it drifted northward.

The Antarctic element includes species of *Drimys, Nothofagus, Araucaria, Podocarpus,* and *Cupressus,* all of which occupy the lower altitudes and latitudes of the southeast. *Nothofagus* and *Araucaria* also extend to New Guinea and New Caledonia, which suggests that these islands may have been connected with Australia at one time through Cape York Peninsula and the Queensland Plateau.

There are few points of agreement, and many of disagreement, over the development of Australia's flora. Mostly, the points of agreement are well-established conclusions drawn from repeated observations of the present vegetation and its fossil record. A few of the major conclusions are given below.

Evolution, diversification, and radiation of major angiosperm orders and families took place during the Cretaceous, probably in the tropics between 45° north and south latitude (Axelrod, 1959). Australia was apparently not a center of development at these taxonomic levels.

Isolation of Australia, through whatever mechanism (geographic, ecological, or reproductive), has been of long duration; periods of contact with other regions have been brief and generally dominated by one-way (inward) migrations.

Australia's vegetation, taken as a whole, is highly endemic—although just how much so is not certain. Sussenguth (1950) has estimated as high as 87.5 percent.

Extra-Australian elements from such diverse biogeographic regions as the Antarctic, southwest Pacific, and Indo-Malaysia are present in the flora and are not uniformly distributed over all parts of Australia.

Australian endemics are overwhelmingly sclerophyllous or in some other manner adapted to xeric environments.

Plio-Pleistocene vegetation over the whole of Australia has fluctuated greatly in response to climatic, edaphic, and physiographic alterations.

According to both Smith-White (1959), and Herbert (1967), the Austral flora began its evolution immediately following a Mid-Cretaceous invasion of angiosperm families and orders across Malaysia. By late Aptian time, according to evidence recoverable from the Winton Flora, tropical Australia, as far as about 22° south latitude, already contained from 10 to 15 percent angiosperm species (Axelrod, 1959). Within this body of newly arriving plants was a complete range of flexible and rigid genetic systems. The more viable segments apparently underwent rapid evolution in pre- or very early Tertiary times to produce such distinctive paleo-Australian genera as *Eucalyptus, Melaleuca, Banksia,* and *Grevillea,* some of which later entered Indo-Malaysia.

Concurrent with this surge in the pace of evolution, the Australian landscape was, by all indications, being reduced to an extensive, lateritized peneplain. It is generally agreed that by Eocene or Miocene time, mild tropical climates prevailed over a featureless topography, and that these conditions favored the development of a pan-Australian vegetation. The significant feature of the flora from this period is that it contained not only the previously evolved paleo-Australian element, but a newly arrived Indo-Malaysian segment as well.

There is marked disagreement about the sequence of events toward the end of the pan-Australian episode. As Wood states (1959, 293), ". . . the Miocene is considered an extremely important period in Australian plant geography. It marks the end of the period of stability and peneplanation, of uniform soil conditions, of mild climate, and of lack of barriers." It is clear that from the late Miocene to the present, the major influences on Australia's vegetation have been (1) climatic fluctuations that not only stimulated the migration of taxa and whole communities, but also inaugurated the immigration of a Southern Hemisphere cold-tolerant element; (2) tectonic instability that resulted in the creation of habitats as well as their migration avenues, that were previously unavailable during the pan-Australian episode; (3) evolution, adaptive radiation, and diversification primarily at the species level; and, (4) edaphic alterations. It is this latter surge in evolutionary rate that resulted in the now distinctively Australian element.

The Fauna

According to Keast (1972), less than half (8 out of 17) of the native mammalian families now living in Australia and New Guinea are placentals (Table 12.2). They are all recent arrivals, and all of them except the dingo (Canidae) are either rats or bats (Rodentia or Chiroptera). The rodents belong to the family Muridae, of which there are 31 genera (77-percent endemic). There are 17 species of the genus *Rattus* endemic to the region—9 of them in Australia, 5 in New Guinea, and 3 in both. From this evidence, Simpson (1961) believes the *Rattus* could have arrived no later than early Pleistocene. "Old Papuan" forms of rat consist of the genera

Table 12.2. Mammalian Families of Australia and Their Relative Proportions in the Total Fauna

		No. of genera	No. of species	% of total fauna
Monotremes				
Tachyglossidae	spiny anteaters	2	3	1.0%
Ornithorhynchidae	platypuses	1	1	.3
Marsupials				
Dasyuridae	"mice," "cats," marsupial anteater	15	42	11.5
Thylacinidae	marsupial wolf	1	1	.3
Peramelidae	bandicoots	8	19	5.2
Phalangeridae	phalangers, honey possum, koala, ringtail possum	17	36	9.8
Macropodidae	wallabies, kangaroos	16–18	42	11.7
Vombatidae	wombats	2	3	1.0
Notoryctidae	marsupial mole	1	2	.5
Placentals				
Muridae	water rats, rats, mice	31	124	33.5
Canidae	dingo	1	1	.3
Pteropodidae	fruit and blossom bats	5	25	6.9
Emballonuridae	all families comprising microchiroptera (small bats)	2	2	.6
Megadermatidae		1	1	.3
Rhinolophidae		1	14	3.8
Vespertilionidae		11	37	10.2
Molossidae		1	10	2.7

Source: A. Keast, 1972, *"Australian Mammals: Zoogeography and Evolution,"* in *Evolution, Mammals and Southern Continents,* A. Keast, F. C. Erk, and B. Glass, eds., State University of New York Press, Albany, N.Y., pp. 204, 205.

Uromys, Melomys, Xenuromys, and *Pogonomelomys.* Indeed, all of the modern forms may have arisen from a single member of *Uromys.* Five other genera—*Mallomys, Hyomys, Anisomys, Pogonomys,* and *Macruromys*—are endemic to New Guinea only. They are estimated to be Miocene arrivals representing several different invasions from the East Indies, and appear to have no relatives outside New Guinea.

"Old Australian" forms of rat are mostly in the *Pseudomys* group. They are endemic and highly diversified, and may have arrived in Australia during the Miocene. Arrival of Muridae is believed to have been along an island arc from Southeast Asia perhaps as early as the Oligocene. Differentiation and sifting of species probably occurred all along this island arc with the final filter occurring at the Torres Strait between New Guinea and Australia.

For the bats, seven families are represented. All are widespread in Eurasia and are relatively old, dating back to the Oligocene of Europe. Endemism is low in Australia. The bats show no particular attenuation series along the island chain and cannot be divided into waves, as can the Muridae.

Though taxonomically less diverse, the marsupials of Australia are all endemic through the superfamily level. Their major radiation took place in Australia, and the forms are surely old. Despite the absence of marsupial fossils in Asia, Simpson (1961), has argued for their arrival by rafting from nearby islands. This proposal, of course, denies the whole notion of continental drift, since rafting suggests that connections between Australia and Asia were then as they are now, and that the island archipelago dates from the Late Cretaceous.

Adaptive radiation of marsupials in Australia and South America is believed to have begun near the end of the Late Cretaceous or early in the Tertiary with stock originating in North America. In the early Tertiary, marsupials were numerous in lowland forested areas of the Nearctic, but they became increasingly scarce as the inland seas dried up. Fossil records from the mid- to late Tertiary (Miocene and Pliocene) of North America lack marsupial remains.

Notwithstanding strong migration barriers, Clemens (1968) feels that many of the early Nearctic marsupials could have crossed an ecological bridge to the Neotropical Realm. He apparently has not ruled out the possibility of a North American land connection to South America at the close of the Cretaceous. If there was one, however, it must have included an almost perfect barrier to all but the marsupial line. Subsequently, he suggests the opossum (*Didelphis*) immigrated back to North America from South America during the Pleistocene. Similarly, European marsupials are believed to have crossed from North America in the early Eocene by an uncertain route. If Fooden (1972) and Kurtén (1968) are correct, it is possible that the Nearctic and western Europe were connected via Greenland until sometime in the early Tertiary.

For Australia, Clemens (1968) reasons that if the time of origin of marsupials is approximately correct, they must have reached Australia through rafting and island hopping along a sweepstakes route that ran along an archipelago between Asia and Australia. Then, according to his reasoning, they died out in southeastern Asia. These marsupials are believed to have reached coastal and insular Asia along the Pacific rim from North America. Unfortunately, no marsupial fossils have ever been found in Asia.

Figure 12.4 shows the limit of the marsupials' northern extension into the New Guinea archipelago and the southern extension of native placentals, other than the bats and rats, toward Australia. A zone of overlap is clear in the area that roughly coincides with the New Guinea uplift. This fact, then, can be taken as zoological evidence that Asia and Australia have only been sharing species since the plates came together.

NEW ZEALAND

The modern flora and fauna of New Zealand can be divided into five elements: (1) a small, ancient, largely endemic group of uncertain origin; (2) a strong Australian element; (3) a large Indo-Malaysian element; (4) a small, Holarctic (cosmopolitan) group; and (5) a large, but heterogeneous, "Austral" group. The origins and dispersal routes of these groups suggest mutually exclusive histories, and some believe that the entire pattern can be explained as the result of chance dispersal, by either wind or water, over stepping stones. Others argue that land connections had to exist, because most biotic forms are ill-adapted to long-distance dispersal over water. One has difficulty, for example,

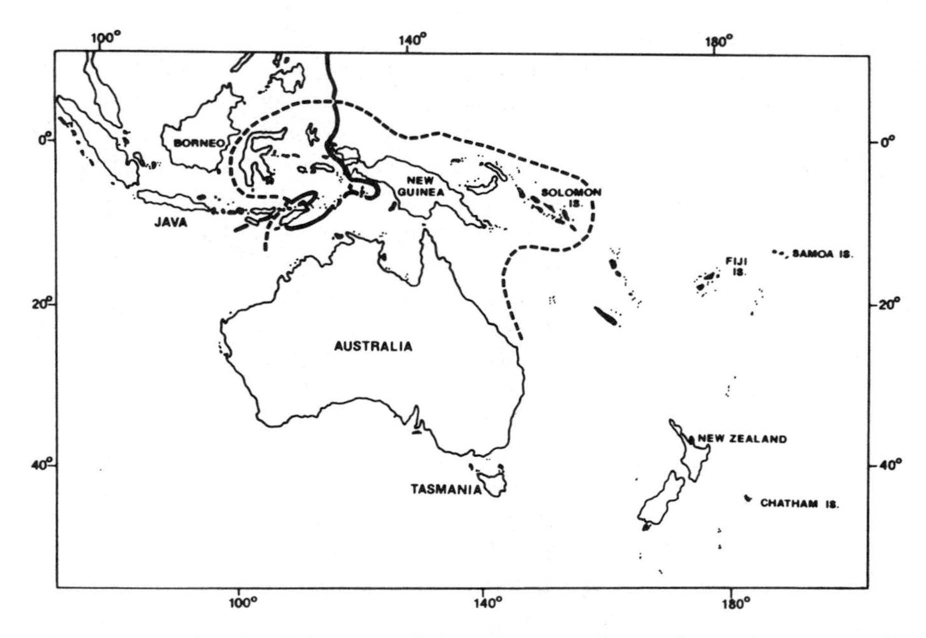

Figure 12.4. Islands of the Australian Region, with numbers of genera and limits reached by marsupials and by native placentals other than bats, murids, and probable late human introductions of pigs and deer (solid line). *(From G. G. Simpson, 1961, Historical Zoogeography of Australian Mammals, Evolution* **15:**439.)

explaining the presence of land snails on Fiji or *Nothofagus* (southern beech) in tropical latitudes. In the case of seeds from southern beech, it has been demonstrated that wind dispersal does not occur over distances of more than a few kilometers. Moreover, when wet, the seeds sink in sea water.

There is even an argument over whether the land connections were direct, as during predrift Gondwanaland, or temporary and discontinuous, as suggested by the land bridge theory. In fact, none of the ideas necessarily excludes the others. It is quite possible that at one time or another in the past 100 to 200 million years, all of the suggested mechanisms have been operational for certain segments of the biota. The question is, which ones were most important in molding the observed modern patterns? Three major dispersal routes can be visualized: (1) a route across the Tasman Sea from eastern and southern Australia; (2) a route along the Melanesian Arc connecting New Guinea with Indo-Malaysia; and (3) a route through Antarctica and South America.

The south Australian connection is interesting because, although there are marked affinities between the biotas of Australia and New Zealand, there are some notable absences. Land snakes (which arose during the Cretaceous), marsupials, and fossil dinosaurs, for example, are totally absent from New Zealand and New Caledonia. This absence would argue that these areas have been separated from Australia since at least Mid-Cretaceous time. A review of New Zealand's angiosperm and gymnosperm flora reveals that 80 percent of the total genera are common to Australia. But, nine out of ten of these genera are widespread beyond Australia. In fact, the number of genera common only to Australia and New Zealand is quite small, suggesting that the overall floras of the two were received simultaneously from yet another source. A connection through the southern end of Australia was not that viable, because such characteristic genera as *Eucalyptus* (500 species, more or less), and *Acacia* never reached New Zealand. The absence of these genera leads one to consider a northern connection through New Guinea and New Caledonia.

Reference to Figures 12.1 and 12.2 shows that a more or less continuous island arc may once have existed between modern New Guinea and New Zealand. As the Australian Plate moved northward from Cretaceous through mid-Tertiary time, the microcontinents referred to as the Fiji Plateau, Queensland Plateau, Lord Howe Rise, and Campbell Plateau apparently foundered, leaving behind island archipelagos. While the microcontinents existed, however, they provided a means for dispersal along an arcuate zone east of present-day Australia.

Gaskin (1969), in fact, refers to two arcs—an inner arc that corresponds roughly to the Queensland Plateau, Lord Howe Rise (remnants of which include Lord Howe Island, Norfolk Island, and New Caledonia), and the

Campbell Plateau; and an outer arc, consisting of the Solomon Islands and the Fiji Plateau (remnants of which include the New Hebrides and the Fiji Islands). Both of these arcs apparently merged at New Guinea.

Detailed studies of the angiosperm flora of the whole region show strong links between New Guinea, New Caledonia, and northern Australia. Many genera, too, show affinities with Indo-Malaysia, but not with Australia. One can speculate that as New Guinea was thrust upward by contact with the Asian Plate, the final "missing link" in a long stepping-stone network was established, along which biotic forms immediately began to filter. It is known that a large segment of Indo-Malaysian rain-forest species has migrated to the east coast of Australia and pushed as far south as Brisbane. Insects and other invertebrate fauna show similar patterns, but placental mammals have, by and large, been excluded.

It is in the context of a Micronesian dispersal route that the *Nothofagus* problem should be addressed. *Nothofagus* can be found in subalpine forests of New Guinea and New Caledonia. Although the genus is believed to have arisen along with the Northern Hemisphere beech (*Fagus*) during the Cretaceous, it apparently did not reach New Caledonia and New Guinea until Pliocene time—after the Australian Plate made contact with the Asian Plate. Some believe the form spread northward from New Zealand, but the New Guinea forms more closely resemble those from western Australia. Their occurrence in New Caledonia is even more difficult to explain. There was no known land connection between this island and either Australia or New Zealand at any time during the Tertiary. Consequently, the New Caledonian *Nothofagus* is suspected of having arrived from the northwest out of New Guinea, which in turn acquired its stock from the south by an as yet unknown route.

The third dispersal route was through the Antarctic. The separation of Gondwanaland apparently began during the Jurassic, with Africa rifting first, followed by New Zealand, and then by Australia. A connection persisted between South America and Antarctica into the early Tertiary via the South Orkney, South Sandwich, and South Georgia islands (referred to as the Scotia Arc). Initial spread of the angiosperms must have occurred while Gondwanaland was still intact, but separation of Africa and New Zealand occurred before the spread of marsupials was complete. This sequence appears to be the only reasonable explanation for the absence of marsupials in these two regions and the presence of marsupials in South America and Australia. One is left bewildered, however, by the apparently paradoxical absence of dinosaurs in New Zealand. If speciation took place as late as the Cretaceous, and if there were residual connections to allow the later spread of *Nothofagus* to New Zealand, why are there no fossil dinosaurs there?

New Zealand emerges in our thinking as an island archipelago sufficiently isolated to be oceanic. Nevertheless, its area was large enough to allow prolific speciation and evolution at the generic level. Although there are some ancient endemics of uncertain affinity, most of the ancestors of the present biota came from outside New Zealand, either by direct overland dispersal or by long-distance island hopping. Although North and South islands have become increasingly isolated during the past 80 million years, the creation of new environments has been a continual stimulus for speciation and extinction. In answer to the dinosaur question, it is conceivable that the fossil remains are to be found in the now submerged Campbell Plateau and Lord Howe Rise. The area that is now New Zealand was apparently submerged when the microcontinents departed Gondwanaland. One would therefore not expect to find fossils from the Mesozoic, but instead would expect North and South islands to become refugia for post-Mesozoic species.

SOUTH POLAR REGION

Little is known of the Cretaceous and pre-Cretaceous terrestrial biota on Antarctica, partly because of the paucity of organisms living there today and the difficulties of fossil exploration. Glaciation apparently commenced in early Tertiary time, some 60 million years ago. Sediments indicate that by Eocene time, at least some glaciers had reached sea level. Prior to 60 million years ago there is evidence of warmer, tropical and subtropical conditions.

Brundin (1970) indicates that most of the present nonmigratory fauna consists of arthropods. There are, of course, migratory seals, penguins, and other seafaring forms. Obviously, these forms have adapted to cold conditions, and for the penguin in particular, a study of phylogeny should prove interesting. Brundin raises three basic questions regarding the nonmigratory fauna. First, is the fauna recent or relict? Evidence suggests that most of the fauna is a preglacial relict, since it appears to be highly endemic (90 percent of the species and 12 percent of the genera). Moreover, for such primitive groups as are present, immigration and speciation in the last million years or so seems unrealistic. Lastly, the fauna is most diverse and has higher populations on ice-free mountains and nunataks than near the coast, where recent immigrants would normally tend to congregate.

Second, why is there such a faunistic difference between eastern and western Antarctica? Only two species of mites are common to both. Brundin believes that eastern Antarctica may contain a relict mountain fauna that evolved in isolation, perhaps since the Mid-Cretaceous. That of western Antarctica differs, perhaps because of links to South America through the Scotia Arc.

Third, what are the trans-Antarctic relations? It appears that Antarctica formed separate connections between the two Australasian realms (Australia and New Zealand) and South America. The phylogenies of Antarctic elements of South America are apparently closely connected to those of Australia-Tasmania via eastern Antarctica and to those of New Zealand via western Antarctica.

The Circum-Antarctic Current

The circum-Antarctic current could not develop until all Gondwanaland masses moved sufficiently north to permit an uninterrupted flow of water. According to Kennett (1974), Australia was the last to move out of the region. Bathymetric and marine stratigraphic data have led to the scenario presented in map form in Figure 12.5. During the Paleocene, the separation of land was narrow with a constriction between Tasmania and Antarctica. The water flowed sluggishly and was shallow. High organic matter in the sediments suggests poor aeration. As separation continued during Eocene and early Oligocene time, the flow of water south of Tasmania increased. Submarine erosion began in the late Oligocene. Today there is a flow of more than 200 million cubic meters of water per second, making the circum-Antarctic current the largest-volume transport of any ocean current. It also circulates completely around the continent, thereby mixing waters of all oceans of the Southern Hemisphere. This changing pattern of ocean currents between the early and late Oligocene may also help to explain the evidence for Miocene climatic change in Australia.

SUMMARY

The Australian Realm is a most intriguing-region in terms of the biological development of land plants and animals. No theory of continental drift or of land bridges, and no summaries of world climatic or oceanic trends, can be considered viable unless it adequately accounts for the peculiar history and composition of Austral lands. A comparison of the floristic elements and affinities of New Zealand, Australia, New Guinea, and New Caledonia suggests relatively late separation from Antarctica. The Campbell Plateau, the Lord Howe Rise, and the Fiji and the Queensland plateaus may all have been bathed by a common northerly trending current before the advent of the West Wind Drift; and this possibility, together with a more extensive land area at the time, may

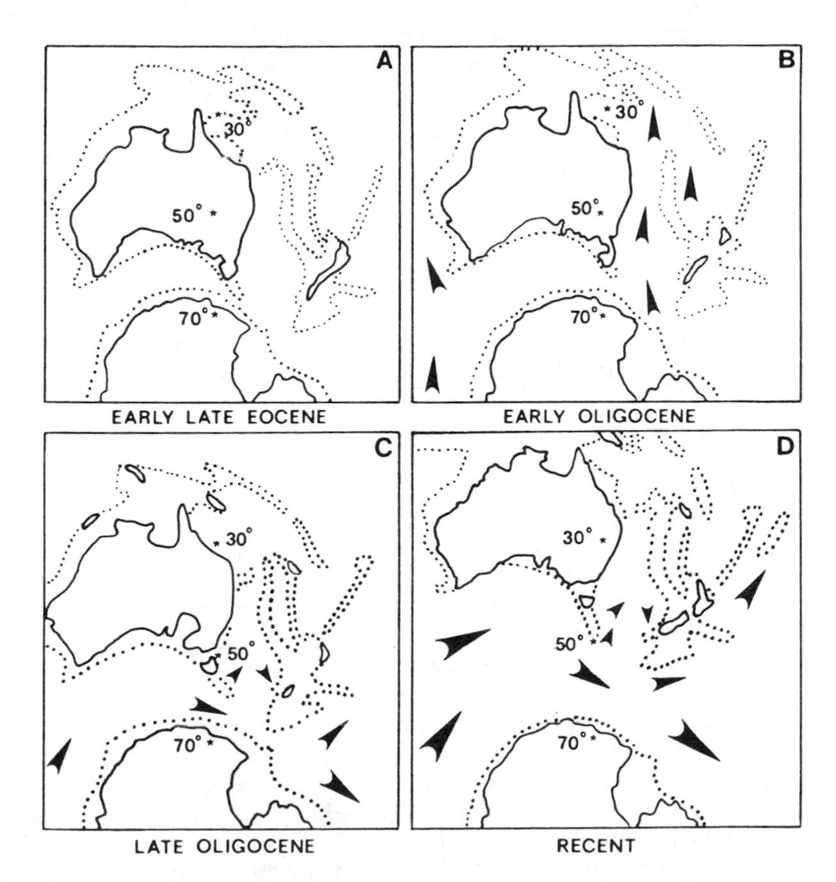

Figure 12.5. Successive Cenozoic reconstructions of Australia, Antarctica, New Zealand, and associated ridges during *(A)* early-late Eocene (45 million years ago); *(B)* Early Oligocene (37 million years ago); *(C)* late Oligocene (30 million years ago); and *(D)* recent time. Directions of bottom-water circulation are shown by arrows. Land areas and shallow ridges and rises are marked. During the early Oligocene *(B)* extensive erosive bottom currents flowed northward through the Tasman and Coral seas. The circum-Antarctic Current did not develop until about the late Oligocene; *(C)*, and a strong northward-flowing western boundary current formed to the east of New Zealand. These directions have been retained to the present day. *(Redrawn from J. P. Kennett, 1974, Development of the Circum-Antarctic Current, Science* **186:***146; copyright © 1974 by the American Association for the Advancement of Science. Reprinted with permission.)*

account for the greater floristic similarity among these islands than any of them has to Australia. Over the last 50 million years, New Guinea has not only emerged as an essential stepping stone to New Caledonia and beyond, but it appears to have served as a hinge for the gradual eastward separation of New Zealand and New Caledonia from Australia.

REFERENCES

Axelrod, D. I., 1959, Poleward Migration of Early Angiosperm Flora, *Science* **130**:203-207.

Brundin, L., 1970, Antarctic Land Faunas and their History, *Antarct. Ecol.* **1**:41-64.

Burbidge, N. T., 1960, The Phytogeography of the Australian Region, *Aust. J. Bot.* **8**(2):75-212.

Clemens, W. A., 1968, Origin and Early Evolution of Marsupials, *Evolution* **22**(1):1-18.

Fooden, J., 1972, Breakup of Pangaea and Isolation of Relict Mammals in Australia, South America, and Madagascar, *Science* **175**:894-898.

Francis, W. D., 1951, *Australian Rain-Forest Trees,* Forestry and Timber Bureau, Canberra, Commonwealth of Australia, 469p.

Gaskin, D. E., 1969, The Origins of the New Zealand Fauna and Flora, *Geogr. Rev.* **59**:414-434.

Good, R., 1960, On the Geographical Relationships of the Angiosperm Flora of New Guinea, *Br. Mus. Nat. Hist. Bull. (Botany)* **2**(8):205-226.

Keast, A., 1972, Australian Mammals: Zoogeography and Evolution, in *Evolution, Mammals, and Southern Continents,* A. Keast, F. C. Erk, and B. Glass, eds., State University of New York Press, Albany, N.Y., pp. 195-246.

Kennett, J. P., 1974, Development of the Circum-Antarctic Current, *Science* **186**:144-147.

Kurtén, B., 1968, *Pleistocene Mammals of Europe,* Aldine, Chicago, 317p.

Morain, S. A., 1970, *Ecological Segregation of Phytogeographic Elements, Upper Burdekin Valley, North Queensland,* Ph.D. dissertation, Kansas University, Lawrence, Kans., 203p.

Raven, P. H., and D.I. Axelrod, 1972, Plate Tectonics and Australasian Paleobiogeography, *Science* **176**:1379-1386.

Simpson, G. G., 1961, Historical Zoogeography of Australian Mammals, *Evolution* **15**:431-446.

Smith-White, S., 1959, Cytological Evolution in the Australian Flora, *Cold Spring Harbor Symp. Quant. Biol.* **24**:273-289.

Specht, R. L., and C. P. Mountford, eds., 1958, *Records of the American-Australian Scientific Expedition to Arnhem Land 3: Botany and Plant Ecology,* Melbourne University Press, Melbourne, 503p.

Suessenguth, K., 1950, The Flora of Australia as a Measure of the Antiquity of the Angiosperms, *Pac. Sci.* **4**:287-308.
Wood, J. G., 1959, The Phytogeography of Australia (in Relation to Radiation of Eucalyptus, Acacia, etc.), in *Biogeography and Ecology in Australia,* A. Keast, R. L. Crocker, and C. S. Christian, eds., W. Junk, The Hague, pp. 291-302.

FURTHER READING

Balgooy, M. M. J., 1960, Preliminary Plant Geographical Analysis of the Pacific, *Blumea* **10**(2):385-443.
Diels, L., 1936, The Genetic Phytogeography of the Southwestern Pacific Area, with Particular Reference to Australia, in *Essays in Geobotany in Honor of William Albert Setchell,* University of California Press, Berkeley, Calif., pp. 189-194.
Fleming, C. A., 1975, The Geological History of New Zealand and its Biota, in *Biogeography and Ecology in New Zealand,* G. Kuschel, ed., W. Junk, The Hague, pp. 1-86.
Herbert, D. A., 1933, The Relationships of the Queensland Flora, *R. Soc. Queensland Proc.* **78**:101-112.
Herbert, D. A., 1967, Ecological Segregation and Australian Phytogeographic Elements, *R. Soc. Queensland Proc.* **78**:101-112.
Hooker, J. D., 1867, Insular Floras, *Gard. Chron. and Agric. Gaz.,* Jan., pp. 6-7, 27, 50-51, 75-76.
Keast, A., ed., 1981, Ecological Biogeography of Australia, W. Junk, The Hague, 805p.
Lam, H. J., 1934, Materials Towards a Study of the Flora of the Island of New Guinea, *Blumea* **1**(1):115-159.
Morain, S. A., 1972, Phytogeographic Sequences on Australian Tropical Laterites, in *International Geography,* vol. 1, Proceedings of the 22nd International Geographical Congress, University of Toronto Press, Montreal, pp. 271-273.
Veevers, J. J., 1969, Palaeogeography of the Timor Sea Region, *Palaeogeogr. Palaeoclimatol., Palaeoecol.* **6**:125-140.
Wardle, P., 1963, Evolution and Distribution of the New Zealand Flora as Affected by Quaternary Climates, *N. Z. J. Bot.* **1**(1):3-17.

Chapter 13

Man: Instigator of Change

Now, in the last quarter of the twentieth century, humankind's role in changing the patterns and processes of biological evolution and extinction are becoming alarmingly clear. Barely six score years after Charles Darwin suggested a mechanism for the gradual increase of species, the world faces a possible wholesale extermination. Over the decades since Darwin's book, scientists have described vast numbers of then unknown species, but human populations and their need for resources have also soared. To trace the larger problem of three or four centuries of deliberate and accidental landscape alterations could easily occupy volumes, and even then would be incomplete. During the millenia preceding the Age of Discovery, early *Homo* was just as active, although to lesser degree, in the modification of plant and animal patterns, so that in many instances no one can tell which forms are truly native to an area and which represent unrecorded introductions. What is certain is that wherever humans have been present in numbers exceeding the densities of other megafaunal species, they have left their indelible mark on plant and animal distributions.

Most of humanity's global impact has been gradual and sinister in its disruptiveness. The study of natural history is scarcely more than 200 years old, and organized concern over humankind's malpractices less than two decades old. Population density and per capita increases in consumerism are unquestionably the root of human influence. Our needs for food, clothing, and shelter have spelled systematic ruin for those elements of ecosystems most easily harvested; and as our desires for the rare and unusual increase, destruction spreads to more delicate linkages in those systems. In a biological sense, cultural evolution is a human adaptation to utilize and exploit natural resources more efficiently. As Eiseley has warned, however, ". . . modern man, for all his developed powers and his imagined insulation in his cities, still lives at the mercy of those giant forces that created him and can equally decree his departure" (Eisley, 1969, 98, 99).

Most of the natural processes to which humanity is inextricably linked, and to which it must also answer, proceed either on a time scale of geologic millenia or biological generations. But human impact is now, in "real time." It is severe and widespread over every biogeographic realm. No ecosystem is spared, and no geographic locale unaffected, whether it be a small remote island like Palau in the far Pacific, or a wheat field in Kansas (see Walsh, 1976). Avenues of land migration have been severed, plant and animal populations have been partitioned into smaller, more isolated ranges, and gene flows have, in many instances, been reduced to functional zero. Selective harvesting has led to loss of species diversity and severe erosion of genetic diversity. What is worse, our socioeconomic and political systems have institutionalized the continuation of these destructive trends (see Regenstein, 1975, and Myers, 1979*a*, 1979*b*). The elimination of natural habitats has forced many species into narrower niches and imperiled their existence, and as the more desirable forms become endangered, their value in wildlife trade adds yet another threat to their tenure on earth.

On the positive side, steps have been initiated, however small and tentative, to restore or prolong the health of some ecosystems and their inhabitants. The need for wildlife protection is well understood, even if the manner of legal enforcement is not. Mitigation of migration barriers is at least being considered in larger developmental projects, even though the mechanisms intended to counteract barriers may in fact create sweepstakes, rather than corridor routes. And most critically, the absolute need to preserve genetic diversity is widely recognized by the creation of seed banks and similar genetic storehouses. When the day arrives that humanity appreciates and understands the intricacies of nature, at least some forms that might otherwise have become extinct can be revitalized.

HABITAT ALTERATION

The rate of loss in existing habitats far exceeds the creation of new territory by humans. For example, tropical moist forests, which

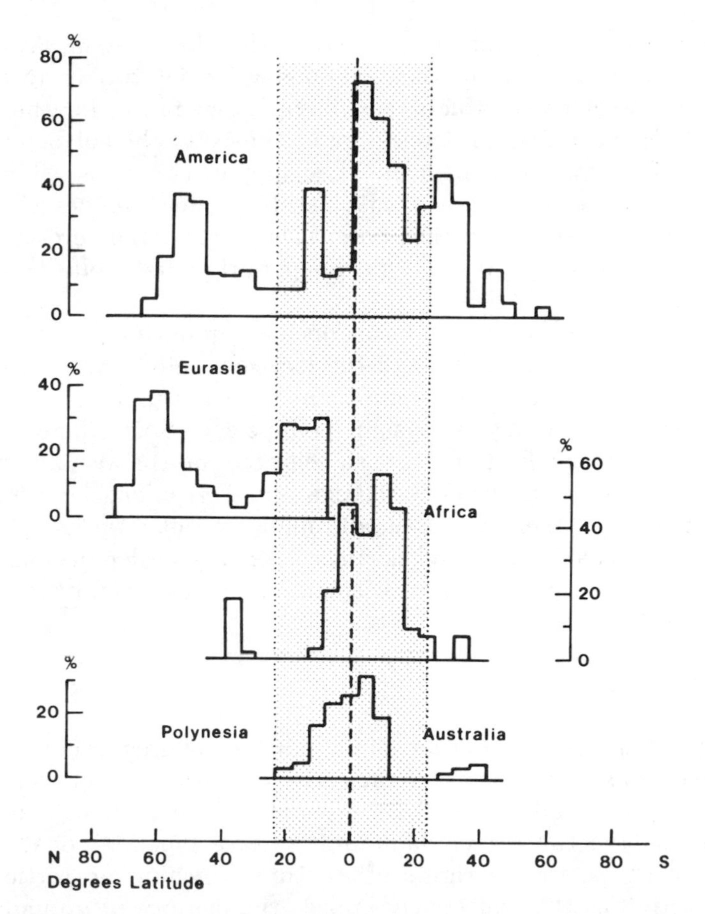

Figure 13.1. Allocation of the world's forests as a percent of land area. Horizontal scale is in increments of 5° latitude. The proportion of tropical moist forests, which have served for millions of years as the incubator of species, is shown in stipple.

contain an estimated 40 to 50 percent of the earth's plant and animal species, are being depleted at the rate of some 23 hectares per minute.* The most recent consensus is that at this rate, the earth will lose most, if not all, of its rain forest by the turn of the century (UNESCO, 1978; see Holden, 1980). Destruction is not even primarily the result of timber harvesting, but of uncontrolled agricultural expansion. Figure 13.1 shows that Central America, South America, and Africa, support most of the

* Myers (1979*b*) reports 20-40 ha/min; the U.S. Interagency Task Force (1980) believes the worst case is 38 ha/min; and Lanly and Clements (1979) estimate 14 ha/min. The most frequently cited value is 23 ha/min.

world's reserve of rain forest, followed by Southeast Asia, Australia, and Polynesia. Aside from their obvious value for timber resources, fuel wood, medicinal products, and oxygen generation for the global atmospheric cycle, tropical forests are the basis for the subsistence life styles of their indigenous peoples. They support countless wildlife and plant species, the larger share of which still have not been described and named, let alone studied for their potential uses; and, rain forests are the natural gene pools and sites for continuing, fairly rapid evolution (Poore, 1978; Eckholm, 1979).

Just how little is known of tropical forest environments is conveyed by the data in Table 13.1. Although the locations cited are all in South America, similar, if not worse, conditions exist throughout the tropical world, especially in the Malaysian Archipelago. Somewhere between one-third and one-half of the species collected, on the average, remain unnamed at the species level. Significant numbers of species are totally unknown at any taxonomic level, and many do not even have local or vernacular names. Humankind assumes a terrible burden in eliminating large numbers of these life forms without making every effort to at least assess their utility.

Barriers

Accompanying the elimination of habitats through agricultural and cultural expansion has been the progressive partitioning of ecosystems into biological islands. This situation is mostly the result of road and railroad construction, river damming, surface pipeline construction, fencing along power-line rights of way and ranching properties, and similar activities that selectively impede the passage of ground-based mammals. The United States, and perhaps Western Europe, are extreme examples of such partitioning, and very little is known about the negative effects these barriers pose.

In pre-European North America, vast herds of megafaunal wildlife roamed freely over their ranges. Only a few Indian trails were present and could hardly have prevented the migrations of larger life forms. Throughout the years of the westward movement, and those of rapid railroad expansion, there was still only a minimal impact on migration—except that it was during this time that millions of bison were eliminated. The advent of sealed roads, following the surge of automobiles in the early 1900s, began the real disruption of ecosystems; and in the 1950s this process took on a significant new complexion with the commencement of an interstate highway network.

Figure 13.2 shows the interstate network as it looked in 1980, when it was roughly 96-percent completed. This network is superimposed on a

Table 13.1. Relative Numbers of Identified and Unidentified Plants from Locales in South America

Country/region	*Size of study area*	*No. of plants identified*	*No. of plants not identified*	*Percent not specifically known*	*Source*
Colombia, S.A.	?	55	75[a]	58%	Espinal and Montenegro, 1963
Surinam, S.A.	?	49	19	28	Schulz, 1960
Amazonia	3.5 hectares	117	62[b]	35	Cain and Castro, 1959
Amazonia (Manaus)	?	30[c]	76[d]	72	Aubréville, 1961
Amazonia (Manaus)	1,850 square meters	76	49[e]	39	Takeuchi, 1961
Amazon River	1,300 kilometer transect parallel to equator	240	134[f]	36	Heinsdijk, 1960
			Average unknown	45%	

[a]Some identified to genus only
[b]46 were identified as to genus, 16 as to family
[c]From a total of 1,652 woody and herbaceous species
[d]31 were identified as to genus, 45 as to family; over 100 were entirely unknown, even in the vernacular
[e]20 were completely unknown, 10 identified as to family, 19 as to genus
[f]93 were identified as to genus, 41 as to family

Figure 13.2. Interstate transportation system in the United States. The biological islands created by this network severely hamper the movement of animals across their natural range. Such restrictions could lead to increased genetic differences between populations of the same species.

much more dense system of state highways and county roads, which themselves represent barriers of lesser magnitude. Each piece in this vastly complex puzzle can be viewed as an island, the boundaries of which serve to reduce gene flow from neighboring finite populations of mammal species. These barriers are obviously not absolute, especially for smaller forms uninhibited by fences. The megafauna, however, have literally been forced to funnel their movements through culverts and other narrow passages.

HUMAN POPULATION DENSITIES

In 1980 the estimated world population was 4.471 billion persons spread over a land area of 151.9 million square kilometers (Hill and Cook, 1980). These figures are equivalent to a density of 29 people per square kilometer. By grouping countries' population totals into their respective biogeographical realms, approximate densities can be calculated, as shown in Table 13.2. These densities range from almost 275 persons per square kilometer in the Oriental Realm to fewer than 3 per square kilometer in the Australian Realm. The Western Hemisphere currently has about 15 persons per square kilometer, while the Palearctic exceeds 20. Africotropical and Neotropical areas are not significantly different, although it is precisely in these regions that 90 percent of future population increases are expected to occur.

Early hominid population densities were obviously much lower than current ones. The success of *Homo sapiens* can be attributed to fire, tool making, language, and related cultural adaptive zones, each one of which reduced the competitive edge of larger, more numerous species. Near Omo, Ethiopia, Boaz (1979) has calculated that 1.9 to 2.8 million years ago, *Australopithecus africanus* and *A. boisei* densities ranged from .001 to 2.58 persons per square kilometer. These values are based on proportional faunal occurrences in excavated fossil beds of the Shungura Formation. The higher number is unbelievably high, since the current population density of Ethiopia is only about 34 persons per square kilometer, with an estimated doubling rate of 28 years. The lower value (.001) is more realistic from the point of view of logistical growth curves. If this number is anywhere near correct, there has been more than a thousandfold increase in human population density since the Pleistocene, while other animal densities have either remained static or declined.

Even steeper increases in human density are being experienced around the world. Eckholm (1975,764) has described the impact of this runaway rate on the world's mountain environments. The highlands linking Europe and the Mediterranean regions with Asia, for example, are experiencing

Table 13.2. Approximate Population Densities by Realm

	Land area (km²)	*1980 population*	*Density*
World	151,906,000	4,471,000,000	29.43
Nearctic	24,531,000	344,400,000	14.04
Neotropical	17,800,000	271,800,000	15.26
Palearctic	59,525,000	1,228,000,000	20.63
Africotropical	33,600,000	406,900,000	12.11
Oriental	8,000,000	2,197,100,000	274.63
Australian	8,450,000	22,800,000	2.69

Source: D. Hill and R. C. Cook, 1980, *1980 World Population Estimates,* The Environmental Fund, Washington, D.C., folio.

Note: These numbers are only approximate because national boundaries do not conform to biogeographical realms. The worst case is represented by the People's Republic of China, with a 1980 estimated population of 1.027 billion people. This country is unevenly distributed between the Palearctic and Oriental realms but the population has been included here entirely within the Oriental Realm. If the population were evenly split between the two realms, their adjusted densities would be 210.51 for the Oriental Realm and 29.25 for the Palearctic Realm. The latter is, coincidentally, extremely close to the world average.

land-use conversions that effectively sever their historical genetic links. "The Himalayan arc . . . forms an ecological Gibraltar whose fate will affect the well-being of hundreds of millions." Afghanistan and Pakistan are most affected at present, but deterioration is spreading rapidly eastward through Nepal to Bhutan. The natural avenues of migration for many life forms have already been cut, creating peripheral populations out of what were once more or less continuous ranges. In Pakistan, only about 3 percent of the land can still be classified as forest and, as in Nepal, evidence of large-scale deforestation is widespread.

The situation in the Andes Mountains of South America has followed a different human scenario than in the Himalayas, Caucasus, Rockies, and Pyranees, but the results are the same. The northern and central Andes historically supported larger populations than their bordering lowlands to the east (rain forest) and west (desert). The Incan civilization reached its apex and climax in the sixteenth century, by which time human numbers already exceeded the carrying capacity during adverse years. "Even before this time, the central Andes were largely deforested and most mountain residents were dependent on the dung of llamas for cooking fuel" (Eckholm, 1975,766). In the century following the arrival of the conquistadores, the population, estimated at anywhere from a few to 16 million, was reduced by 75 percent. Today, the population explosions of Peru, Colombia, and Bolivia far surpass any environmental pressures ever before experienced. The population doubling rates are roughly 25 years, and the current density is estimated at 14 persons per square kilometer. If the Incan resource base was tenuous, the modern situation is several orders of magnitude worse and declining.

WILDLIFE TRADE

The destruction of habitat and the pressures of human populations are dire enough without the aggravation of deliberate extermination of wildlife for trade. In *The Sinking Ark,* Myers (1979*a*) conjectures that at least one species per day is becoming extinct in the tropical forest alone, and that in a few years this pace may accelerate to one per hour. Based on current rates of population increase, land-use conversion, pollution, and other human activity, some 500,000 species may be exterminated by the year 2000 (Lovejoy, 1978). A disappearing plant can take 10 to 30 dependent species with it, including insects, higher animals, and other plants. No current rate of extinction is known for the world in general, but somewhere between 20,000 and 30,000 flowering plants and 1,000 mammals and birds are already considered to be in jeopardy of extinction (Eckholm, 1978).

The Convention on International Trade in Endangered Species (CITES) now prohibits international commercial trade in the rarest 600 or so species of animals and plants. According to Inskipp and Wells (1979), 51 nations had ratified the convention as of 1979, and an additional 30 or so countries had participated in conferences. Figure 13.3 shows those countries that have either ratified or participated in CITES activities. Absences in the European community are probably due to commercial pressure, while those in the developing world probably reflect inadequate scientific or logistical support to carry out the aims of the convention. Table 13.3 lists the mammals and flora considered to be actually threatened with extinction unless trade, contributing to their destruction, is halted. Commercial trade in animals focuses on their uses for meat, fur, hides, and horn; or on their value as zoo specimens, for medical experiments, and as personal pets. (Some of the most bizarre and senseless products include gorilla skulls and gorilla hands for use as ashtrays.) The flora are almost always collected for their horticultural value, which is heightened by the fact that they generally have restricted geographic ranges and habitats.

As with all products flowing through international trade, there are countries that are producers and others that are consumers. Producing countries are concentrated in the tropical regions of Asia, Africa, and Latin America. The consumers, expectedly, are located in the industrial regions of Europe, North America, and the Far East. Control of illicit trade requires sophisticated technologies at both ends of this flow, and it

Table 13.3. Mammals and Flora Protected from International Trade by the Convention on International Trade in Endangered Species (CITES)

MAMMALS

Bettongia spp (Rat-kangaroos)
Caloprymnus campestris(Desert rat-kangaroo)
Lagorchestes hirsutus (Western hare-wallaby)
Lagostrophus fasciatus (Banded hare-wallaby)
Onychogalea frenata (Merrin)
Onychogalea lunata (Wurrung)
Lasiorhinus krefftii (Queensland hairy-nosed wombat)
Chaeropus ecaudatus (Pig-footed bandicoot)
Macrotis lagotis (Bilby)
Macrotis leucura (Yallara)
Perameles bougainville (Marl)
Sminthopsis longicaudata (Long-tailed dunnart)
Sminthopsis psammophila (Sandhill dunnart)
Thylacinus cynocephalus (Thylacine)
Allocebus spp (Hairy-eared dwarf lemurs)
Cheirogaleus spp (Fat-tailed dwarf lemurs)
Hapalemur spp (Gentle lemurs)
Lemur spp (Lemurs)
Lepilemur spp (Sportive lemurs)
Microcebus spp (Mouse lemurs)
Phaner spp (Fork-marked mouse lemurs)
Avahi spp (Woolly indris)
Indri spp (Indris)
Propithecus spp (Sifakas)
Daubentonia madagascariensis (Aye-aye)
Callimico goeldii (Goeldi's marmoset)
Axis (Hyelaphus) porcinus annamiticus (Ganges hog deer)
Blastocerus dichotomus (Marsh deer)
Cervus duvauceli (Swamp deer)
Cervus elaphus hanglu (Kashmir red deer)
Cervus eldi (Brow-antlered deer)
Dama mesopotamica (Persian fallow deer)
Hippocamelus antisensis (North Andean huemal)
Hippocamelus bisulcus (South Andean huemal)
Moschus moschiferus (Himalayan musk deer)
Ozotoceros bezoarticus (Pampas deer)
Pudu pudu (Southern pudu)
Antilocapra americana peninsularis (Sonoran pronghorn)
Antilocapra americana sonoriensis (Lower California pronghorn)
Bison bison athabascae (Wood bison)
Bos gaurus (Gaur)
Bos (grunniens) mutus (Wild yak)
Pygathrix nemaus (Douc langur)
Simias concolor (Pig-tailed langur)
Hylobates spp (Gibbons)
Symphalangus syndactylus (Siamang)
Pongidae spp (Gorilla, orang-utan and chimpanzees)
Priodontes giganteus (Giant armadillo)
Ursus arctos
Ursus arctos isabellinus (Himalayan brown bear)
Ursus arctos nelsoni (Mexican grizzly bear)
Ursus arctos pruinosus (Tibetan brown bear)
Aonyx microdon (Cameroon clawless otter)
Enhydra lutris nereis (Southern sea otter)
Lutra felina (Marine otter)
Lutra longicaudis (platensis/annectens)
Lutra lutra (European otter)
Lutra provocax (Southern river otter)
Mustela nigripes (Black-footed ferret)
Pteronura brasiliensis (Giant otter)
Bubalus (Anoa) depressicornis (Lowland anoa)
Bubalus (Anoa) mindorensis (Tamaraw)
Bubalus (Anoa) quarlesi (Mountain anoa)
Capra falconeri chiltanensis (Chiltan markhor)
Capra falconeri jerdoni (Straight-horned markhor)
Capra falconeri megaceros (Kabul markhor)
Capricornis sumatraensis (Serow)
Damaliscus dorcas dorcas (Bontebok)
Hippotragus niger variani (Giant sable antelope)
Nemorhaedus goral (Goral)
Novibos (Bos) sauveli (Kouprey)
Oryx leucoryx (Arabian oryx)
Ovis ammon hodgsoni (Nyan)

Callithrix aurita (White-eared marmoset)
Callithrix flaviceps (Buff-headed marmoset)
Leontopithecus (Leontideus) spp (Golden tamarins)
Saguinus bicolor (Bare-faced tamarin)
Saguinus leucopus (White-footed tamarin)
Saguinus oedipus (Cotton-headed tamarin)
Alouatta palliata (Mantled howler)
Ateles geoffroyi frontatus (Black-browed spider monkey)
Ateles geoffroyi panamensis (Red spider monkey)
Brachyteles arachnoides (Wooly spider monkey)
Cacajao spp (Uakaris)
Chiropotes albinasus (White-nosed saki)
Saimiri oerstedii (Red-backed squirrel monkey)
Cercocebus galeritus galeritus (Tana River mangabey)
Colobus badius kirkii (Zanzibar red colobus)
Colobus badius rufomitratus (Tana River red colobus)
Macaca silenus (Lion-tailed macaque)
Nasalis larvatus (Proboscis monkey)
Presbytis entellus (Entellus langur)
Presbytis geei (Golden langur)
Presbytis pileatus (Capped langur)
Presbytis potenziani (Mentawai leaf monkey)
Axis (Hyelaphus) calamianensis (Calamian deer)
Axis (Hyelaphus) khuli (Kuhl's deer)
Manis temmincki (Cape pangolin)
Caprolagus hispidus (Hispid hare)
Romerolagus diazi (Volcano rabbit)
Cynomys mexicanus (Mexican prairie dog)
Leporillus conditor (Stick-nest rat)
Pseudomys fumeus (Smokey mouse)
Pseudomys praeconis (Shark Bay mouse)
Xeromys myoides (False water rat)
Zyzomys pedunculatus (Macdonnell Range rock-rat)
Chinchilla spp (Chinchillas)
Lipotes vexillifer (White flag dolphin)
Platanista gangetica (Ganges susu)
Platanista indi (Indus susu)
Sotalia spp (South American river dolphins)
Sousa spp (Hump-backed dolphins)
Neophocaena phocaenoides (Finless porpoise)
Phocoena sinus (Cochito)
Eschrichtius robustus (Grey Whale)
Balaenoptera borealis (Sei whale)
Balaenoptera musculus (Blue whale)
Balaenoptera physalus (Fin whale)
Megaptera novaeangliae (Humpback whale)
Balaena mysticetus (Bowhead whale)
Eubalaena spp (Right whales)
Canis lupus (Grey wolf)
Speothos venaticus (Bush dog)
Vulpes velox hebes (Northern swift fox)
Helarctos malayanus (Sun bear)
Selenarctos thibetanus (Asiatic black bear)
Tremarctos ornatus (Spectacled bear)
Ovis orientalis ophion (Cyprus mouflon)
Ovis vignei (Urial)
Pantholops hodgsoni (Chiru)
Rupicapra rupicapra ornata (Abruzzi chamois)
Prionodon pardicolor (Spotted linsang)
Hyaena brunnea (Brown hyaena)
Acinonyx jubatus (Cheetah)
Felis bengalensis bengalensis (Indian leopard cat)
Felis caracal (Caracal)
Felis concolor coryi (Florida puma)
Felis concolor costaricensis (Costa Rican puma)
Felis concolor cougar (Eastern puma)
Felis jacobita (Mountain cat)
Felis marmorata (Marbled cat)
Felis nigripes (Black-footed cat)
Felis pardalis mearnsi (Costa Rican ocelot)
Felis pardalis mitis (Brazilian ocelot)
Felis planiceps (Flat-heated cat)
Felis rubiginosa (Rusty-spotted cat)
Felis (Lynx) rufa escuinapae (Mexican bobcat)
Felis temmincki (Asiatic golden cat)
Felis tigrina oncilla (Costa Rican tiger cat)
Felis wiedii nicaraguae (Nicaraguan margay)
Felis wiedii salvinia (Guatemalan margay)
Felis yagouaroundi cacomitli (Tamaulipas jaguarundi)
Felis yagouaroundi fossata (Yucatan jaguarundi)
Felis yagouaroundi panamensis (Panama jaguarundi)

Table 13.3. *(continued)*

MAMMALS

Felis yagouaroundi tolteca (Sinaloa jaguarundi)
Neofelis nebulosa (Clouded leopard)
Panthera leo persica (Asiatic lion)
Panthera onca (Jaguar)
Panthera pardus (Leopard)
Panthera tigris (Tiger)
Panthera uncia (Snow leopard)
Arctocephalus townsendi (Guadalupe fur seal)
Monachus spp (Monk seals)
Elephas maximus (Asian elephant)
Dugong dugon (Dugong)
Trichechus inunguis (South American manatee)
Trichechus manatus (North American manatee)
Equus grevyi (Grevy's zebra)
Equus hemionus hemionus (Mongolian wild ass)
Equus hemionus khur (Indian wild ass)
Equus przewalskii (Przewalski's horse)
Equus zebra zebra (Cape mountain zebra)
Tapirus bairdii (Central American tapir)
Tapirus indicus (Malayan tapir)
Tapirus pinchaque (Mountain tapir)
Rhinocerotidae spp (Rhinoceroses)
Babyrousa babyrussa (Babirusa)
Sus salvanius (Pygmy hog)
Vicugna vicugna (Vicuna)

FLORA

Alocasia sanderana (an arum)
Alocasia zebrina (an arum)
Araucaria araucana (Monkey-puzzle tree)
Caryocar costaricense
Gymnocarpos przewalskii (a pink)
Melandrium mongolicus (a pink)
Silene mongolica (a pink)
Stellaria pulvinate (a chickweed)
Fritzroya cupressoides (Fitzroy's cypress)
Pilgerodendron uviferum (a cypress)
Microcycas calocoma (Palma corcho)
Prepusa hookeriana (a gentian)
Vantanea barbourii
Engelhardtia pterocarpa (a hickory)
Ammopiptanthus mongolicum
Cynometra hemitomophylla
Platymiscium pleiostachyum
Tachigalia versicolor
Aloe albida (an aloe)
Aloe pillansii (an aloe)
Aloe polyphylla (Spiral aloe [Kharetsal])
Aloe thorncropftii (an aloe)
Aloe vossii (an aloe)
Lavoisiera itambana
Guarea longipetiola (a mahogany)
Batocarpus costaricensis (a mulberry)
Cattleya skinneri (White nun)
Cattleya trianae
Didiciea cunninghamii
Laelia jongheana
Laelia lobata
Lycaste virginalis var alba
Peristeria elata (Holy ghost orchid)
Renanthera imschootiana (Red vanda)
Vanda coerulea (Blue vanda orchid)
Abies guatemalensis (a fir)
Abies nebrodensis (a fir)
Podocarpus costalis
Podocarpus parlatorei
Orothamnus zeyheri (Marsh rose)
Protea odorata (a protea)
Balmea stormae
Ribes sardoum (a currant)
Stangeria eriopus (a cycad)
Celtis aetnensis (a hackberry)
Welwitschia bainesii
Encephalartos spp (bread palm)
Hedychium phillippinense (ginger lily)

is here that efforts to eliminate the problem generally fail. One of the biggest failures is that customs officials often do not recognize products by their scientific names, and hence cannot apprehend traders of illegal goods listed in Table 13.3. Another major obstacle is the sheer number of political entities that must be coordinated. More than 60 sovereign tropical countries would have to coordinate and cooperate to effectively halt tropical exploitation. Under current conditions, the efforts of individual nations are easily circumvented by dealers smuggling their products out of neighboring ports. West German statistics, for example, show that in 1976, 75 percent of their imported cat skins originated in South America, mainly from Brazil. In 1977 because of new regulations, they came mainly from Paraguay. Brazil showed a decrease in the numbers of exported skins of 21 percent, while Paraguay showed an increase of 360 percent (Inskipp and Wells, 1979).

Zoos

Zoos, including aquaria, have proliferated from a sparse 125 locations worldwide in 1925 to 822 in 1980 (Figure 13.4). Of these, 176 (more than 20 percent) are located in the United States and its territories, and if those in Canada are added, almost a quarter of all world zoos are located in the Nearctic Realm. The *International Zoo Yearbook* (Olney, 1980) contains a wealth of data on the physical size of facilities, the diversity of species they house, and the specialties of each collection. Aside from the

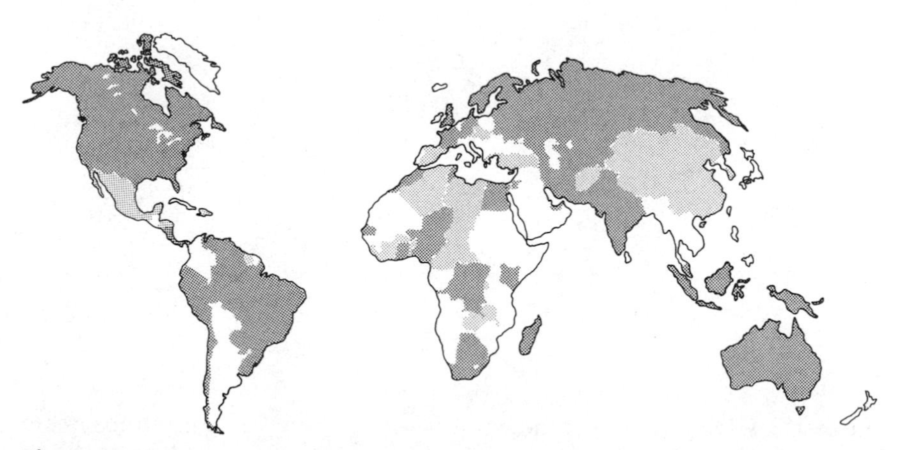

Figure 13.3. Countries that have either ratified the CITES convention articles (dark stipple) or attended conferences to regulate world trade in endangered species (light stipple). Some of the most critical areas to control are the tropical countries shown in white.

immediate attraction of zoos for public viewing of the world's fauna, rapidly growing concern for the populations of rare and endangered species has stimulated active cooperation among zookeepers to preserve and enhance the genetic variability of certain stock. By 1978, 47 studbooks were officially authorized for the purpose of tracing pedigrees and lineages of zoo populations. Table 13.4 provides a listing of the species for which such data are maintained.

The role of zoos in maintaining a storehouse of genetic material appears to be vital in view of other deleterious processes affecting natural habitats. Few zoos, however, maintain viable breeding populations of any species, so to avoid inbreeding, it has become necessary to transport individuals between zoos to maintain gene flow and enhance the prospects for genetic recombination. Without this human intervention, it is safe to assume that each of the species listed in Table 13.4 would be functionally extinct, because their isolated population sizes would be smaller than the threshold of genetic variability. Unfortunately, it sometimes happens that in their desire to acquire a stake in the business, zookeepers themselves contribute to the trade of illegal wildlife (see Holden, 1979).

PRESERVES AND NATURAL RESOURCE AREAS

Worldwide, the number of jeopardized species is too great, and time too short, for standard protection measures to be applied. The

Figure 13.4. Distribution of the world's zoos. Heavy concentrations occur in the United States, United Kingdom, Germany, the Netherlands, Denmark, Czechoslovakia, and Japan. The day may dawn when the African megafauna can be found in only a few natural preserves and in middle latitude cages.

Table 13.4. Species for Which Studbooks Are Being Maintained, as of 1978

Common name	*Scientific name*
Kiwi	*Apteryx* spp.
Red-cheeked ibis or waldrapp	*Geronticus eremita*
Cabot's tragopan	*Tragopan caboti*
Edwards' pheasant	*Lophura edwardsi*
White-eared pheasant	*Crossoptilon crossoptilon*
Manchurian crane	*Grus japonensis*
Rare amazon parrots	*Amazona* spp.
Malachite parrot	*Triclaria malachitacea*
Golden lion tamarin	*Leontopithecus rosalia rosalia*
Douc langur	*Pygathrix nemaeus*
Orangutan	*Pongo pygmaeus*
Gorilla	*Gorilla gorilla*
Bonobo	*Pan paniscus*
Giant anteater	*Myrmecophaga tridactyla*
Mexican wolf	*Canis lupus baileyi*
Maned wolf	*Chrysocyon brachyurus*
Bushdog	*Speothos venaticus*
Spectacled bear	*Tremarctos ornatus*
Lesser or red panda	*Ailurus fulgens*
Giant panda	*Ailuropoda melanoleuca*
Brown hyaena	*Hyaena brunnea*
Pakistan sand cat	*Felis margarita scheffeli*
Clouded leopard	*Neofelis nebulosa*
Rare tigers	*Panthera tigris,* spp.
Rare leopards	*Panthera pardus,* spp.
Snow leopard	*Panthera uncia*
Przewalski horse	*Equus przewalskii*
Asiatic wild ass	*Equus hemionus*
African wild ass	*Equus africanus*
Grevy's zebra	*Equus grevyi*
Indian rhinoceros	*Rhinoceros unicornis*
Black rhinoceros	*Diceros bicornis*
White rhinoceros	*Ceratotherium simum*
Pygmy hippopotamus	*Choeropsis liberiensis*
Vicuna	*Lama vicugna*
Eld's deer	*Cervus eldi*
Père David's deer	*Elaphurus davidianus*
Pudu	*Pudu pudu*
Pampas deer	*Ozotoceros bezoarticus*
Okapi	*Okapia johnstoni*
Gaur	*Bos gaurus*
European bison	*Bison bonasus*
Wood bison	*Bison bison athabascae*
Lechwe waterbuck	*Kobus leche*
Arabian oryx	*Oryx leucoryx*
Japanese serow	*Capricornis crispus*

destruction of habitats supporting large numbers of interdependent species cannot be analyzed or halted using a species-by-species approach. The overriding conservation need of the next few decades, according to Eckholm (1978), is the protection of as many varied habitats as possible. These efforts should preserve a representative cross section of the world's ecosystems, especially those rich in life forms.

To more systematically ensure the futures of plant and animal species, a network of reserves has been established in the United States that considers the number, size, and location of linkages between the biological islands created by cultural and agricultural disruption. These areas will hopefully serve to maintain or improve the viability of normal gene flow for many, if not most, of the species comprising our major biotic provinces. As of 1977, 27 reserves had been established under the Biosphere Reserve Program, and these are shown in Figure 13.5 (Franklin, 1977). In addition, by 1972 the U.S. federal government had preserved over 500 research natural areas, and the Nature Conservancy had acquired some 100,000 hectares and set them aside in 650 natural areas (Franklin, Jenkins, and Romancier, 1972).

On a global basis, it has been suggested by Sullivan and Shaffer (1975) that reserves be stratified on the basis of watershed size in such a way that larger areas support stable populations of large carnivores, and that reserves be distributed in a manner that would include most of the world's biological diversity. Smaller reserves might then serve as stepping stones and avenues of migration and dispersal.

CONCLUSION

Species are biological units whose evolution and distribution reflect eons of progressive change in response to the environment. Unlike that of individuals, whose passing goes unnoticed in the grand scheme of nature, the genetic material of species, once lost, is lost forever. Life is truly a nonrenewable resource. The ebb and flow of evolutionary rates in geologic time has led to a vast array of life forms long since vanished. The "worldeaters," as Eiseley refers to human beings, may in a matter of a few years experience a loss of biotic diversity unprecedented in any past natural catastrophe. Since the loss of the European lion in 80 A.D., human impact on plant and animal life has accelerated at such an alarming rate that by the year 2000, according to estimates provided by Lovejoy (1978), the total extinction of species may approach the staggering figure of 70 per day. In the remaining days of the twentieth century, the world could lose as many as 500,000 species, based on a global extrapolation from estimates of tropical moist forest clearing.

Figure 13.5. Location of biotic provinces (capital letters) and established biosphere reserves in the conterminous United States. *(Redrawn from J. F. Franklin, 1977, The Biosphere Program in the United States, Science* **195:**263; *copyright © 1977 by the American Association for the Advancement of Science. Reprinted with permission.)*

To reverse the process of human impact is clearly impossible, but it may not be too late to stem the tide. The development of an "evolutionary ethic," the beginnings of which are seen in genetic storehouses and improved environmental planning, would indicate that humans are trying to keep their own evolutionary options open. Any other ethic, says Eckholm, places humanity in the untenable position of being arbiter over the fate of all habitats and species that do not seem to have an immediate use. The course of future evolution would thus be set on the basis of short-term considerations, and a great deal of ignorance.

REFERENCES

Aubréville, A., 1961, *Etude ecologique des principales formations vegetales du Brésil et contribution à la connaissance des forêts de l'Amazonie bresilienne,* Centre Technique Forestier Tropical (CTFT), Nogent-sur-Marne, 268p.

Boaz, N. T., 1979, Early Hominid Population Densities: New Estimates, *Science* **206**:592-595.

Cain, S. A., and G. M. De Oliveira Castro, 1959, *Manual of Vegetation Analysis,* Harper Bros., New York, 325p.

Eckholm, E., 1975, The Deterioration of Mountain Environments, *Science* **189**:764-770.

Eckholm, E., 1978, *Disappearing Species: The Social Challenge,* Worldwatch Paper 22, Worldwatch Institute, Washington, D.C., 38p.

Eckholm, E., 1979, *Planting for the Future: Forestry for Human Needs,* Worldwatch Paper 26, Worldwatch Institute, Washington, D.C., 64p.

Eiseley, L., 1969, *The Unexpected Universe,* Harcourt Brace, New York, 239p.

Espinal, L. S., and E. Montenegro, 1963, *Formaciones vegetales de Colombia. Memoria Explicativa Sobre el Mapa Ecologico (1/1,000,000),* Bogota, 201p.

Franklin, J. F., 1977, The Biosphere Reserve Program in the United States, *Science* **195**:262-267.

Franklin, J. F., R. E. Jenkins, and R. M. Romancier, 1972, Research Natural Areas: Contributors to Environmental Quality Programs, *J. Environ. Qual.* **1**(2):133-139.

Heinsdijk, D., 1960, *Interim Report to the Government of Brazil on the Dry Land Forests on the Tertiary and Quaternary South of the Amazon River,* FAO Report No. 1284, Rome.

Hill, D., and R. C. Cook, 1980, *1980 World Population Estimates,* The Environmental Fund, Washington, D.C., folio.

Holden, C., 1979, Cracking Down on Illegal Wildlife Trade, *Science* **206**:801-802.

Holden, C., 1980, Rain Forests Vanishing, *Science* **208**:378.

Inskipp, T., and S. Wells, 1979, *International Trade in Wildlife,* EarthScan Publications, London, 104p.

Lanly, J. P., and J. Clement, 1979, *Present and Future Forest and Plantation Areas in the Tropics,* United Nations Food and Agriculture Organization, Rome, 13p.

Lovejoy, T. E., 1978, Global Changes in Biological Diversity, *The Dodo* **15**:7-12.
Myers, N., 1979*a*, *The Sinking Ark,* Pergamon Press, Oxford, England, 307p.
Myers, N., 1979*b*, Tropical Rain Forests: Whose Hand is on the Axe? *Nat. Parks and Conserv. Mag.* **53**(11).
Olney, P. J. S., ed., 1980, *International Zoo Yearbook,* vol. 20, Zoological Society of London, Dorset Press, Dorchester, 521p.
Poore, D., 1978, Values of Tropical Moist Forests, paper presented at *Improved Utilization of Tropical Forests Conference,* U.S. Forest Service Agency for International Development, Madison, Wis., May 21-26.
Regenstein, L., 1975, *The Politics of Extinction,* Macmillan, New York, 280p.
Schulz, J. P., 1960, *Ecological Studies on Rain Forest in Northern Surinam,* Amsterdam, 267p.
Sullivan, A. L., and M. L. Shaffer, 1975, Biogeography of the Megazoo, *Science* **189**:13-17.
Takeuchi, M., 1961, 1962, *The Structure of the Amazonian Vegetation, Fac. Sci. Univ. Tokyo, J.,* Sect. III., Bot 8, 1961, pp. 1-26, 27-35; 1962, pp. 279-288, 289-296, 297-304.
UNESCO, 1978, *Tropical Forest Ecosystems,* Natural Resources Research Series #14, Presses Universitaires de France, Vendôme, France, 683p.
U.S. Interagency Task Force, 1980, *The World's Tropical Forests: A Policy, Strategy and Program for the United States,* Department of State Publication 9117. International Organization and Conference Series #145. U.S. Government Printing Office, Washington, D.C., 53p.
Walsh, J., 1976, Superport for Palau Debated: Ecopolitics in the Far East, *Science* **194**:919-921.

FURTHER READING

Brown, L. R., P. L. McGrath, and B. Stokes, 1976, *Twenty-two Dimensions of the Population Problem, Worldwatch Paper 5.,* Worldwatch Institute, Washington, D.C., 83p.
Butzer, K. W., 1980, Civilizations: Organisms or Systems? *American Scientist* **68**(5):517-523.
Chesser, P. K., M. H. Smith, and I. L. Brisbin, Jr., 1980, Management and Maintenance of Genetic Variability in Endangered Species, in *International Zoo Yearbook,* vol. 20, P. J. S. Olney, ed., Zoological Society of London, Dorset Press, Dorchester, pp. 146-154.
Cole, M. M., 1965, Biogeography in the Service of Man, Inaugural Lecture, University of London School of Geography, Bedford College, 63p.
Davids, R. C., 1978, Managing America's Man-Made Islands, *Exxon Quarterly Report* **17**(4):16-21.
Duplaix, N., and L. Grady, 1980, Is the International Trade Convention For or Against Wildlife? in *International Zoo Yearbook,* P. J. S. Olney, ed., Zoological Society of London, Dorset Press, Dorchester, pp. 171-177.
Frankel, O. H., and M. E. Soulé, 1981, *Conservation and Evolution,* Cambridge University Press, New York, 328p.

Jackson, J. A., 1979, Highways and Wildlife—Some Challenges and Opportunities for Management, in *The Mitigation Symposium,* General Technical Report RM-65, Rocky Mountain Forest and Range Experiment Station, Fort Collins, Colo., pp. 566-571.

McNeill, W. H., 1976, *Plagues and Peoples,* Anchor Press/Doubleday, New York, 368p.

Raup, H. M., 1942, Trends in the Development of Geographic Botany, *Assoc. Am. Geogr. Ann.* **32**(4):319-354.

Sagan, C., O. B. Toon, and J. B. Pollack, 1979, Anthropogenic Albedo Changes and the Earth's Climate, *Science* **206**:1363-1366.

Simberloff, D. S., and L. G. Abele, 1976, Island Biogeography: Theory and Conservation Practice, *Science* **191**:285-286.

Terborgh, J., 1976, Island Biogeography and Conservation: Strategy and Limitations, *Science* **193**:1027-1030.

U.S. Fish and Wildlife Service, 1975, Threatened or Endangered Fauna or Flora, *Federal Register* **40**(127):27824-27924.

Glossary

Adaptability: capability of an organism to make changes that fit it better to its environmental conditions.

Adaptation: the process or processes by which an organism becomes apparently better suited to its environment or for particular functions. Also, the structures or activities of an organism, or of one or more of its parts, that tend to fit it better to life in its environment.

Adaptedness: the sum of genetic characters by which an organism is suited to its environment.

Adaptive Divergence: a process that occurs when a population is split into two subpopulations by some extrinsic barrier to gene flow, providing an opportunity for the two gene pools to evolve independently.

Adaptive Radiation: the evolution of taxa as they become adapted to new habitats; applicable also to the development of a new community.

Allopatric: refers to organisms originating in, or occupying, different geographic areas; cf. **Sympatric.**

Alpine: belonging to the Alps or, alternatively, to the higher regions of a mountain system. More strictly, the term refers to the mountainous region lying above the coniferous forests and below the permanent snow; that is, between the *timberline* and the *snowline.* The climate of this region is often known as the *alpine climate.*

Anemochore: an organism that is normally disseminated by the wind; cf. **Anthropochore, Hydrochore, Zoochore.**

Angiosperm: the subdivision of *spermatophytes* (seedplants) in which seeds are produced within the ovary; includes *monocotyledons* and *dicotyledons;* cf. **Gymnosperm.**

Anthropochore: a species that is regularly disseminated by man, for example, weeds and crop plants; cf. **Anemochore, Hydrochore, Zoochore.**

Barrier: (1) a topographic feature, or a physical or biological condition, that restricts or prevents migration of organisms or prevents establishment of organisms that have migrated; (2) a condition that prevents, or appreciably reduces, crossbreeding of organisms.

Bight: an indentation in the sea coast similar to a bay but either larger or with a gentler curvature, for example, the Great Australian Bight.

Binomial: a name of an organism consisting of two words, for instance, *Quercus alba,* the white oak; the first name is the genus, the second the species.

Bioclimatology: the study of climate in relation to life and health; one object of such study is to determine the climatic conditions most favorable to human habitation.

Biogeography: the branches of biology or of geography that deal with the geographic distribution of plants and animals; cf. **Plant Geography, Phytogeography, Zoogeography.**

Biome: a major biotic community composed of all the plants and animals and communities, including the successional stages of an area; the communities possess certain similarities in physiognomy and in environmental conditions; similar to *formation,* for example, North American grassland; cf. **Biotic province.**

Biosphere: a term sometimes applied to that portion of the earth occupied by the various forms of life, which is additional, for purposes of classification, to the three main physical zones, the *lithosphere,* the *hydrosphere,* and the *atmosphere.*

Biota: all of the species of plants and animals occurring within a certain area or region.

Biotic Province: a major ecological portion of a continent, occupying a continuous geographic area and containing one or more regional communities of plants and animals; for example, the Hudsonian biotic province, which occupies most of Canada and Alaska; cf. **Biome.**

Bipolar Distribution: discontinuous distribution of a taxon in the Northern and Southern Hemispheres.

Boreal Forest: the forest, consisting chiefly of conifers, that extends across northern North America, from Newfoundland to Alaska, and across Siberia.

Capacity, Adaptive: the genetically determined range (or plasticity) of reactions of an organism that enable it to respond in different ways to a variety of conditions.

Carrying Capacity: the maximum number of a wildlife species that a certain territory will support through the most critical period of the year. Also, the maximum quantity of a standing crop that can be maintained indefinitely on an area.

Cenozoic Era: the "New Life" era—the fourth of the five major subdivisions of the geologic scale of time; that period in the earth's history extending from the close of the *Mesozoic era* to the beginning of the *Quaternary era.* Some geologists, however, omitting the Quaternary era from the geologic scale of time, consider the Cenozoic era to extend to the present. The Cenozoic era is the era when mammals were widely distributed over the earth and the vegetation was beginning to resemble that of today.

Chordate: an animal in the phylum Chordata, characterized by a notochord, a dorsal central nervous system, and gill slits, for example, the *vertebrates.*

Chromosome: the thread-like or rod-like bodies bearing genes in the cells of plants and animals; they are formed from chromatin during the process of cell division.

Class: a unit of classification of organisms, composed of orders, for example, *Monocotyledoneae, Mammalia.*

Coadaptation: the correlated modification of two or more mutually dependent organs or organisms, for instance, the structure of a flower and the proboscis of an insect.

Competition: the condition that exists when the requirements of one or more of the organisms living in a community cannot be obtained from the available supply of resources.

Coniferous Forest: a forest of evergreen coniferous or cone-bearing trees carrying needle-shaped leaves; the most extensive coniferous forests stretch in a belt across northern Canada and Eurasia, with "tongues" extending southwards on the higher land, for instance, along the Rocky Mountains.

Continental Bridge Theory: the hypothesis that the present-day continents were once connected by isthmuses, or other areas of land.

Continental Drift Theory: the hypothesis that the present-day continents were displaced horizontally from an original mass, called Pangaea, to their present positions.

Continental Shelf: the seabed bordering the continents, which is covered by shallow water, in general, 100 fathoms or less in depth; it thus takes the form of a ledge sloping gently downwards from the coast and is approximately outlined by the isobath of 100 fathoms. It varies considerably in width, reaching 200 kilometers in some places. Where it is widest, the angle of slope is usually least and may be less than 1°; where the coast is mountainous, the shelf is usually narrow, and there is a quick transition from high land to deep water. Beyond the continental shelf there is a sudden drop in the seabed to a depth of 1,000 fathoms or more, so that the shelf has a steep, cliff-like edge. As the land slopes continuously downwards on the continental shelf, with no change of angle at the waterline, it is widely held that the edge of the shelf represents the former boundary of the continent; the shelf may have been formed by a rise in sea level or a fall in the level of the land, by the denudation of the fringe of the land by the sea, or by the deposition of solid materials by rivers or geographic events.

Convergent Evolution: the development of similarity in characteristics of organisms that were originally more different.

Corridor: a broad, continuous land connection enduring a long time and thus permitting the extensive interchange of organisms by migration, as for example, between Asia and Europe; cf. **Filter, Sweepstakes.**

Cretaceous: the most recent geological period of the *Mesozoic era,* which began about 135 million years ago and lasted for about 60 million years.

Diploid: refers to a cell in which there are two copies of each type of chromosome; cf. **Haploid, Polyploid.**

Dispersal: the actual transfer or movement of disseminules or organisms from one place to another. Also, the history of the movement of a group of organisms. cf. **Vicariance.**

Dominance: the condition in communities in which one or more species, by means of their number, coverage, or size, have considerable influence or control upon the conditions of existence of associated species. Also, the influence exerted by a dominant character or allele; for example, redness of petals in certain flowers is dominant over white. Also, a species that manifests ecological or social dominance.

Ecology: the science that treats organisms in relation to their environment; it is frequently subdivided into human ecology, animal ecology, plant ecology, and bioecology, the last-named dealing with the interrelationship between animal life and plant life. Ecology lies on the frontiers of so many other subjects, including various branches of geography, that its limits have not yet been precisely defined.

Ecotype: the smallest taxon or group of similar biotypes within an ecospecies, each one adapted to a certain combination of environmental conditions. Differences between ecotypes may be morphological, or only physiological.

Emigration: the migration of an organism out of an area, or the spread of a population out of previously occupied territory; cf. **Immigration.**

Endemic: a taxon confined to a certain country or region and with a comparatively restricted distribution; cf. **Exotic.**

Endemism: the occurrence of endemics in an area.

Environment: the sum total of all the external conditions that may influence organisms; cf. **Habitat.**

Eocene: the second geological epoch in the *Cenozoic era,* Tertiary period; it began about 58 million years ago and lasted for about 19 million years.

Equatorial Forest or **Tropical Rain Forest:** the equatorial region's hot, wet, evergreen forest, where rainfall is very heavy and where there is no dry season; it extends in parts into the typical monsoon areas. Owing to the extreme heat and moisture, the growth is dense and luxuriant. Many of the trees grow to tremendous heights; lianes and epiphytes are also abundant. Individual species of trees are scattered, but among them are such valuable tropical hardwoods as mahogany and ebony; the wild rubber tree is also a native of the equatorial forest. Typical equatorial forest regions include the basins of the Amazon and Zaire, and much of the East Indies.

Eucaryotes: a complex cell characterized by a nuclear membrane, mitochondria, and numerous chromosomes.

Evolution: the process of natural or artificial selection acting upon genetic diversity in organisms.

Evolutionary Rate: the rate at which new genetic characters arise in a breeding population, or the rate at which new phylogenetic forms can be recognized.

Exotic: refers to any organism that is not native in the area where it occurs; introduced; cf. **Endemic.**

Family: in the classification of organisms a group of one or more related genera, for example, the rose family, in which the roses, strawberries, and so on are classified. In ecology, it has been used for a community comprising individuals of a single species.

Fauna: a collective term to include all the kinds of animals in an area or in a geologic period.

Filter: a land connection that is temporary in duration and restricted in extent, limiting the kinds of organisms that can migrate over it; an example is the Bering Land Bridge in the Pleistocene; cf. **Corridor, Sweepstakes.**

Fitness: the degree of adaptation to the environment that an organism possesses.

Flora: the plant life of a region or of a geologic period, corresponding to the term **Fauna** for animal life.

Floristic Composition: the kinds of plant species, in the aggregate, that occur in a community or in an area.

Floristic Element: species that are characteristic of a certain territory but occur also in a different one; for instance, an arctic species growing in the high Rocky Mountains is an arctic element in the flora of the Rockies.

Gamete: a sex cell—a sperm or egg; in some of the simplest organisms the gametes are not differentiated into egg and sperm.

Gene: a localized unit of genetic material with a specific function in transmitting characters from one generation to the following one.

Gene Flow: the geographic dispersal of genes in a population or between populations.

Gene Pool: in a narrow sense, the genetic material of a local interbreeding population at the present time; in a broad sense, the total genetic resources or materials of a species throughout its geographical range.

Genetic Drift: accidental fluctuations in the proportions of a particular allele so that exact Mendelian ratios do not occur, which may result in the fixation or loss of certain genes in small populations without reference to selective value; cf. **Natural Selection.**

Genome: a reproductive unit having a haploid set of chromosomes.

Genotype: the entire genetic constitution, or the sum total of genes, of an organism, in contrast to **Phenotype.**

Genus: a group of related species, or occasionally only one species, used in the classification of organisms, for example, the white and Scotch pines belong to the genus *Pinus.*

Geography: the subject that describes the earth's physical features, climates, products, peoples, and the like, along with their distribution. Four main subdivisions of the subject are generally recognized: mathematical geography, which deals with the shape, size, and movements of the earth; physical geography, which is synonymous with physiography; and biological geography (including Biogeography); and Human Geography, which is sometimes regarded as synonymous with Anthropogeography.

Grasslands: those regions of the world where the natural vegetation consists of grass; the rainfall is too light to permit forest growth but is less scanty and irregular than that of the deserts. The grasslands are thus normally situated between the forest belts and the arid regions. Two main subdivisions are recognized, the tropical grasslands and the intermediate grasslands, of which the steppes, prairies, and pampas are examples.

Gymnosperm: a plant in the class Gymnospermae of the seed plants, spermatophytes, in which the seeds are not enclosed within an ovary, for instance, pine or spruce; cf. **Angiosperm.**

Habitat: the sum total of environmental conditions of a specific place that is occupied by an organism, a population, or a community; cf. **Environment, Niche, Microhabitat.**

Habitat Divergence: a process that occurs when parts of a single breeding population become progressively more adapted to different niches or habitats in the absence of isolation by distance.

Haploid: refers to an organism in which the nuclei contain a single set of chromosomes; cf. **Diploid, Polyploid.**

Hemisphere: one half of the earth's surface, formed when a plane through its center bisects the earth. The earth is usually divided into the Northern Hemisphere and Southern Hemisphere, the former being that half of the earth's surface north of the equator, the latter that half south of the equator.

Heterozygous: refers to an organism that originated from the fusion of gametes containing unlike genomes.

Holarctic: refers to the combined Palearctic and Nearctic regions of the faunal realm.

Homeostatic: the maintenance of constancy under changing conditions because of the capabilities of organisms to make adjustments.

Hydrochore: a plant whose diaspores are disseminated primarily by water, for instance, water lilies; cf. **Anemochore.**

Hypsodont: in teeth, having high or deep crowns and short roots, as in the molar teeth of horses.

Ice Age: a geologic period in which ice sheets and glaciers covered large areas of the continents, reaching the sea in places and lowering the temperature of the oceans. The latest Ice Age began early in the Quaternary period when ice covered much of Europe and North America; the present ice sheets of Greenland and Antarctica are relicts of the Ice Age.

Ice Sheet: a mass of ice and snow that covers large land areas in the polar regions; its surface is almost flat. The ice sheets of Greenland and Antarctica are the only large ones now in existence.

Immigration: the migration of an organism into an area where it did not previously occur; cf. **Emigration.**

Insectivore: an animal in the order Insectivora, a primitive group in the class Mammalia, whose diet consists chiefly of insects; examples are the mole and the shrew.

Invasion: the migration and establishment of an organism in a new location.

Irruption: an abrupt, irregular increase in a population's number or size, and its rapid emigration into new territory.

Isolation: the separation of populations from other populations of the same species by geographic, ecological, climatic, physiologic, or other barriers.

Isthmus: a narrow strip of land joining two larger land areas, such as two continental land masses, or joining a peninsula to the mainland; examples are the Isthmus of Panama, the Isthmus of Suez, and the Isthmus of Kra.

Land Bridge: a land connection between two bodies of land over which migration of organisms has occurred.

Latitude: the angular distance of a point on the earth's surface north or south of the equator, as measured from the center of the earth. Parallels of latitude are lines drawn around the earth parallel to the equator and may thus be described as approximate circles with the two poles as centers; the circles become smaller with increasing proximity to the poles. Since one-quarter of the complete circumference of the earth is traversed in moving from the equator to the pole, parallels of latitude are marked off in ninety divisions or degrees from the equator to each of the poles; the equator represents 0° latitude, while the North Pole is 90°N, and the South Pole is 90°S. Each degree of latitude is subdivided into 60 miles everywhere on the earth's surface.

Life form: the characteristic form or appearance of a species at maturity, as in tree, herb, worm, or fish.

Linnaean: refers to the work or the concepts of Carolus Linnaeus; the Linnean, or binomial, system of classification segregates species as to genus and species.

Logistic Curve: a graph that represents the growth of an individual or a population, typically it is S-shaped.

Longitude: the angular distance, measured along the equator, between a given point and a standard or prime meridian; since 1884, the meridian through the Royal Astronomical Observatory at Greenwich, near London, has been accepted by international agreement as the prime meridian. Longitude is thus measured in degrees west or east of Greenwich, the Greenwich meridian representing longitude 0°; the meridian 180°W coincides with the meridian 180°E and is roughly equal to the international date line. The complete angular distance around the earth is 360°, so if 360 meridians are drawn from pole to pole at equal intervals, they will lie 1° of longitude apart. Since the parallels of latitude become shorter with increasing nearness to the pole, it follows that a degree of longitude also decreases toward the poles. It is longest at the equator, where it approximates to a degree of latitude, about 69 miles; at 30°N or S the length of a degree of longitude is about 60 miles; at 60°N or S it is 34.6 miles; at 80°N or S it is 12 miles; and at the poles it is zero. The longitude of a place may be found from the difference between local time and Greenwich mean time or GMT, as shown by chronometers.

Microhabitat: a small habitat, such as a tree stump or a space between clumps of grass. There is no precise size range included in the definition and in some discussions forest clearings and coppices are considered to be microhabitats in their larger forest or grassland habitats, respectively.

Miocene: a geologic epoch of the Tertiary period that began about 28 million years ago and lasted about 16 million years.

Monotreme: an animal in the primitive order Monotremata, egg-laying mammals, restricted to the Australian Realm; an example is the platypus.

Monsoon: a wind system, ocurring mostly in Southeast Asia, that reverses its direction with the season.

Mutant: an organism, characteristic, or gene resulting from mutation; an animal or plant with inheritable characteristics that differ from those of the parents.

Mutation: a sudden, inheritable variation in an organism resulting from changes in a gene or in alterations of the structure or number of chromosomes.

Natural Selection: the agent of evolutionary change by which the organisms possessing certain characteristics in a given environment give rise to more offspring than those lacking such characteristics; cf. **Genetic Drift, Mutation.**

Nearctic Realm: the faunal region of the earth including North America except the tropical part of Mexico.

Neotropical Realm: the faunal region of the earth, including Central and South America.

Niche: the ecological position of an organism—the organism's "occupation" within the biological community.

Oligocene: the geologic epoch near the middle of the Tertiary period, that began about 39 million years ago and lasted for about 11 million years.

Oriental Realm: the faunal region that includes India, tropical Asia, and the Malaysian archipelago.

Paleogeography: the study of the distribution of land, water, and related features during earlier periods of the earth's history.

Palearctic: the faunal region that includes Eurasia north of the tropics and the northernmost part of Africa.

Paleocene: the earliest geological epoch in the Tertiary period; it began about 75 million years ago and lasted for about 11 million years.

Panmixis: the wide interbreeding of individuals of a population, where each individual has the potential capacity of mating with any other individual.

Paramo: the bleak, barren lands extending above the puna of the Andes, in South America, as far as the snowline, and invaded in parts by glaciers. Although flowering plants grow in sheltered spots, the vegetation over much of the area is stunted, as on the tundra, and consists largely of lichens and mosses.

Peripheral Population: a breeding population of plants or animals located near the margins of the species' geographic range.

Phenology: the study of the periodic phenomena of animal and plant life, for example, the time of flowering in plants, and their relations to the weather and climate.

Phenotype: the expression of the characteristics of an organism as determined by the interaction of its genetic constitution and the environment; cf. **Genotype.**

Phylum: one of the major subdivisions used in classifying plants and animals, for example, Tracheophyta (vascular plants) and Arthropoda (arthropods).

Phytogeography: the study of the distribution of plants on the earth in relation to their geographical environment, cf. **Plant Geography.**

Pioneer: plant, animal, or community that first invades a bare area; an example would be willows on a newly formed sandbar.

Plant Geography: the science that deals with the geographic distribution of plants and the causes of their distribution and dispersal; cf. **Phytogeography.**

Pleistocene: the geologic epoch preceding the Recent, in the Quaternary period, that began about 1.5 million years ago and lasted for about 1.5 million years.

Pliocene: the latest geological epoch in the Tertiary period; it began about 12 million years ago and lasted for about 10.5 million years.

Ploidy: refers to the number or sets of chromosomes in a cell—diploid, polyploid, haploid.

Polymorphism: the presence of several distinct forms in a species, particularly within a certain habitat or population.

Polyploid: an organism or a cell that contains more than the diploid number of chromosomes per cell; cf. **Haploid.**

Population Density: the number of individuals in a population per unit of area.

Prairies: the gently undulating, almost flat, generally treeless, grassy plains of North America, covering the southern regions of Alberta, Saskatchewan, and Manitoba in Canada, and the central United States from the foothills of the Rocky Mountains about as far east as the longitude of Lake Michigan. The light summer rains, with local droughts and high summer temperatures, permit a rich growth of natural grasses, but not, in general, of trees. The prairies form the North American counterpart of the pampas of South America, the steppes of Eurasia, and the veld of South Africa.

Predation: the behavior of animals, predators, in killing other animals, prey.

Predator: an animal that attacks other animals, prey, for example, a fox that kills mice or other prey.

Prey: an animal that is attacked and killed by a predator, for instance, a ground squirrel killed by a coyote.

Puna: the higher and bleaker parts of the plateau of the Andes, lying between about 3,000 and 4,000 meters above sea level.

Quaternary: the latest geologic period of the Cenozoic era; it began about 1.5 million years ago and includes the Recent and Pleistocene epochs.

Range: the extent of the geographic area over which a plant or animal breeds.

Realm: one of the major divisions in the classification of continental faunas.

Relict: a remnant or fragment of a flora or fauna; a relict remains from a former period when it was more completely developed. Alternatively, a relict may be a remnant of the population of a species that was formerly more widespread.

Reproductive Isolation: the separation of populations or organisms so that interbreeding cannot occur.

Rift Valley: a valley that has been formed by the sinking of land between two roughly parallel faults; such a valley is long in proportion to its width. The best-known example is the rift valley that extends through Syria, Palestine, and East Africa. Over 4,800 kilometers in length, it includes the Sea of Galilee, the valley of the River Jordan and the Dead Sea, the Gulf of Akaba, and the Red Sea, and runs through Lakes Tanganyika, Edward, and Albert. Examples of smaller rift valleys are the Rhine Valley between the Vosges Mountains and the Black Forest, and the central lowlands of Scotland.

Savanna: the region that borders the equatorial forest in each hemisphere, and thus lies between the rain forest and the hot deserts; the natural vegetation is mainly grass with scattered trees, for there are distinct wet and dry seasons. The rains occur during the hot season, when the belt of equatorial calms move to the savannas, but in the opposite season the savanna is under the influence of the drying trade winds. The llanos and campos of South America are examples of savannas, but the most extensive savanna is in Africa.

Sclerophyll: plants with stiff, leathery, evergreen leaves. Most often associated with Mediterranean-type climates dominated by dry summers and moist winters. Sclerophylly refers to the adaptive traits in broadleaved plants that reduce moisture loss during the dry season.

Selection: any process, whether natural or artificial, by which certain organisms or characteristics are permitted or favored to survive and reproduce in, or as if in, preference to others.

Specialized: refers to an organism, or part thereof, that is adapted to a particular kind of life or to a certain combination of environmental conditions; more limited than an unspecialized organism.

Speciation: the processes in evolution by which new species are formed; cf. **Mutation, Natural Selection.**

Species: a unit of classification of plants and animals, consisting of the largest and most inclusive array of sexually reproducing and cross-fertilizing individuals that share a common gene pool.

Steppes: the temperate grasslands, consisting of the level, generally treeless plains of Eurasia that extend over the lower regions of the Danube and in a broad belt over the southern European U.S.S.R. and southwest Siberia. The term is sometimes applied to the corresponding temperate grasslands on other continents, for example, in North America, and sometimes to the semiarid regions bordering the hot deserts.

Subspecies: a taxon of distinct, geographically separated complexes of genes, immediately below species and above variety (if varieties are recognized in a species); sometimes considered as synonymous with variety, or as an incipient species.

Superspecies: a group of related species that are geographically isolated, without any implication of natural hybridization among them.

Sweepstakes: describes the emigration of organisms across a barrier, usually where no land connection exists; an example would be the migration of a few kinds of animals from Africa to Madagascar. The term implies that the chances of successful migration are very low and that only a few of many forms actually breach the barrier; cf. **Corridor, Filter.**

Sympatric: refers to the origin or area of occupation of two or more closely related species in the same geographical area; cf. **Allopatric.**

Taiga: the coniferous forest land of Siberia, bordered on the north by the treeless, inhospitable tundra and on the south by the steppes. The principal species of trees are pine, fir, spruce, and larch. The taiga contains many swampy areas, and during the spring much of the land is flooded from the upper reaches of northward-flowing rivers whose lower stretches are still frozen. The name is often applied to other coniferous forests of the Northern Hemisphere.

Taxon: any taxonomic category, for example, species, genus, variety.

Taxonomy: the science of classification of organisms; the arrangement of organisms into systematic groupings such as species, genus, family, and order.

Terrestrial: refers to land as opposed to rivers, lakes, oceans, or seas.

Territory: the area occupied by an individual or group of organisms; the area that an animal defends against intruders.

Tertiary: the first of two geologic periods in the Cenozoic era; it comprized the Paleocene, Eocene, Oligocene, Miocene, and Pliocene epochs.

Timberline: the boundary line above which trees do not grow. It is usually clearly marked on high ground in low and middle latitudes, but it is not uniform in any region, for its height depends on local as well as general conditions of climate and soil. It is lower on the shady side of mountains than on the sunny side, and it is highest on those slopes that provide the best protection from winds and the longest exposure to the sun.

Tolerance: the capacity of an organism to live under a given set of conditions within its range of ecological amplitude.

Tundra: the treeless plains of northern North America and northern Eurasia, lying principally along the Arctic Circle and on the northern side of the coniferous forests. There is no corresponding region in the Southern Hemisphere. For most of the year the mean monthly temperature is below freezing, and winters are long and severe. The summers are short and warm, but even in July, the mean monthly temperature does not rise above 50°F (10°C). Relatively high temperatures may be reached during the daytime, but the subsoil, about a foot below the surface, is perpetually frozen. This condition, and the strong, intensely cold winds of winter, make tree growth impossible. In summer, mosses, lichens, and some flowering plants appear in abundance. Much of the flat ground, where drainage is poor, then becomes swampy.

Variation: divergences in the characteristics of organisms, or other objects, of the same kind caused either by the environment or by differences in the genetic constitutions of the organisms. cf. **Phenotype, Genotype.**

Vicariance: a philosophy, within the field of biogeography, that recognizes that and seeks to explain how organisms become dispersed through biotic (essentially geologic) events or disjunctions, such as by ruptures or separations of tectonic plates; cf. **Dispersal.**

Wallace's Line: the line established and described by A. R. Wallace in 1860 (On the Zoological Geography of the Malay Archipelago, *Linn. Soc. London J.* **14**:172-184) as a boundary between the Oriental and the Australian faunal regions.

Zoochore: an organism that is normally disseminated by an animal; cf. **Anemochore, Antropochore, Hydrochore.**

Zoogeography: the science that deals with the geographic distribution of animals; cf. **Biogeography, Plant Geography.**

PLANT AND ANIMAL INDEX

A

B

C

D

E

M

N

O

P

Q

R

S

T

U

V

W

X

Y

Z

SUBJECT INDEX

E

K

L

M

R

T

V

W

Z

About the Author

STANLEY A. MORAIN received his degree in geography from the University of Kansas in 1970 and since that time has taught biogeography, soils geography, and remote sensing of natural resources, both at KU and at the University of New Mexico. As a consulting scientist for the U. S. Agency for International Development and the United Nations, he has traveled widely in the past fifteen years, researching the historical biogeography of Australia, Southeast Asia, Central America, parts of Africa, and, more recently, China. Plant and animal evolution and processes of speciation remain an abiding pursuit in a career that includes designing biogeographic projects for satellite survey and the use of remote sensing techniques for resource inventories.